Lecture Notes in Computer Science 10064

More information about this series at http://www.springer.com/series/7407

Sylvain Duquesne · Svetla Petkova-Nikova (Eds.)

Arithmetic of Finite Fields

6th International Workshop, WAIFI 2016
Ghent, Belgium, July 13–15, 2016
Revised Selected Papers

Springer

Editors
Sylvain Duquesne
University of Rennes
Rennes
France

Svetla Petkova-Nikova
KU Leuven
Leuven
Belgium

ISSN 0302-9743 ISSN 1611-3349 (electronic)
Lecture Notes in Computer Science
ISBN 978-3-319-55226-2 ISBN 978-3-319-55227-9 (eBook)
DOI 10.1007/978-3-319-55227-9

Library of Congress Control Number: 2017933869

LNCS Sublibrary: SL1 – Theoretical Computer Science and General Issues

Printed on acid-free paper

This Springer imprint is published by Springer Nature
The registered company is Springer International Publishing AG
The registered company address is: Gewerbestrasse 11, 6330 Cham, Switzerland

Preface

These are the proceedings of WAIFI 2016, the 6th International Workshop on the Arithmetic of Finite Fields, held in Ghent, Belgium, during July 13–15, 2016. The five previous editions of this workshop were held in Madrid, Spain (WAIFI 2007), Siena, Italy (WAIFI 2008), Istanbul, Turkey (WAIFI 2010), Bochum, Germany (WAIFI 2012), and Gebze, Turkey (WAIFI 2014). Springer has published all previous volumes of the WAIFI proceedings in the LNCS series.

Since 2008, WAIFI has been held every even year, bringing together mathematicians, computer scientists, engineers, and physicists who conduct research in different areas of finite field arithmetic.

The program consisted of three invited talks and 17 contributed papers. The invited speakers were Swastik Kopparty (Rutgers University, USA), Simeon Ball (Universitat Politècnica de Catalunya, Spain) and Razvan Barbulescu (CNRS, Paris 6 and 7, France). The papers supporting the two last invited talks were also included in the proceedings. The contributed talks were selected from 38 submissions, each of which was assigned to at least three committee members or external reviewers chosen by the members. Additionally, the Program Committee had a significant online discussion phase for several days. Three additional presentations were made during the workshop but are not part of these proceedings.

We are very grateful to the members of the Program Committee for their dedication, professionalism, and careful work with the review and selection process. We also sincerely thank the external reviewers who contributed with their special expertise to review papers for this workshop.

We deeply thank the general co-chairs, Vincent Rijmen and Leo Storme, for their support of the Program Committee and their hard work in leading the overall organization of the workshop helped by the Organizing Committee. We would also like to sincerely thank members of the Steering Committee of the workshop series for their constant support and encouragement in our efforts to create a stimulating scientific program leading to this volume. Furthermore, we thank Jean-Jacques Quisquater for his valuable help in publicity and we are also very grateful to José Luis Imaña and Jan de Beule for diligently maintaining the workshop website. As with the previous volumes, Springer agreed to publish the revised and expanded versions of the WAIFI 2016 papers as an LNCS volume. We thank Alfred Hoffman and Anna Kramer from Springer for making this possible.

The submission and selection of papers were done using the EasyChair conference management system. Hence, thank you EasyChair! We would also like to acknowledge the Foundation Compositio Mathematica and FWO for being sponsors of the workshop.

Finally, but most importantly, we deeply thank all the authors who submitted their papers to the workshop and the participants all over the world who chose to honor us with their attendance.

February 2017

Sylvain Duquesne
Svetla Petkova-Nikova

Organization

Steering Committee

Berk Sunar	Worcester Polytechnic Institute, USA
Anwar Hasan	University of Waterloo, Canada
Çetin Kaya Koç	University of California Santa Barbara, USA
Jean-Jacques Quisquater	Université Catholique de Louvain, Belgium
Christof Paar	Ruhr-Universität Bochum, Germany
Gustavo Sutter	Autonomous University of Madrid, Spain
José Luis Imaña	Complutense University of Madrid, Spain
Francisco Rodriguez-Henriquez	CINVESTAV-IPN, Mexico
Ferruh Ozbudak	Middle East Technical University, Turkey
Sihem Mesnager	University of Paris 8, France
Erkay Savaş	Sabanci University, Turkey
Claude Carlet	University of Paris 8, France

General Co-chairs

Vincent Rijmen	KU Leuven, Belgium
Leo Storme	Ghent University, Belgium

Local Organizing Committee

Daniele Bartoli	Ghent University, Belgium
Wouter Castryck	Ghent University, Belgium
Maarten De Boeck	Ghent University, Belgium
John Sheekey	Ghent University, Belgium
Leo Storme	Ghent University, Belgium
Peter Vandendriessche	Ghent University, Belgium
Geertrui Van de Voorde	Ghent University, Belgium
Jan De Beule	Vrije Universiteit Brussel, Belgium
Vincent Rijmen	KU Leuven, Belgium
Bart Preneel	KU Leuven, Belgium
Jan Tuitman	KU Leuven, Belgium
Jean-Jacques Quisquater	Université Catholique de Louvain, Belgium
Joost Vercruysse	Université Libre de Bruxelles, Belgium

Program Co-chairs

Sylvain Duquesne	University of Rennes 1, France
Svetla Petkova-Nikova	KU Leuven, Belgium

Publicity Chair

Jean-Jacques Quisquater	Université Catholique de Louvain, Belgium

Program Committee

Tsonka Baicheva	Bulgarian Academy of Sciences, Bulgaria
Jean-Claude Bajard	University Pierre et Marie Curie, France
Josep Balasch	KU Leuven, Belgium
Anne Canteaut	Inria Rocquencourt, France
Claude Carlet	University of Paris 8, France
Luca De Feo	University of Versailles-Saint Quentin, France
Sylvain Duquesne	University Rennes 1, France
Tor Helleseth	University of Bergen, Norway
Sophie Huczynska	University of St. Andrews, UK
Alexander Kholosha	University of Bergen, Norway
Miroslav Knezevic	KU Leuven and NXP Semiconductors, Belgium
Gohar Kyureghyan	University of Magdeburg, Germany
Ivan Landzhev	New Bulgarian University, Bulgaria
Gregor Leander	Ruhr University Bochum, Germany
Sihem Mesnager	University of Paris 8, France
Amir Moradi	Ruhr University Bochum, Germany
Gary Mullen	Penn State University, USA
Svetla Petkova-Nikova	KU Leuven, Belgium
Daniel Panario	Carleton University, Canada
Ruud Pellikaan	Technical University Eindhoven, The Netherlands
Alexander Pott	Otto von Guericke University, Germany
Christophe Ritzenthaler	University of Rennes 1, France
Leo Storme	Ghent University, Belgium
Arnaud Tisserand	CNRS, University of Rennes 1, France
Frederik Vercauteren	KU Leuven, Belgium
Paul Zimmermann	Inria Nancy - Grand Est, France

Additional Reviewers

Domingo Gomez	University of Cantabria, Spain
Sujoy Sinha Roy	KU Leuven, Belgium
Valentin Suder	University of Versailles Saint-Quentin, France
Nicolas Estibals	University of Rennes 1, France
Kanat Abdukhalikov	United Arab Emirates University, UAE
Alina Ostafe	University of New South Wales, Australia
Karim Bigou	University of Brest, France
Wilfried Meidl	Sabanci University, Turkey
Yi Lu	University of Bergen, Norway
Omran Ahmadi	Institute for Research in Fundamental Sciences, Iran
Zhixiong Chen	Mercy College, USA

Wouter Castryck	Ghent University, Belgium
Qiang Wang	Carleton University, Canada
Aurore Guillevic	Inria Nancy, Grand Est, France
Audrey Lucas	University of Rennes 1, France
Ayoub Otmani	University of Rouen, France
Ayça ÇeŞmelioglu	Istanbul Kemerburgaz University, Turkey
Nicolas Sendrier	Inria Paris, France
Gregoire Lecerf	Ecole Polytechnique, France
Gabriel Gallin	University of Rennes 1, France
Arthur Beckers	KU Leuven, Belgium
Tobias Oder	Ruhr-Universität Bochum, Germany

Contents

Invited Talk I

A Brief History of Pairings

Razvan Barbulescu(✉)

CNRS, Univ. Paris 6, Univ. Paris 7, Paris, France
razvan.barbulescu@imj-prg.fr

Abstract. Pairings are a relatively new tool in cryptography. Recent progress on the attack algorithms have changed the security estimations. We make a list of pairing families and explain their advantages but also their weaknesses.

1 Introduction

Pairings are a mathematical tool which has been known to cryptographers for a long time and which switched sides during its history. If in the early 90's it was on the attacker's side, it is now used to create secure cryptologic protocols.

Let E be an elliptic curve defined over a finite field $\mathbb{F}_q$, r an integer number, P a point of order r and μ an rth root of unity in the algebraic closure $\overline{\mathbb{F}_q}$. The Weil pairing (restricted to the subgroup generated by P) is the map

$$\forall (a,b) \in (\mathbb{Z}/r\mathbb{Z})^2 \qquad \begin{array}{rl} e: \mathbb{Z}/r\mathbb{Z}P \times \mathbb{Z}/r\mathbb{Z}P & \rightarrow \mu^{\mathbb{Z}/r\mathbb{Z}} \\ ([a]P,[b]P) & \mapsto \mu^{ab}. \end{array} \tag{1}$$

Two properties of the Weil pairing are direct:

– bilinearity: for all a, a', b, b' we have

$$\begin{aligned} e([a+a']P,[b]P) &= e([a]P,[b]P) \cdot e([a']P,[b]P) \\ e([a]P,[b+b']P) &= e([a]P,[b]P) \cdot e([a]P,[b']P) \end{aligned}$$

– non-degeneracy: for all $a \neq 0$ there exists b so that

$$e([a]P,[b]P) \neq 1,$$

and similarly with the roles of a and b exchanged.

The Weil pairing owes its name to André Weil who gave an equivalent definition in 1940 [Wei40]. More precisely, Weil defined the map

$$e_W(S,T) = \frac{g(X+S)}{g(X)}, \tag{2}$$

where g is a function so that $\operatorname{div}(g) = rT - r\mathcal{O}_E$ and $X \in E \backslash E[r^2]$. This map is bilinear and non-degenerate (see Proposition III.8.1 in [Sil07]) and, since there is a unique map with these two properties (up to a multiplicative constant),

S. Duquesne and S. Petkova-Nikova (Eds.): WAIFI 2016, LNCS 10064, pp. 3–17, 2016.
DOI: 10.1007/978-3-319-55227-9_1

we conclude that Eqs. (2) and (1) are alternative definitions of the same object. In 1985 Miller [Mil04] invented an algorithm based on this equivalent definition to compute e in polynomial time with respect to the bit sizes of q and r. Frey and Rück [FR94] created an alternative manner to compute Weil pairings using results of Tate and Lichtenbaum.

From attacker's point of view, pairings are a tool to reduce hard problems to easier ones. Given a cyclic group G of known order, a generator P of G and an other point $[a]P$ for $a \in \{0, 1, \ldots, \#G - 1\}$, the discrete logarithm problem (DLP) consists in finding a. In 1992 Menezes, Okamoto and Vanstone [MOV93] showed that the Weil pairing associated to an elliptic curve over $\mathbb{F}_q$, an integer r, a point P of order r and an rth root of unity in $\overline{\mathbb{F}_q}$ allows to reduce the DLP on E to the DLP in the multiplicative group of $\mathbb{F}_q(\mu)$, the smallest subfield of $\overline{\mathbb{F}_q}$ which contains μ. The embedding degree of E with respect to r is the degree of $\mathbb{F}_q(\mu)$.

From a constructive point of view, pairings are a tool to combine two encrypted secrets into a common encrypted secret, without decrypting them at any time. In 2001 Joux [Jou00] proposed a three-party Diffie-Hellman key exchange which requires only one round of communications. If Alice, Bob and Carol want to agree on a common key they need to agree on an elliptic curve and on a point P of order r. Then they proceed in two steps

1. each participant generates a random integer, raises P to that power and broadcasts the result:
   - Alice generates a and computes $[a]P$ and broadcasts it,
   - Bob generates b and computes $[b]P$ and broadcasts it,
   - Carol generates c and computes $[c]P$ and broadcasts it;
2. each participant computes the Weil pairing of the received points and raises it to its own secret number:
   - Alice computes $e([b]P, [c]P)^a$,
   - Bob computes $e([c]P, [a]P)^b$,
   - Carol computes $e([a]P, [b]P)^c$.

Due to Eq. (1) all participants have computed μ^{abc}.

This protocol has inspired alternative solutions which are based on lattices and therefore belong to the exponential cryptography [GGH13].

The three party Diffie-Hellman protocol can be broken by solving the DLP in the subgroup of E generated by P or by solving the DLP in the multiplicative group of $\mathbb{F}_{q^k}$. This is true for other applications of pairings but we stick to this example for simplicity.

2 Known Attacks Against Pairings

2.1 Attacks on the Curve Side

Pollard Rho. In the three-party Diffie-Hellman protocol an attacker can compute the discrete logarithm of $[a]P$ and obtain the secret information a. The state-of-the-art algorithm to solve DLP in elliptic curves over prime fields is Pollard's rho [Pol78] which has a cost of $O(\sqrt{r})$ operations. Hence, for a given security level one has to set $\log_2 r = 2s$ and therefore $\log_2 \#E(\mathbb{F}_q) \geq 2s$. Due to Hasse's theorem, q and $\#E(\mathbb{F}_q)$ have the same bit size up to an error of 3 bits, so we have $\log_2 q \geq \log_2 r = 2s$.

Faults on the Twist Curve. Biehl, Meyer and Müller [BMM00] explained that, since some implementations of the scalar multiplication use only the x coordinate of the points on the elliptic curve $E : y^2 = x^3 + ax + b$, by error injection one can transfer the DLP from E to its twisted curve $E' : \epsilon y^2 = x^3 + ax + b$, where ϵ is a non-square of $\mathbb{F}_q$. As a counter-measure we require that the elliptic curves used in cryptography are twist-safe, i.e. that both $\#E(\mathbb{F}_q)$ and $2(q+1) - \#E(\mathbb{F}_q)$ have large prime factors.

Faults in Miller's Algorithm. Page and Vercauteren [PV06] studied the fault attacks which concern precisely the evaluation of the pairings and are independent on the protocol in which this primitive is used.

2.2 Attacks on the Finite Field Side

In the three-party Diffie-Hellman protocol an attacker, who has access to the public information $[a]P$, can compute $\mu^a = e([a]P, P)$ using solely public information. By solving the DLP in the group generated by μ one can obtain the secret information a. Hence a safe pairing requires that the DLP in the multiplicative group of $\mathbb{F}_{q^k}$ is hard.

The best algorithms to solve DLP in finite fields inherited the main traits from Index Calculus [Adl79] and have a complexity inferior to any exponential function. A suitable notation to express their complexity is

$$L_Q(\alpha, c) = \exp((c + o(1))(\log Q)^\alpha (\log\log Q)^{1-\alpha}),$$

where Q is the cardinality of the target finite field and α and c are two constants such that $0 < \alpha < 1$. When the constant c is not important we simply write $L_Q(\alpha)$. By extension we use a similar notation when α is a function.

The state-of-the-art algorithms depend on the size of the characteristic p with respect to $Q = p^n$ (we switch notations from q^k to p^n to show that p is not necessarily prime). When $p = L_Q(l_p, c_p)$ we have the following complexities:

- $L_Q(\frac{1}{3}, \sqrt[3]{\frac{64}{9}})$ when the field has large characteristic, i.e. if $l_p > \frac{2}{3}$, [JLSV06];
- $L_Q(\frac{1}{3}, c)$ with $c \in [\sqrt[3]{\frac{48}{9}}, \sqrt[3]{\frac{96}{9}}]$ in the boundary case, i.e. if $l_p = \frac{2}{3}$; the constant $c = \sqrt[3]{\frac{48}{9}}$ is obtained when $c_p = 12^{\frac{1}{3}}$, [SS16];
- $L_Q(\frac{1}{3}, \sqrt[3]{\frac{48}{9}})$ when the field has medium characteristic, i.e. if $\frac{1}{3} < l_p < \frac{2}{3}$, and n has a factor of size $12^{-\frac{1}{3}}(\frac{\log Q}{\log\log Q})^{\frac{1}{3}}$; and $L_Q(\frac{1}{3}, \sqrt[3]{\frac{96}{9}})$ if n has no factor of the suitable size (e.g. if n is prime) [BGGM15b];

- $L_Q(\frac{1}{3}, c)$ with $c \in [\sqrt[3]{\frac{8}{9}}, \sqrt[3]{\frac{96}{9}}]$ when the field has a characteristic at the boundary between medium and small, i.e. if $l_p = \frac{1}{3}$; the complexity $c = \sqrt[3]{\frac{8}{9}}$ is obtained when $c_p = 3^{-\frac{1}{3}}$ [Jou13]; one has a better complexity in the case of Kummer extensions;
- $L_Q(l_p + o(1))$ when the field has small characteristic, i.e. $l_p < \frac{1}{3}$; the best complexity corresponds to $\exp(O(1)(\log\log Q)^2) = L_Q(o(1))$ when $p = (\log Q)^{O(1)}$ [BGJT14].

When the characteristic is non-small, i.e. $l_p \geq 1/3$, the best complexities are all obtained with the same algorithm, presented below.

Number Field Sieve (NFS). The main steps of NFS [JL03] are similar to those of Index calculus and the key ingredient is smoothness: an integer is B-smooth if all its prime factors are less than B.

Polynomial Selection. One selects two polynomials f and g with integer coefficients which, when seen as elements of $\mathbb{F}_p[x]$, have a common factor φ which has degree n and is irreducible. The performance of the algorithm depends strongly on the degrees of the two polynomials as well as on their norms, i.e. larges coefficient in absolute value.

Relation Collection. Given two polynomials $f = \sum_{i=0}^{\deg f} f_i x^i$ and $g = \sum_{i=0}^{\deg g} g_i x^i$ we collect all the pairs (a, b) of integers (or equivalently linear polynomials $a - bx \in \mathbb{Z}[x]$) such that $\max(|a|, |b|) \leq E$ for a parameter E, $\gcd(a, b) = 1$ and the two norms $N_f(a, b) = \sum_{i=0}^{\deg f} f_i a^i b^{\deg f - i}$ and $N_g(a, b) = \sum_{i=0}^{\deg g} g_i a^i b^{\deg g - i}$ are B-smooth. This stage is usually done using a technique called sieve.

Linear Algebra. For each pair (a, b) yielded by the sieve one can write a linear equation whose unknowns are in bijection with set of prime ideals of degree one in the number fields of f and g of norm less than B. The square matrix has less than B unknowns and less than $\log_2 p^n$ non-zero entry per row so that one can use sparse-matrix algorithms like Wiedemann [Wie86].

Individual Logarithm. The unknowns obtained after the linear algebra stage, called virtual logarithms, allow to compute any discrete logarithm. This stage takes a negligible amount of time compared to the other stages.

When p has a special form, e.g. a low Hamming weight, a variant of NFS has a better asymptotic complexity.

The Special Number Field Sieve (SNFS). Given an integer d, an integer p is d-SNFS if there exists a polynomial $P \in \mathbb{Z}[x]$ and an integer u so that $\|P\| \leq 50$ (or other absolute constant) and $p = P(u)$. Semaev [Sem02] proved that the DLP is easier in prime finite fields $\mathbb{F}_p$ when p is d-SNFS with $d = (\frac{9}{2})^{\frac{1}{3}}(\frac{\log p}{\log\log p})^{\frac{1}{3}}$. One doesn't have to change anything in the NFS algorithm except for the choice of polynomials: $f = P(x)$ and $g = x - u$. In practice d is the value of $\deg f$ in the record computations using NFS and goes from 5 for fields of about 500 bits to 8 for fields of about 1200 bits. Experiments conducted with SNFS in the case of discrete logarithm [HT11] as well as of factorization [KBL14] confirm the efficiency of the algorithm for d-SNFS numbers with $d \geq 3$.

2.3 The LogJam Attack

A simple remark about the algorithms of the Index Calculus family is that they have two types of input data: a group G and a generator g of G which are used in the costly stages of the algorithm, relation collection and linear algebra, and an element h of G which isn't used before the individual logarithm stage. An attacker can therefore perform the expensive computations which depend on G and g once for all and use then to compute many secrete keys by solving many instances of individual logarithm with respect to that group.

Adrian et al. [ABD+15] conducted real life attacks in this manner. They estimated that 82% of the scanned servers use the same group and therefore can be attacked with one stone. One can easily imagine a situation where this is unacceptable: 80 bits of security are enough to protect bits of one minute for a pay-TV channel whereas it might be unacceptable for the whole program.

Consequences. Whenever the security of a cryptosystem is measured using Index Calculus attacks, as NFS, one is vulnerable to the LogJam attack. In this case one might either use a stronger level of security or generate on-the-fly the group used in the cryptosystem. For example in the case of pairings one should be able to generate on-the-fly pairing-friendly curves. However in the case of hardware implementation of cryptosystems, where parameters have to be hard-coded, the only option is to use larger key sizes.

3 Recent Progress of the NFS Attack

The first estimations of security of pairings have been done at a time when NFS could only be used for prime fields, and one had to make the hypothesis that the DLP in the general case is as hard as in prime fields [Len01]. Since then the NFS was adapted to the case $\mathbb{F}_{p^n}$ of non-small characteristic and in some cases the complexity is smaller than in the prime case, as we present below.

3.1 New Methods of Polynomial Selection

The first manner to go from $\mathbb{F}_p$ to $\mathbb{F}_{p^n}$ is to create new methods of polynomial selection whose result is a pair $(f, g) \in \mathbb{Z}[x]$ not necessarily irreducible which have a common irreducible factor φ in $\mathbb{F}_p[x]$.

For any pair (p, φ) formed of a prime p and a monic polynomial with integer coefficients φ which is irreducible in $\mathbb{F}_p[x]$ and any parameter $D \geq \deg \varphi$ one defines the lattice

$$\mathcal{L}(p,\varphi,D) = \{(a_0,\ldots,a_D) \in \mathbb{Z}^{D+1} \mid \sum_{i=0}^{D} a_i x^i \in p\mathbb{Z}[x] + \varphi\mathbb{Z}[x])\}.$$

A naive method of polynomial selection would be to pick a random monic irreducible $\varphi \in \mathbb{F}_p[x]$ of degree n and to make f and g from the shortest two vectors b_1 and b_2 in an LLL-reduced basis of $\mathcal{L}(p,\varphi,D)$. By the Lenstra-Lenstra-Lovasz theorem [LLL82] we know that $\|b_1\|_2 \leq c_1 \mathrm{Vol}(\mathcal{L})^{\frac{1}{\dim \mathcal{L}}}$ where $c_1 = 2^{\frac{\dim \mathcal{L}}{4}}$. Heuristically we expect b_1 and b_2 to have no non-zero coordinates (random vectors) so that $\deg f = \deg g = D$ and $\|b_1\| \approx \|b_2\| \approx \mathrm{Vol}(\mathcal{L})^{\frac{1}{\dim \mathcal{L}}}$.

JLSV$_2$. In [JLSV06] Joux, Lercier, Smart and Vercauteren take φ of degree $n < D$ such that $\|\varphi\|_2 = 1 + c_1 \mathrm{Vol}(\mathcal{L})^{\frac{1}{\dim \mathcal{L}}}$. Then one can make f from the coordinates of the shortest vector of $\mathcal{L}(p,\varphi,D)$ and set $g = \varphi$. By the Lenstra-Lenstra-Lovasz theorem $\|f\| \leq c_1 \mathrm{Vol}(\mathcal{L})^{\frac{1}{\dim \mathcal{L}}} < \|g\|$ so the two polynomials are distinct. The advantage is that $\deg g = n$ which is smaller than D whereas $\deg f$, $\|f\|$ and $\|g\|$ are the same as in the naive method.

GJL. In [JL03] Joux and Lercier proposed a method of polynomial for $\mathbb{F}_p$ which was generalized [Mat06,BGGM15b] to $\mathbb{F}_{p^n}$ with $n > 1$ (generalized Joux Lercier). One takes f to be a polynomial of degree $D+1$ with $\|f\| = 1$ which has an irreducible factor $\varphi \in \mathbb{F}_p[x]$ of degree n, and then one makes g from the shortest vector of $\mathcal{L}(p,\varphi,D)$. The advantage in this case is that f has coefficients of size $O(1)$ instead of $c_1(p^n)^{\frac{1}{D+1}}$ for the small cost of increasing the degree of f from D to $D+1$.

JLSV$_1$. Also in [JLSV06] Joux, Lercier, Smart and Vercauteren proposed to take f equal to a polynomial of degree n which is irreducible in $\mathbb{F}_p[x]$ with $\|f\| \leq 1$ and to set $g = f + p$. An additional improvement, which doesn't change the asymptotic complexity, consists in selecting polynomials such that $\deg f = \deg g$ and $\|f\| = \|g\|$. We can obtain this if we apply the JLSV$_2$ method with $D = 2n$, when $\|f\| \approx \|g\| \approx c_1(p^n)^{\frac{1}{2n}} = c_1\sqrt{p}$. However, one can obtain polynomials of the same characteristics by reducing a lattice of dimension 2 instead of $2n$. Indeed one takes two polynomials $f_0, f_1 \in \mathbb{Z}[x]$ of degree n respectively $\leq n-1$ so that, for all integers a, $f_0 + af_1$ has degree n. Next one LLL-reduces the lattice generated by $M(a,p) = \begin{pmatrix} 0 & p \\ 1 & a \end{pmatrix}$ and obtains a vector (u,v) of norm $\leq 2^{\frac{1}{4}}\sqrt{p}$. Finally one sets $f = f_0 + af_1$ and set $g = vf_0 + uf_1$, which is a multiple of f in $\mathbb{F}_p[x]$.

Conjugation Method. This method, presented in [BGGM15b], is similar to JLSV$_1$. First we select f_0 and f_1 so that, for all integer a, $f_0 + af_1$ has degree n. Next we select m as small as possible so that $x^2 - m$ has a root $a \in \mathbb{Z}$ modulo p and $f_0 + af_1$ is irreducible in $\mathbb{F}_p[x]$. We finish as in JLSV$_1$ by reducing $M(a,p)$ and setting $g = vf_0 + uf_1$.

At this point one would like to set $f = f_0 + \sqrt{m} f_1$ but this polynomial belongs to $\mathbb{Z}[\sqrt{m}][x]$ instead of $\mathbb{Z}[x]$. We overcome this difficulty by setting $f =$

$(f_0 + \sqrt{m}f_1)(f_0 - \sqrt{m}f_1) = f_0^2 - mf_1^2$ which has integer coefficients and is a multiple of g in $\mathbb{F}_p[x]$.

Methods for Composite n. Sarkar and Singh [SS16] proposed a method which improves the asymptotic complexity of NFS when $p = L_Q(2/3, c_p)$ with $c_p \in [1.12, 1.45] \bigcup [3.15, 20.91]$. The authors made a precise estimation of efficiency in the case of finite fields of cryptographic sizes $n = 4$ and $n = 6$.

Practical Efficiency of the New Methods. The new methods have been tested in practice and one concluded that the DLP in non-prime finite fields can be easier than in the prime case. In Table 1 we compare the cases $n = 2$ and $n = 3$ using the Conjugation method (Conj) to the prime case ($n = 1$). For this we converted the computation time into GIPS years (1GIPS year = the number of instructions done in one year by a CPU core of 1GHz) and made the convention that 1 GPU hour = 10 CPU hours.

Table 1. Time of discrete logarithms computations in $\mathbb{F}_{p^n}$ measured in GIPS years.

Bit size of p^n	160 dd (≈532 bits)	180 dd (≈600 bits)
n = 1	55.5 [Kle07]	260 [BGI+14]
n = 2 (Conj)	0.5 [BGGM14]	1 [BGGM15b]
n = 3 (Conj)	34 [BGGM15a]	46 [GGM16]

3.2 The Tower Number Field Sieve

A second method to go from $\mathbb{F}_p$ to $\mathbb{F}_{p^n}$ with $n > 1$ has been proposed by Schirokauer [Sch00] and revised in [BGK15]. One selects $h \in \mathbb{Z}[x]$ of degree n which is irreducible in $\mathbb{F}_p[x]$ and call ι a root of h in its number field. Then one selects f and g in $\mathbb{Z}[x]$ which have a common root in $\mathbb{F}_p$ using one of the methods for $\mathbb{F}_p$ and calls α_f (resp. α_g) a root of f (resp. g) in its number field and set $K_f = \mathbb{Q}(\iota, \alpha_f)$ (resp. $K_g = \mathbb{Q}(\iota, \alpha_g)$) and compute θ_f (resp. θ_g) a primitive element of K_f (resp. K_g).

One sets the parameters E, B and d at the same value as when computing discrete logarithms in a prime field of same bit size as $\mathbb{F}_{p^n}$. The factor base is formed of the prime ideals of K_f and K_g whose norm is less than B and whose inertia degree over $\mathbb{Q}(\iota)$ is one, together with all the prime ideals dividing the leading coefficients of f and g. The algorithm continues as follows.

1. Enumerate all pairs $a, b \in \mathbb{Z}[t]$ of degree $n-1$ with $\|a\|, \|b\| \leq E^{\frac{1}{n}}$ and collect those such that $\mathrm{Res}_t(F(a,b), h(t))$ and $\mathrm{Res}_t(G(a,b), h(t))$ are B-smooth.
2. Consider each element $a(\iota) + \alpha_f b(\iota)$ (resp. $a(\iota) + \alpha_g b(\iota)$) and compute the corresponding linear equations, as in the case of the classical version of NFS. Then solve the linear system to obtain the virtual logarithms of the factor base.

3. Compute the desired discrete logarithm in a similar manner to the classical case.

The practical efficiency of the TNFS has not been tested. Indeed, the relation collection consists of sieving on pairs $(a, b) \in \mathbb{Z}[t]$ of degree less than n which is equivalent to sieving on pairs of $2n$-tuples of integers. Several teams [Zaj10, HAKT15, GGV16] made experiments in the case of 3-tuples and concluded that this does not represent a major practical obstacle. This might be a starting point for future experiments in the case of 4-tuples so that TNFS in $\mathbb{F}_{p^2}$ can be tested.

3.3 The Extended Tower Number Field Sieve

The extended number field sieve (exTNFS), presented in [KB16], consists in combining the two ideas of the previous sections: new methods of polynomial selection and tower number fields. One writes $n = \eta\kappa$ with $\eta, \kappa \in \mathbb{Z}$ but not necessarily different from 1 and n and selects polynomials:

1. f and g as in Sect. 3.1 with κ instead of n;
2. h as in Sect. 3.2 with η instead of n.

When $\eta = 1$ we obtain the variant of NFS in Sect. 3.1, when $\eta = n$ we obtain TNFS (Sect. 3.2), but when n is composite and η is a proper factor of n we obtain a new algorithm. When $\gcd(\eta, \kappa) \neq 1$ one has to use a special method of polynomial selection which is due to Jeong and Kim [JK16]. The advantage of exTNFS is that, in a similar manner in which in TNFS one has the same size of norms as in classical NFS, in exTNFS one has the same size of the norms when attacking $\mathbb{F}_{p^{\eta\kappa}}$ as when attacking $\mathbb{F}_{P^\kappa}$ for a prime P of the same bit size as p^η.

The Case of General Primes. In order to analyze the efficiency of exTNFS we estimate the bit size of the norms product. Using Lemma 1 in [KB16] we find that, when the Conjugation method is used to select f and g, the two upper bound on the norms bit size is:

$$\text{norms bit size(exTNFS-Conj)} \leq 3\kappa \log_2 E + \frac{1}{2\kappa} \log_2 Q + o(1). \tag{3}$$

where $o(1)$ is a negligible term when $\log_2 Q$ goes to infinity. The $o(1)$ term is indeed negligible in cryptographic examples, e.g. Example 1 in [KB16]. Hence exTNFS has the same efficiency as NFS with the difference that now we can tune the parameter κ and make it equal to any factor of n.

The right hand member of Eq. (3) has its minimum when $\kappa \approx \sqrt{\frac{\log_2 Q}{6 \log_2 E}}$. Although the bit size of the parameter E depends on the size of the norms it doesn't vary of more than a factor 2 among variants of NFS when one attacks the same size of finite fields. In [KDL+16] one has $\log_2 Q = 768$ and $\log_2 E \approx 43$ so that the optimal value of $\kappa \approx 1.72$. We conclude that if one selects f and g using the Conjugation method then for target fields of approximatively 1000 bits with $n \leq 24$ composite the best options are $\kappa = 2$ if n is even and $\kappa = 3$ if n is odd. This would allow to obtain similar practical results as in Table 1.

The Case of Primes of Special Form. The exTNFS variant for SNFS numbers, abbreviated SexTNFS, consists in writing $n = \eta\kappa$ for two integers κ and η not necessarily different from 1 and n, in selecting h as in Sect. 3.2 with η instead of n and in selecting f and g using the Joux-Pierrot method [JP13], that we describe below, with κ instead of n.

One selects a monic polynomial $S \in \mathbb{Z}[x]$ of degree n such that $f = P(S(x))$ is irreducible in $\mathbb{F}_p[x]$ and then sets $g = S(x) - u$. The method is correct due to the following equation:

$$f(x) - p = P(S(x)) - P(u) \equiv 0 \bmod (S(x) - u) \text{ in } \mathbb{F}_p[x].$$

Once again we evaluate the practical efficiency using the estimation of the bit size of the norms product, which from [KB16, Sect. 5.2] is:

$$\text{norms bit size(SexTNFS)} \leq (d+1)\kappa \log_2 E + \frac{1}{\kappa d} \log_2 Q + o(1),$$

where $o(1)$ is negligible when Q goes to infinity. The advantage of SexTNFS is that we have the possibility to set κ equal to any divisors of n.

4 Pairings Families and Their Security

In the light of the recent progress, a perfect pairing family needs to contain a large number of curves for each security level that can be rapidly generated. Each curve of a perfect family has an embedding degree k which can be set as desired to any prime of desired size. The characteristic p is large and is not d-SNFS with $d \geq 3$. Finally for efficiency reasons the parameter r has the same bit size as q.

Freeman, Scott and Teske [FST10] made a taxonomy of known pairing-friendly families of elliptic curves. Given a bit size and an embedding degree k, most of them are constructed in two steps:

(i) one selects a prime power q of prescribed bit size and an integer t so that any elliptic curve over $\mathbb{F}_q$ of trace t has embedding degree k and its cardinality has a large prime factor r;
(ii) one uses the CM method [Mor91, AM93], which, given a prime power q and an integer t, allows to construct elliptic curves over $\mathbb{F}_q$ of trace t.

The CM method has complexity $O(D^{1+\epsilon})$ where D is the unique integer so that $(4q-t^2)/D$ is a perfect square. This imposes that we fix D in advance: it will be either small or will have common factors with q. By definition $\#E(\mathbb{F}_q) = q+1-t$ so we ask the existence of a prime r so that $q+1-t \equiv 0 \bmod r$. Finally, the property that k is the embedding degree of the curve is equivalent to $\Phi_k(q) \equiv 0 \pmod r$. We summarize the conditions on the output of the first step as follows:

CM-1. $\Phi_k(t-1) \equiv 0 \pmod r$
CM-2. $q+1-t \equiv 0 \pmod r$
CM-3. $\exists y,\ 4q = Dy^2 + t^2$

4.1 Supersingular Curves

When $k = 2$ there is a value of D for which the system is easy to solve. Indeed we set $t = 0$ so that we have $\Phi_2(t-1) = 0$ and therefore the first equation is satisfied independently on r. In Equation CM-3, we take $D = q$ and $y = 2$ so that there is no condition on q. Finally Equation CM-2, states that $q+1$ has a prime factor r which is easy to fulfill by enumerating primes q. Bröker [Brö06] presented the CM method in the case $D = q$, which is fast although D is large.

A natural question is whether this method can be extended to other values of k. The answer is given by the following classical result.

Proposition 1. *If $p \geq 5$ is a prime then any supersingular elliptic curve over $\mathbb{F}_p$ has embedding degree $k = 2$.*

Proof. By the definition of supersingular curves we have $\gcd(t, p) \neq 1$ so p divides t and therefore $t = 0$ or $|t| \geq p$. By Hasse's theorem $|t| \leq 2\sqrt{p}$ which is less than p and therefore $t = 0$. Then $q \equiv t - 1 \equiv -1 \pmod r$ and $q^2 \equiv 1 \pmod r$ which shows that $k = \mathrm{ord}_r(q) = 2$.

Drawback. Due to the quasi-polynomial algorithm the cases $p = 2$ and $p = 3$ are forbidden. When $p \geq 5$ the embedding degree $k = 2$ is fixed to a value which is far from the optimal value and has made the object of recent computation records which were faster than the prime case.

4.2 Pinch-Cocks [CP01]

One starts by replacing Equation CM-2, with

$$\text{CM-2}'. \qquad Dy^2 + (t-2)^2 \equiv 0 \pmod r$$

so that we obtain an equivalent system. Then we select r so that $r \equiv 1 \bmod k$ and $(\frac{-D}{r}) = 1$. Then Equation CM-2', is factorized into

$$(\sqrt{-D}y + (t-2))(\sqrt{-D}y - (t-2)) \equiv 0 \pmod r.$$

The choice of r allows to set t equal to a root of the polynomial $\Phi_k(X-1) \in \mathbb{F}_r[X]$. The same choice allows to solve this Equation CM-2$'$, for y: $y = (t-2)/\sqrt{-D} \pmod r$. Finally q is set to $(Dy^2 + t^2)/4$. Heuristically this is integer in a constant proportion of the cases and has the same probability to be prime as a random integer of the same size, i.e. one succeeds on average after $O(\log q)$ trials.

Drawback. With high probability the integer y has the same bit size as r so that $\log_2 q \approx 2\log_2 r$ which affects the efficiency of pairings.

4.3 Dupont-Enge-Morain [DEM05]

Once again we start by replacing Equation CM-2, by Equation CM-2′. Then we see Equations CM-1 and CM-2′ as a system which has to be solved with $y, t \in \mathbb{F}_r$:

$$\begin{cases} \Phi_k(t-1) = 0 \\ Dy^2 + (t-2)^2 = 0. \end{cases}$$

We solve the system (for a given D and bit size b of q) as follows:

```
R(y) ← Res_t(Φ_k(t − 2), Dy^2 + (t − 2)^2);
for y ≤ 2^(b/(2φ(k))) do
    r ← the largest prime factor of R(y)
    t = 2 + sqrt(−Dy^2) (if it exists)
    q ← q = (Dy^2 + t^2)/4
    q' ← q + 1 + t (cardinality of the twisted curve to be tested)
    if q and q' are integer primes and log_2 r ≥ b/2 then return y
    end if
end for
```

For example we ran the algorithm for the bit size $b = 256$, embedding degree $k = 16$ and the parameter $D = 3$. The output list was: $y \in \{39193, 61815\}$.

Drawback. The total number of curves which can be constructed for cryptographic sizes is very small if we restrict to twist-safe curves so that this family is vulnerable to the LogJam attack.

4.4 Sparse Families (e.g. MNT [MNT01])

The following construction is possible for all integers k so that $\varphi(k) = 2$, i.e. $k = 3, 4$ and 6, but for simplicity we present only the case $k = 3$. We set $r = \Phi_k(t-1)$ so that Equation CM-1 is satisfied. Next we set $q = r + t - 1$, which satisfies CM-2. The method was generalized by Freeman when $\varphi(k)$ but cannot be generalized further.

Proposition 2. *If $\varphi(k) > 4$ then the system CM-1, 2, 3 has a finite set of solutions.*

Proof. When we set $r = \Phi_k(t-1)$ Equation CM-3 becomes

$$y^2 = f(t) \text{ where } f(t) = \frac{1}{D}(4q - t^2) = \frac{1}{D}(4(\Phi_k(t-1) + t - 1) - t^2).$$

By the Riemann-Hurwitz formula the genus of the curve is $\lfloor \frac{\deg f - 1}{2} \rfloor = \lfloor \frac{\varphi(k)-1}{2} \rfloor \geq 2$. By Faltings' theorem the equation has a finitely many solutions in $\mathbb{Q}$

The integer solutions obtained when setting t equal to a linear polynomial in an additional variable are a subset of the rational solutions, so we have a finite number in total.

Drawback. The embedding degree k has a very small set of possibilities all of which are divisible by 2 or 3.

4.5 Complete Families (e.g. BN [BN05])

Once again we replace Equation CM-2 by CM-2′. Then we set r equal to a polynomial $r(x)$ whose number field contains $\mathbb{Q}(\sqrt{-D}, \zeta_k)$ for a kth root of unity ζ_k. This translates into

1. Φ_k is totally split modulo $r(x)$;
2. $x^2 + D$ is totally split modulo $r(x)$.

Next we take t to be a polynomial $t(x)$ so that $\Phi_k(t(x)) \equiv 0 \bmod r(x)$. Since Equation CM-2′ factors we can set $y(x) = t(x) \cdot \frac{t(x)}{\sqrt{-D}}$ where $\frac{1}{\sqrt{-D}}$ is a polynomial $z(x)$ in $\mathbb{Q}[x]$ so that $Dz^2+1 \equiv 0 \bmod r(x)$. Finally set $q(x) = \frac{1}{4}(Dy(x)^2+t(x)^2)$. The advantage of this method is that pairing-friendly curves can be generated on the fly by evaluating r and q at integer values x.

Drawback. The primes constructed by this method are $2\varphi(k)$-SNFS and therefore the NFS attacks have a smaller asymptotic complexity.

4.6 Menezes-Köblitz [KM05]

Not all the pairing constructions are obtained using the CM method. Menezes and Köblitz proposed a family which is not affected by the recent progress: p is not d-SNFS with $d \geq 3$ so that the SNFS attack has no consequences and $k = 1$ so that the security on the finite field side is the same as that of DSA.

Drawback. The embedding degree k cannot be tuned as desired.

5 Conclusion

We have identified a list of properties that a perfect pairing family should have and, by a thorough examination, concluded that in the present state of the art there is no perfect pairing family. In particular there is no clear champion because the Barreto-Naehrig family, long believed to be perfect for 128 bits of security, has a characteristic of a special form and is target to the SNFS attack.

Pairings are subject to two contradictory trends. On the one hand they require more time before being standardized because no perfect family has been proposed. On the other hand, time is running against pairings as they are subject to the NFS attack and therefore belong to the sub-exponential cryptography as RSA and DSA whereas there exist alternative primitives which are based on lattices and belong to the exponential cryptography.

References

[ABD+15] Adrian, D., Bhargavan, K., Durumeric, Z., Gaudry, P., Green, M., Halderman, J.A., Heninger, N., Springall, D., Thomé, E., Valenta, L., VanderSloot, B., Wustrow, E., Zanella-Béguelin, S., Zimmermann, P.: Imperfect forward secrecy: how Diffie-Hellman fails in practice. In: Proceedings of the 22nd ACM SIGSAC Conference on Computer and Communications Security-CCS 2015, pp. 5–17. ACM, New York (2015)

[Adl79] Adleman, L.M.: A subexponential algorithm for the discrete logarithm problem with applications to cryptography. In: 20th Annual Symposium on Foundations of Computer Science, pp. 55–60. IEEE (1979)

[AM93] Oliver, A., Atkin, L., Morain, F.: Elliptic curves and primality proving. Math. Comput. **61**(203), 29–68 (1993)

[BGGM14] Barbulescu, R., Gaudry, P., Guillevic, A., Morain, F.: Discrete logarithms in GF(p^2) – 160 digits (2014). Announcement available at the NMBRTHRY archives, item 004706

[BGGM15a] Barbulescu, R., Gaudry, P., Guillevic, A., Morain, F.: New record in $\mathbb{F}_{p^3}$ (2015). https://webusers.imj-prg.fr/~razvan.barbaud/p3dd52.pdf

[BGGM15b] Barbulescu, R., Gaudry, P., Guillevic, A., Morain, F.: Improving NFS for the discrete logarithm problem in non-prime finite fields. In: Oswald, E., Fischlin, M. (eds.) EUROCRYPT 2015. LNCS, vol. 9056, pp. 129–155. Springer, Heidelberg (2015). doi:10.1007/978-3-662-46800-5_6

[BGI+14] Bouvier, C., Gaudry, P., Imbert, L., Jeljeli, H., Thomé, E.: Discrete logarithms in GF(p) – 180 digits (2014). Announcement available at the NMBRTHRY archives, item 004703

[BGJT14] Barbulescu, R., Gaudry, P., Joux, A., Thomé, E.: A heuristic quasi-polynomial algorithm for discrete logarithm in finite fields of small characteristic. In: Nguyen, P.Q., Oswald, E. (eds.) EUROCRYPT 2014. LNCS, vol. 8441, pp. 1–16. Springer, Heidelberg (2014). doi:10.1007/978-3-642-55220-5_1

[BGK15] Barbulescu, R., Gaudry, P., Kleinjung, T.: The tower number field sieve. In: Iwata, T., Cheon, J.H. (eds.) ASIACRYPT 2015. LNCS, vol. 9453, pp. 31–55. Springer, Heidelberg (2015). doi:10.1007/978-3-662-48800-3_2

[BMM00] Biehl, I., Meyer, B., Müller, V.: Differential fault attacks on elliptic curve cryptosystems. In: Bellare, M. (ed.) CRYPTO 2000. LNCS, vol. 1880, pp. 131–146. Springer, Heidelberg (2000). doi:10.1007/3-540-44598-6_8

[BN05] Barreto, P.S.L.M., Naehrig, M.: Pairing-friendly elliptic curves of prime order. In: Preneel, B., Tavares, S. (eds.) SAC 2005. LNCS, vol. 3897, pp. 319–331. Springer, Heidelberg (2006). doi:10.1007/11693383_22

[Brö06] Bröker, R.: Constructing elliptic curves of prescribed order. Ph.D. thesis, Leiden University (2006). http://www.math.leidenuniv.nl/~reinier/thesis.pdf

[CP01] Cocks, C., Pinch, R.G.E.: Identity-based cryptosystems based on the Weil pairing. Unpublished manuscript, 170 (2001)

[DEM05] Dupont, R., Enge, A., Morain, F.: Building curves with arbitrary small mov degree over finite prime fields. J. Cryptol. **18**(2), 79–89 (2005)

[FR94] Frey, G., Rück, H.-G.: A remark concerning m-divisibility and the discrete logarithm in the divisor class group of curves. Math. Comput. **62**(206), 865–874 (1994)

[FST10] Freeman, D., Scott, M., Teske, E.: A taxonomy of pairing-friendly elliptic curves. J. Cryptol. **23**(2), 224–280 (2010)

[GGH13] Garg, S., Gentry, C., Halevi, S.: Candidate multilinear maps from ideal lattices. In: Johansson, T., Nguyen, P.Q. (eds.) EUROCRYPT 2013. LNCS, vol. 7881, pp. 1–17. Springer, Heidelberg (2013). doi:10.1007/978-3-642-38348-9_1

[GGM16] Gaudry, P., Guillevic, A., Morain, F.: Discrete logarithm record in GF(p^3) of 592 bits (180 decimal digits) (2016). Announcement available at the NMBRTHRY archives, item 004706

[GGV16] Gaudry, P., Grémy, L., Videau, M.: Collecting relations for the number field sieve in GF(p^6) (2016). Accepted for publication at ANTS-XII, Kaiserslautern

[HAKT15] Hayasaka, K., Aoki, K., Kobayashi, T., Takagi, T.: A construction of 3-dimensional lattice sieve for number field sieve over GF(p^n). Cryptology ePrint Archive, Report 2015/1179 (2015). http://eprint.iacr.org/2014/300

[HT11] Hayasaka, K., Takagi, T.: An experiment of number field sieve over GF(p) of low hamming weight characteristic. In: Chee, Y.M., Guo, Z., Ling, S., Shao, F., Tang, Y., Wang, H., Xing, C. (eds.) IWCC 2011. LNCS, vol. 6639, pp. 191–200. Springer, Heidelberg (2011). doi:10.1007/978-3-642-20901-7_11

[JK16] Jeong, J., Kim, T.: Extended tower number field sieve with application to finite fields of arbitrary composite extension degree. Cryptology ePrint Archive, Report 2016/526 (2016). http://eprint.iacr.org/2016/526

[JL03] Joux, A., Lercier, R.: Improvements to the general number field for discrete logarithms in prime fields. Math. Comput. **72**(242), 953–967 (2003)

[JLSV06] Joux, A., Lercier, R., Smart, N., Vercauteren, F.: The number field sieve in the medium prime case. In: Dwork, C. (ed.) CRYPTO 2006. LNCS, vol. 4117, pp. 326–344. Springer, Heidelberg (2006). doi:10.1007/11818175_19

[Jou00] Joux, A.: A one round protocol for tripartite Diffie–Hellman. In: Bosma, W. (ed.) ANTS 2000. LNCS, vol. 1838, pp. 385–393. Springer, Heidelberg (2000). doi:10.1007/10722028_23

[Jou13] Joux, A.: Faster index calculus for the medium prime case application to 1175-bit and 1425-bit finite fields. In: Johansson, T., Nguyen, P.Q. (eds.) EUROCRYPT 2013. LNCS, vol. 7881, pp. 177–193. Springer, Heidelberg (2013). doi:10.1007/978-3-642-38348-9_11

[JP13] Joux, A., Pierrot, C.: The special number field sieve in $\mathbb{F}_{p^n}$ - application to pairing-friendly constructions. In: Cao, Z., Zhang, F. (eds.) Pairing 2013. LNCS, vol. 8365, pp. 45–61. Springer, Heidelberg (2014). doi:10.1007/978-3-319-04873-4_3

[KB16] Kim, T., Barbulescu, R.: Extended tower number field sieve: a new complexity for the medium prime case. In: Robshaw, M., Katz, J. (eds.) CRYPTO 2016. LNCS, vol. 9814, pp. 543–571. Springer, Heidelberg (2016). doi:10.1007/978-3-662-53018-4_20

[KBL14] Kleinjung, T., Bos, J.W., Lenstra, A.K.: Mersenne factorization factory. In: Sarkar, P., Iwata, T. (eds.) ASIACRYPT 2014. LNCS, vol. 8873, pp. 358–377. Springer, Heidelberg (2014). doi:10.1007/978-3-662-45611-8_19

[KDL+16] Kleinjung, T., Diem, C., Lenstra, A.K., Priplata, C., Stahlke, C.: Discrete logarithms in GF(p) – 768 bits (2016). Announcement available at the NMBRTHRY archives, item 004917

[Kle07] Kleinjung, T.: Discrete logarithms in GF(p) – 160 digits (2007). Announcement available at the NMBRTHRY archives, item 003269

[KM05] Koblitz, N., Menezes, A.: Pairing-based cryptography at high security levels. In: Smart, N.P. (ed.) Cryptography and Coding 2005. LNCS, vol. 3796, pp. 13–36. Springer, Heidelberg (2005). doi:10.1007/11586821_2

[Len01] Lenstra, A.K.: Unbelievable security matching AES security using public key systems. In: Boyd, C. (ed.) ASIACRYPT 2001. LNCS, vol. 2248, pp. 67–86. Springer, Heidelberg (2001). doi:10.1007/3-540-45682-1_5

[LLL82] Lenstra, A.K., Lenstra, H.W., Lovász, L.: Factoring polynomials with rational coefficients. Math. Ann. **261**(4), 515–534 (1982)

[Mat06] Matyukhin, D.: Effective version of the number field sieve for discrete logarithms in the field GF(p^k). Trudy po Discretnoi Matematike **9**, 121–151 (2006). (in Russian)

[Mil04] Miller, V.S.: The weil pairing, and its efficient calculation. J. Cryptol. **17**(4), 235–261 (2004)

[MNT01] Miyaji, A., Nakabayashi, M., Takano, S.: New explicit conditions of elliptic curve traces for FR-reduction. IEICE Trans. Fundam. Electron. Commun. Comput. Sci. **84**(5), 1234–1243 (2001)

[Mor91] Morain, F.: Building cyclic elliptic curves modulo large primes. In: Davies, D.W. (ed.) EUROCRYPT 1991. LNCS, vol. 547, pp. 328–336. Springer, Heidelberg (1991). doi:10.1007/3-540-46416-6_28

[MOV93] Menezes, A.J., Okamoto, T., Vanstone, S.A.: Reducing elliptic curve logarithms to logarithms in a finite field. IEEE Trans. Inf. Theory **39**(5), 1639–1646 (1993)

[Pol78] Pollard, J.: Monte carlo methods for index computation (mod p). Math. Comput. **32**(143), 918–924 (1978)

[PV06] Page, D., Vercauteren, F.: A fault attack on pairing-based cryptography. IEEE Trans. Comput. **55**(9), 1075–1080 (2006)

[Sch00] Schirokauer, O.: Using number fields to compute logarithms in finite fields. Math. Comput. **69**(231), 1267–1283 (2000)

[Sem02] Semaev, I.: Special prime numbers and discrete logs in finite prime fields. Math. Comput. **71**(237), 363–377 (2002)

[Sil07] Silverman, J.H.: The Arithmetic of Dynamical Systems, vol. 241. Springer Science & Business Media, Heidelberg (2007)

[SS16] Sarkar, P., Singh, S.: New complexity trade-offs for the (multiple) number field sieve algorithm in non-prime fields. In: Fischlin, M., Coron, J.-S. (eds.) EUROCRYPT 2016. LNCS, vol. 9665, pp. 429–458. Springer, Heidelberg (2016). doi:10.1007/978-3-662-49890-3_17

[Wei40] Weil, A.: Sur les fonctions algébriquesa corps de constantes fini. CR Acad. Sci. Paris **210**(1940), 592–594 (1940)

[Wie86] Wiedemann, D.: Solving sparse linear equations over finite fields. IEEE Trans. Inform. Theory **32**(1), 54–62 (1986)

[Zaj10] Zajac, P.: On the use of the lattice sieve in the 3D NFS. Tatra Mt. Math. Publ. **45**(1), 161–172 (2010)

Elliptic Curves

Differential Addition on Binary Elliptic Curves

Reza Rezaeian Farashahi[1,2](✉) and Seyed Gholamhossein Hosseini[1]

[1] Department of Mathematical Sciences, Isfahan University of Technology, Isfahan 84156-83111, Iran
farashahi@cc.iut.ac.ir, g.hoseini@math.iut.ac.ir

[2] School of Mathematics, Institute for Research in Fundamental Sciences (IPM), P.O. Box 19395-5746, Tehran, Iran

Abstract. This paper presents extremely fast differential addition (i.e., the addition of two points with the known difference) and doubling formulas, as the core step in Montgomery scalar multiplication, for various forms of elliptic curves over binary fields. The formulas are provided for *binary Edwards*, *binary Hessian* and *binary Huff* elliptic curves with cost of $5\mathbf{M} + 4\mathbf{S} + 1\mathbf{D}$ when the given difference point is in affine form. Here, **M**, **S**, **D** denote the costs of a field multiplication, a field squaring and a field multiplication by a constant, respectively. This paper also presents, new *complete* differential addition formulas for *binary Edwards* curves with cost of $5\mathbf{M} + 4\mathbf{S} + 2\mathbf{D}$.

Keywords: Elliptic curves · Binary Edwards curves · Hessian curves · Binary Huff curves · Differential addition

1 Introduction

An elliptic curve E over a field $\mathbb{F}$ can be given by the Weiersrasß equation

$$y^2 + a_1xy + a_3y = x^3 + a_2x^2 + a_4x + a_6$$

where coefficients a_1, a_2, a_3, a_4 and a_6 are in $\mathbb{F}$. There are many other ways to represent elliptic curves such as Legendre equation, cubic equations, quartic equations and intersection of two quadratic surfaces [18]. The use of elliptic curves over finite fields based on their finite groups in cryptography (ECC) was independently proposed in the mid 1980s by Koblitz [11] and Miler [14]. Since the introduction of elliptic curve cryptography many proposals have been made to speed up the group arithmetic. Efficient arithmetic (addition, doubling, tripling and scalar multiplication) on elliptic curves over finite fields is the core requiromont of elliptic curve cryptography. Several forms of elliptic curves over finite fields with several coordinate systems have been studied to improve the efficiency and the speed of the arithmetic on the group law.

S. Duquesne and S. Petkova-Nikova (Eds.): WAIFI 2016, LNCS 10064, pp. 21–35, 2016.
DOI: 10.1007/978-3-319-55227-9_2

Elliptic curves over binary finite fields are interesting particularly for hardware implementations. Every ordinary elliptic curve over the binary finite filed $\mathbb{F}_{2^m}$ can be represented in the Weierstraß form

$$y^2 + xy = x^3 + ax^2 + b,$$

where $a, b \in \mathbb{F}_{2^m}$ and $b \neq 0$. There are alternative ways to represent binary elliptic curves such as binary Hessian [1,5,6,17], binary Edwards [3], binary Huff curves [9] and binary μ_4-normal forms [12].

The scalar multiplication is the most important operation of elliptic curve cryptography. That is to compute kP for a given point P on elliptic curve E defined over a finite field $\mathbb{F}_q$ and a given integer k. The scalar multiplication can be performed by a sequence of point additions and point doublings. Speed and efficiency are the main factors to be considered in the correct implementing of scalar multiplication. Moreover, the implementations should be performed in a way to be resistant against passive and active side channel attacks. There are several mathematical countermeasures proposed for preventing these attacks. Simple side-channel attacks get information from a single scalar multiplication when the power trace reveal distinctive key dependent patterns. The main idea of the countermeasure against simple side-channel attacks is to make the computation uniform. And the main solutions are making indistinguishable point addition and point doubling, using double and add always method, using window method or applying the Montgomery technique.

The Montgomery method [15,16] is introduced for scalar multiplication of points for a special type of curve in large characteristic. This method has been extended to other form of elliptic curves and to binary elliptic curves [8]. The Montgomery scalar multiplication is known also as Montgomery ladder. In the Montgomery ladder, for each bit of the scalar both doubling and addition are performed, so this prevents the computation secure against simple power analysis. Also this method is not subject to fault attacks.

The countermeasures for some other passive or active attacks are to insert suitable randomness to the key and also to the base point of the scalar multiplication. Therefore, here the scalar key may be larger than the order of the base point, which makes some exceptional cases like the point at infinity in the computation of the Montgomery ladder. Thus, obtaining *complete* or *almost complete* formulas for addition and doubling makes the ladder performs completely.

In this paper we present fast explicit formulas for differential additions and doublings on well known binary elliptic curves such as binary Edwards, binary Hessian and binary Huff curves.

2 Differential Addition

A Montgomery curve over a field $\mathbb{F}$ of characteristic different from 2 is given by the equation

$$bY^2Z = X^3 + aX^2Z + XZ^2,$$

where a, b are elements of $\mathbb{F}$ with $b(a^2 - 4) \neq 0$. The Montgomery ladder for scalar multiplication is performed by a sequence of simultaneous point addition and doubling, which makes this method interesting against side-channel attacks. In Montgomery curves, the basic computation in a each step is done without the Y coordinate, i.e., the technique involves special formulas for addition and doubling that relies on only the X and Z coordinates of a point in projective form. Also, the Y coordinate of the output point can be derived from the X and Z coordinates.

In general, the basic computation in a each step of the Montgomery ladder is differential addition and doubling. That is for given points P_1, P_2 and $P_1 - P_2$ on elliptic curve E over $\mathbb{F}_q$ to compute $P_1 + P_2$ and $2P_1$. The idea is extended by a suitable rational function on the elliptic curve. Suppose w is a rational function defined over an elliptic curve E over a finite field $\mathbb{F}_q$. The function w is given by fraction of polynomials in the coordinate ring of E over $\mathbb{F}_q$. Let $w(P) = w(-P)$ for any point P on $E(\mathbb{F}_q)$. Then the w-coordinate *differential addition* and *doubling* means to compute $w(2P_1)$ and $w(P_1+P_2)$ from given values $w(P_1)$, $w(P_2)$ and $w(P_1-P_2)$, where P_1, P_2 are points on $E(\mathbb{F}_q)$. For Montgomery curves the function w is x, where $w(P)$ equals the x-coordinate of the point P. Since field inversion is costly, practically computations are performed where points are represented in projective coordinates. Therefore, when w is regular at the point P then $w(P)$ is represented by $(w(P) : 1)$ in the projective line $\mathbb{P}(\mathbb{F}_q)$. Otherwise, it is represented by $(1 : 0)$. The projective w-coordinate differential addition and doubling (dADD) algorithm is given in Algorithm 1. Notice, in Algorithm 1, the given input values $w(P_1)$, $w(P_2)$ and $w(P_0) = w(P_1 - P_2)$ are represented by W_i/Z_i where $i = 1, 2, 0$ respectively. Then $w(P_1 + P_2)$, i.e. the w-coordinate differential addition, is given by $\frac{f_a}{g_a}$ with some homogenous polynomials f_a and g_a in variables W_i, Z_i, where $i = 0, 1, 2$. Also, $w(2P_1)$ is given by $\frac{f_d}{g_d}$, where f_d and g_d are homogenous polynomials with variables W_1, Z_1.

Algorithm 1. Projective w-coordinate dADD

Input : E/F_q, $w : E(\mathbb{F}_q) \to \mathbb{P}(\mathbb{F}_q)$, ▷ The elliptic curve E over $\mathbb{F}_q$
$(W_i : Z_i) = w(P_i)$, $i = 0, 1, 2$. ▷ $w(P_0) = w(P_1 - P_2)$
Output : $(W_i : Z_i) = w(P_i)$, $i = 3, 4$. ▷ $w(P_3) = w(P_1 + P_2)$, $w(P_4) = w(2P_1)$

```
function dADD((W0 : Z0), (W1 : Z1), (W2 : Z2))
    W3 = fa(W0, Z0, W1, Z1, W2, Z2)        ▷ Differential addition computation
    Z3 = ga(W0, Z0, W1, Z1, W2, Z2)
    W4 = fd(W1, Z1)                         ▷ Doubling computation
    Z4 = gd(W1, Z1)
    return ((W4 : Z4), (W3 : Z3))           ▷ The differential addition and doubling
end function
```

The Montgomery scalar multiplication based on a projective w-coordinate dADD is given in Algorithm 2. Notice, the base point P can be considered such that one of the coordinates of $w(P)$ equals 1, which makes less field operation computation in each step of the ladder.

Algorithm 2. The Montgomery scalar multiplication

Input : E/F_q, $w : E(\mathbb{F}_q) \rightarrow \mathbb{P}(\mathbb{F}_q)$, ▷ The elliptic curve E over $\mathbb{F}_q$
Projective w-coordinate dADD funtion,
$P \in E(\mathbb{F}_q)$, $k = (k_{m-1}, \cdots, k_1, k_0)$ ▷ k is a positive integer, $k_{m-1} = 1$
$(W_0 : Z_0) := w(P)$, $(W_1 : Z_1) := w(P)$, $(W_2 : Z_2) := w(2P)$.
Output : $w(kP)$

for $i := m-2$ **down to** 0 **do**
  if $k_i = 0$ **then**
    $((W_1 : Z_1), (W_2 : Z_2)) := dADD((W_0 : Z_0), (W_1 : Z_1), (W_2 : Z_2))$
  else
    $((W_2 : Z_2), (W_1 : Z_1)) := dADD((W_0 : Z_0), (W_2 : Z_2), (W_1 : Z_1))$
  end if
end for
return $(W_1 : Z_1)$, $(W_2 : Z_2)$ ▷ The differential addition and doubling

Note that if there are some exceptional points where the function dADD is not computed correctly, then the Montgomery ladder does not work properly. We say that the differential w-coordinate is *complete* if the Algorithm 1 works for any input without any exception. We also say that the function dADD is *almost complete* if the Algorithm 1 works for all inputs except for the case where $w(P_0)$ equals $w(\mathcal{O})$, where $\mathcal{O}$ is the neutral element of the group of points $E(\mathbb{F}_q)$. Therefore, for the complete function dADD the Montgomery ladder is performed without any problem for any input. Moreover, for the almost complete function dADD the Montgomery ladder works for any base point P except for the points where $w(P)$ equals $w(\mathcal{O})$. Notice, the almost complete function is also suitable for cryptographic application.

In this paper, we concentrate on differential addition on binary elliptic curves. Let E be a binary elliptic curve over $\mathbb{F}_{2^m}$ in Weiersraß form

$$y^2 + xy = x^3 + ax^2 + b,$$

where a, b are in $\mathbb{F}_{2^m}$. Lopez and Dahab [13] presented the projective formulas for the addition and doubling of points on E. And, they generalized the Montgomery's idea to binary curves. Algorithm 3 provides the Lopez and Dahab differential x-coordinate on elliptic curve E over $\mathbb{F}_{2^m}$.

If we assume $Z_0 = 1$, then the Lopez and Dahab formulas are computed using $5\mathbf{M} + 4\mathbf{S} + 1\mathbf{D}$. Here, a multiplication in $\mathbb{F}_q$ costs one $\mathbf{M}$ and a squaring costs one $\mathbf{S}$. Also the cost of field multiplication by a parameter (as a constant) is denoted by $\mathbf{D}$.

We note, that the point at infinity on the binary elliptic curve E over $\mathbb{F}_{2^m}$ is $\mathcal{O} = (0 : 1 : 0)$ and $x(\mathcal{O})$ is represented by $(1 : 0)$. One can easily check that the projective x-coordinate formulas work for all inputs if $Z_0 \neq 0$, that is where $P_0 \neq \mathcal{O}$. In other words the formulas are almost complete and the Montgomery ladder works for all inputs if the base point is not the point at infinity. So, the

Algorithm 3. Lopez and Dahab projective x-coordinate dADD

Input : $E/F_q : y^2 + xy = x^3 + ax^2 + b$ ▷ The elliptic curve E over $\mathbb{F}_{2^m}$
$(X_i : Z_i) = x(P_i),\ i = 0, 1, 2.$ ▷ $x(P_0) = x(P_1 - P_2)$
Output : $(X_i : Z_i) = x(P_i),\ i = 3, 4.$ ▷ $x(P_3) = x(P_1 + P_2),\ x(P_4) = x(2P_1)$

function DADD($(X_0 : Z_0),\ (X_1 : Z_1),\ (X_2 : Z_2)$)
  $X_3 = X_0\,(X_1 Z_2 + X_2 Z_1)^2 + Z_0\,(X_1 Z_1 X_2 Z_2)$
  $Z_3 = Z_0\,(X_1 Z_2 + X_2 Z_1)^2$
  $X_4 = (X_1^4 + bZ_1^4)$
  $Z_4 = X_1^2\, Z_1^2$
  return $((X_4 : Z_4),\ (X_3 : Z_3))$ ▷ The differential addition and doubling
end function

Montgomery ladder can be modified as Algorithm 4. Here there is no need to assume that the bit k_{m-1} of the integer k is equal to '1'. Also, there is no need to precompute $2P$ from the base point P. Moreover, the ladder works properly even if the integer k is bigger than the order of the base point P. So, for Lopez and Dahab formulas, one can use random scalar k as a countermeasure to protect against differential power analysis attack.

Algorithm 4. The modified Montgomery scalar multiplication

Input : $E/F_q : y^2 + xy = x^3 + ax^2 + b$ ▷ The elliptic curve E over $\mathbb{F}_q$
$P = (x : y : z) \in E(\mathbb{F}_q)$ ▷ $P \neq \mathcal{O} = (0 : 1 : 0)$
$k = (k_{m-1}, \cdots, k_1, k_0)$ ▷ $0 \leq k \in \mathbb{Z}$
$(X_0 : Z_0) := (x : z),\ (X_1 : Z_1) := (1 : 0),\ (X_2 : Z_2) := (x : z).$
Output : $w(kP)$

for $i := m - 1$ **down to** 0 **do**
  if $k_i = 0$ **then**
    $((X_1 : Z_1), (X_2 : Z_2)) := dADD((X_0 : Z_0), (X_1 : Z_1), (X_2 : Z_2))$
  else
    $((X_2 : Z_2), (X_1 : Z_1)) := dADD((X_0 : Z_0), (X_2 : Z_2), (X_1 : Z_1))$
  end if
end for
return $(X_1 : Z_1),\ (X_2 : Z_2)$ ▷ The differential addition and doubling

3 Binary Edwards Curves

In this section we review the *Binary Edwards curve* [3] and propose new differential addition and doubling formulas.

Let d_1, d_2 be elements of $\mathbb{F}_{2^m}$ such that $d_1 \neq 0$ and $d_2 \neq d_1(d_1 + 1)$. The binary Edwards curve with parameters d_1 and d_2 is given by the equation

$$E_{B,d_1,d_2} : d_1(x + y) + d_2(x + y)^2 = xy(x + 1)(y + 1). \tag{1}$$

The curve is symmetric in x, y and the negation of (x, y) is (y, x). This curve has two points $(0, 0)$ and $(1, 1)$ which are invariant under the negation law. The point $(0, 0)$ is the neutral element of the addition law and the point $(1, 1)$ has order 2. We denote the point $(0, 0)$ by $\mathcal{O}$.

The binary Edwards curve E_{B,d_1,d_2} is birationally equivalent to the ordinary elliptic curve in Weierstraß form

$$v^2 + uv = u^3 + au^2 + b,$$

where a, b are in $\mathbb{F}_{2^m}$ with $b \neq 0$. The map $(x, y) \longmapsto (u, v)$ defined by

$$u = ((d_1^3 + d_1^2 + d_1d_2)(x + y))/(xy + d_1(x + y))$$

$$v = (d_1^3 + d_1^2 + d_1d_2)(d_1 + 1 + x/(xy + d_1(x + y))$$

is a birational equivalence form E_{B,d_1,d_2} to the elliptic curve

$$v^2 + uv = u^3 + (d_1^2 + d_2)u^2 + d_1^4(d_1^4 + d1^2 + d_2).$$

Affine Addition. The sum of two points (x_1, y_1) and (x_2, y_2) on E_{B,d_1,d_2} is the point (x_3, y_3) defined as follows:

$$x_3 = \frac{d_1(x_1 + x_2) + d_2(x_1 + y_1)(x_2 + y_2) + (x_1 + x_1^2)(x_2(y_1 + y_2 + 1) + y_1y_2)}{d_1 + (x_1 + x_1^2)(x_2 + y_2)}, \tag{2}$$

$$y_3 = \frac{d_1(y_1 + y_2) + d_2(x_1 + y_1)(x_2 + y_2) + (y_1 + y_1^2)(y_2(x_1 + x_2 + 1) + x_1x_2)}{d_1 + (y_1 + y_1^2)(x_2 + y_2)}.$$

Affine Doubling. The doubling of point (x_1, y_1) is the point (x_4, y_4) defined as follows:

$$x_4 = 1 + \frac{d_1 + d_2(x_1^2 + y_1^2) + y_1^2 + y_1^4}{d_1 + (x_1^2 + y_1^2) + d_2/d_1(x_1^4 + y_1^4)}, \tag{3}$$

$$y_4 = 1 + \frac{d_1 + d_2(x_1^2 + y_1^2) + x_1^2 + x_1^4}{d_1 + (x_1^2 + y_1^2) + d_2/d_1(x_1^4 + y_1^4)}.$$

Differential Addition. Bernstein, Lange and Farashahi in [3] proposed the differential addition and doubling formulas for binary Edwards curve. Assume that $P = (x_1, y_1)$, $Q = (x_2, y_2)$ are points on E_{B,d_1,d_2} and $Q - P = (x_0, y_0)$, $Q + P = (x_3, y_3)$ and $2P = (x_4, y_4)$. They considered w-function as $w(x_i, y_i) = x_i + y_i$ and obtained the following complete formulas for differential addition:

$$w_4 = \frac{w_1^2 + w_1^4}{d_1 + w_1^2 + (d_2/d_1)w_1^4},$$

$$w_3 + w_0 = \frac{d_1w_1w_2(1 + w_1)(1 + w_2)}{d_1^2 + w_1w_2(d_1(1 + w_1 + w_2) + d_2(w_1w_2))},$$

$$w_3w_0 = \frac{d_1^2(w_1^2 + w_2^2)}{d_1^2 + w_1w_2(d_1(1 + w_1 + w_2) + d_2(w_1w_2))}.$$

Assume that w_0 is given as a field element, and w_1, w_2 are given as fractions W_1/Z_1, W_2/Z_2 and w_4, w_3 are outputs as fractions W_4/Z_4 and W_3/Z_3. Then, the mixed projective w-coordinate differential addition and doubling formulas are given as follows.

$$A = W_1(W_1 + Z_1), \quad B = W_2(W_2 + Z_2), \quad C = Z_1 Z_2, \quad D = W_1 W_2, \quad E = AB,$$
$$F = E + (\sqrt{d_1}C + \sqrt{d_2/d_1 + 1}D)^2,$$
$$W_4 = A^2, \quad Z_4 = W_4 + ((\sqrt[4]{d_1}Z_1 + \sqrt[4]{d_2/d_1 + 1}W_1)^2)^2,$$
$$Z_3 = F, \; W_3 = E + w_0 F.$$

From above formulas, for the general case $d_1 \neq d_2$, the cost of differential addition is $6\mathbf{M}+1\mathbf{S}+2\mathbf{D}$ and the cost of doubling is $1\mathbf{M}+3\mathbf{S}+2\mathbf{D}$. And the total cost is $6\mathbf{M}+4\mathbf{S}+4\mathbf{D}$. If $d_1 = d_2$ the total cost is $5\mathbf{M}+4\mathbf{S}+2\mathbf{D}$. Recently Kim, Lee and Negre [10] for the case $d_1 = d_2$, by using the co-Z trick improved the differential addition formulas by $1\mathbf{D}$ and obtained almost complete differential addition formulas with cost of $5\mathbf{M}+4\mathbf{S}+1\mathbf{D}$.

New Differential Addition. In this section, we consider *binary Edwards curves* in general form and present two new w-coordinates differential formulas where one of this formulas is complete and the other is almost complete.

Let define the rational function w by $w(x,y) = (x+y)/(d_1(x+y+1))$. The function is well computed for all affine points on a binary Edwards curve except for the points (x,y) where $x+y=1$. Since $-(x,y)=(y,x)$, for all points P on the curve, we have $w(P) = w(-P)$. Also, we have $w(\mathcal{O}) = 0$. As before, for $i = 0,1,2,3,4$, let $w_i = w(P_i)$, where $P_i \in E_{B,d_1,d_2}(\mathbb{F}_{2^m})$ with $w(P_0) = w(P_1 - P_2)$, $w(P_3) = w(P_1 + P_2)$ and $w(P_4) = w(2P_1)$. From the addition formula (2), with a straightforward calculation, we obtain the following differential addition formulas.

$$w_3 + w_0 = \frac{w_1 w_2}{d_1^2(d_1^2 + d_1 + d_2)w_1^2 w_2^2 + 1}, \tag{4}$$

$$w_3 w_0 = \frac{w_1^2 + w_2^2}{d_1^2(d_1^2 + d_1 + d_2)w_1^2 w_2^2 + 1}. \tag{5}$$

Also, from the doubling formula (3) and some calculations we obtain

$$w_4 = \frac{w_1^2}{d_1^2(d_1^2 + d_1 + d_2)w_1^4 + 1}. \tag{6}$$

We recall [3] that the binary Edwards curve E_{B,d_1,d_2} over $\mathbb{F}_{2^m}$ is complete if $\mathrm{Tr}(d_2) = 1$. Here Tr is the trace function from $\mathbb{F}_{2^m}$ to $\mathbb{F}_2$. Moreover, if $\mathrm{Tr}(d_1) = 0$ then there is no point (x,y) on the curve with $x+y+1=0$. Since, if there is a point (x,y) with $x+y+1=0$ on the curve E_{B,d_1,d_2} with $\mathrm{Tr}(d_2)=1$ and $\mathrm{Tr}(d_1)=0$, then by the curve Eq. (1), we have $x^4 + x^2 + d_1 + d_2 = 0$. Then,

$$\begin{aligned}
\mathrm{Tr}(0) &= \mathrm{Tr}(x^4 + x^2 + d_1 + d_2) \\
&= \mathrm{Tr}(x^4) + \mathrm{Tr}(x^2) + \mathrm{Tr}(d_1) + \mathrm{Tr}(d_2) \\
&= \mathrm{Tr}(x^2) + \mathrm{Tr}(x^2) + 0 + 1 = 1,
\end{aligned}$$

which is a contradiction. Therefore, the function w is well defined for all affine points on the complete binary Edwards curve E_{B,d_1,d_2} with $\mathrm{Tr}(d_1) = 0$.

Notice, the set of affine $\mathbb{F}_{2^m}$-rational points of the complete binary Edwards curve E_{B,d_1,d_2} is an abelian group. And, with the condition $\mathrm{Tr}(d_1) = 0$, for any point $P = (x, y)$ on the curve, the value $w(P)$ is well computed and belongs to $\mathbb{F}_{2^m}$. By the Eqs. (4) and (6) we have

$$(w_3 + w_0)(d_1^2(d_1^2 + d_1 + d_2)w_1^2 w_2^2 + 1) = w_1 w_2 ,$$
$$w_4(d_1^2(d_1^2 + d_1 + d_2)w_1^4 + 1) = w_1^2.$$

So, we see that if $\mathrm{Tr}(d_1) = 0$ then the denominators of Eqs. (4) and (6) never equal zero. In other words, above w-coordinates differential addition and doubling formulas for complete binary Edwards curve are complete where $\mathrm{Tr}(d_1) = 0$.

For further speedup, we can divide the Eq. (4) by Eq. (5) and obtain the following faster formula.

$$\frac{1}{w_3} + \frac{1}{w_0} = \frac{w_1 w_2}{(w_1 + w_2)^2}. \tag{7}$$

Cost of Projective w-Coordinates. Using Eqs. (4) and (6), we obtained new and complete differential addition formulas for general *binary Edwards curves* with the total cost of $5\mathbf{M} + 4\mathbf{S} + 2\mathbf{D}$ where the difference of input points is affine. Then, by using the Eqs. (6) and (7) we obtain, new and fast, but almost complete, differential addition formulas in mixed projective coordinates with the total cost of $5\mathbf{M} + 4\mathbf{S} + 1\mathbf{D}$. Thus, the total cost of differential addition and doubling in general binary Edwards curves is reduced from $6\mathbf{M} + 4\mathbf{S} + 4\mathbf{D}$ to $5\mathbf{M} + 4\mathbf{S} + 1\mathbf{D}$.

As before assume that w_0 is given as a field element, and w_1, w_2 are given as fractions W_1/Z_1, W_2/Z_2 and w_4, w_3 are to be output as fraction W_4/Z_4 and W_3/Z_3. From Eq. (6) the explicit doubling formula is given by

$$\frac{W_4}{Z_4} = \frac{W_1^2 Z_1^2}{(d_1^4 + d_1^3 + d_1^2 d_2)W_1^4 + Z_1^4} \tag{8}$$

and from Eq. (4) the explicit addition formula is given by

$$\frac{W_3}{Z_3} = \frac{W_0((d_1^4 + d_1^3 + d_1^2 d_2)\, W_1^2 W_2^2 + Z_1^2 Z_2^2) + Z_0(W_1 W_2 Z_1 Z_2)}{Z_0((d_1^4 + d_1^3 + d_1^2 d_2)\, W_1^2 W_2^2 + Z_1^2 Z_2^2))}. \tag{9}$$

So, from the Eqs. (8) and (9), the cost of projective w-coordinates is $7\mathbf{M} + 4\mathbf{S} + 2\mathbf{D}$. If we set $Z_0 = 1$, then the mixed projective w-coordinates differential addition and doubling formulas have the total cost $5\mathbf{M} + 4\mathbf{S} + 2\mathbf{D}$ as follows.

$$\begin{aligned} &A = W_1 Z_1, \quad B = W_1 W_2, \quad C = Z_1 Z_2, \\ &W_4 = A^2, \quad Z_4 = (\sqrt[4]{(d_1^4 + d_1^3 + d_1^2 d_2)}W_1 + Z_1)^4, \\ &Z_3 = (\sqrt{(d_1^4 + d_1^3 + d_1^2 d_2)}B + C)^2, \quad W_3 = BC + w_0 Z_3. \end{aligned} \tag{10}$$

From Eq. (7), we also obtain the following explicit projective differential addition formulas.

$$\frac{Z_3}{W_3} = \frac{Z_0(W_1Z_2 + W_2Z_1)^2 + W_0(W_1Z_2W_2Z_1)}{W_0(W_1Z_2 + W_2Z_1)^2}. \tag{11}$$

Thus, by Eqs. (8) and (11), the cost of projective w-coordinates is $7\mathbf{M} + 4\mathbf{S} + 2\mathbf{D}$. If we set $W_0 = 1$ and using the mixed projective coordinates we have the following formulas for computing differential addition.

$$\begin{aligned} &A = W_1Z_1, \quad B = W_1Z_2, \quad C = W_2Z_1, \\ &W_4 = A^2, \quad Z_4 = (\sqrt[4]{(d_1^4 + d_1^3 + d_1^2 d_2)}W_1 + Z_1)^4, \\ &W_3 = (B + C)^2, \quad Z_3 = BC + z_0 W_3. \end{aligned} \tag{12}$$

From differential addition and doubling formulas (12), the costs of differential addition and doubling are $4\mathbf{M} + 1\mathbf{S}$, $1\mathbf{M} + 3\mathbf{S} + 1\mathbf{D}$ respectively. And, the total cost is $5\mathbf{M} + 4\mathbf{S} + 1\mathbf{D}$.

The binary Edwards curve E_{B,d_1,d_2}, has the neutral element $\mathcal{O}$ represented by w-coordinate as $(0 : 1)$. For the complete binary Edwards curve E_{B,d_1,d_2} with $\mathrm{Tr}(d_1) = 0$, any point P on the curve can be represented by $(w(P) : 1)$. In other words, for any w-coordinate representation of the point P by $(W : Z)$ we have $Z \neq 0$. So, from the completeness of the affine w-coordinates differential addition and doubling formulas for complete binary Edwards curve with $\mathrm{Tr}(d_1) = 0$, we deduce that the projective w-coordinates differential addition and doubling formulas (8) and (9) are also complete. The mixed projective formulas (10) have the cost of $5\mathbf{M} + 4\mathbf{S} + 2\mathbf{D}$. Furthermore, the projective w-coordinates differential addition and doubling formulas (8) and (11) are almost complete; the exceptional cases are points P_0 where $w(P_0) = w(\mathcal{O})$. The mixed projective formulas (12) have the cost of $5\mathbf{M} + 4\mathbf{S} + 1\mathbf{D}$.

4 Binary Hessian Curve

A *Hessian curve* over a field $\mathbb{F}_{2^m}$ is given by the cubic equation

$$\mathrm{H}_d: \; x^3 + y^3 + 1 + dxy = 0,$$

for some $d \in \mathbb{F}_{2^m}$ with $d^3 \neq 27$ [5]. The family is extended to the family of generalized Hessian [5] or twisted Hessian curves [1]. A generalized Hessian curve $\mathrm{H}_{c,d}$ over $\mathbb{F}_{2^m}$ is defined by the equation

$$\mathrm{H}_{c,d} : x^3 + y^3 + c + dxy = 0,$$

where c, d are elements of $\mathbb{F}_{2^m}$ such that $c \neq 0$ and $d^3 \neq 27c$. The projective closure of the curve $\mathrm{H}_{c,d}$ is

$$\mathbf{H}_{c,d}: \; X^3 + Y^3 + cZ^3 = dXYZ.$$

It has the points $(1 : \omega : 0)$ with $\omega^3 = 1$ at infinity. The neutral element of the group of $\mathbb{F}_{2^m}$-rational points of $\mathbf{H}_{c,d}$ is the point at infinity $(1 : 1 : 0)$ that is denoted by $\mathcal{O}$. And, the negation of point $(X : Y : Z)$ is $(Y : X : Z)$.

Affine Addition. The sum of two different points (x_1, y_1), (x_2, y_2) on $\mathrm{H}_{c,d}$ is the point (x_3, y_3) given by

$$x_3 = \frac{{y_1}^2 x_2 + {y_2}^2 x_1}{x_2 y_2 + x_1 y_1} \quad \text{and} \quad y_3 = \frac{{x_1}^2 y_2 + {x_2}^2 y_1}{x_2 y_2 + x_1 y_1}.$$

Affine Doubling. The doubling of the point (x_1, y_1) on $\mathrm{H}_{c,d}$ is the point (x_4, y_4) given by

$$x_4 = \frac{y_1(c + {x_1}^3)}{{x_1}^3 + {y_1}^3} \quad \text{and} \quad y_4 = \frac{x_1(c + {y_1}^3)}{{x_1}^3 + {y_1}^3}.$$

Differential Addition. Farashahi and Joye in [5] adapted differential addition formulas for the binary curve $\mathbf{H}_{c,d}$. They defined the rational function $w(x, y) = x^3 + y^3$. As before, for $i = 0, 1, 2, 3, 4$, let $w_i = w(P_i)$, where P_i are points of $\mathrm{H}_{c,d}(\mathbb{F}_{2^m})$ with $w(P_0) = w(P_1 - P_2), w(P_3) = w(P_1 + P_2)$ and $w(P_4) = w(2P_1)$. From [5], we have

$$w_4 = \frac{{w_1}^4 + c^3(d^3 + c)}{d^3 {w_1}^2}, \tag{13}$$

$$w_0 + w_3 = \frac{d^3 w_1 w_2}{(w_1 + w_2)^2} \quad \text{and} \quad w_0 w_3 = \frac{{w_1}^2 {w_2}^2 + c^3(d^3 + c)}{(w_1 + w_2)^2}. \tag{14}$$

To have mixed projective formulas, w_i are given by the fractions W_i/Z_i for $i = 0, 1, 2, 3$ where $Z_0 = 1$. The following explicit formulas give the output w_3 defined by W_3/Z_3:

$$A = W_1 Z_2,\ B = W_2 Z_1,\ C = AB,\ U = d^3 C,\ V = (A + B)^2,$$
$$Z_3 = V,\ W_3 = U + w_0 V.$$

Moreover, we write w_4 by the fraction W_4/Z_4. Then, the explicit doubling formulas is

$$A = {W_1}^2,\ B = {Z_1}^2,\ C = A + \sqrt{c^3(d^3 + c)}B,\ D = d^3 B,$$
$$W_4 = C^2,\ Z_4 = AD.$$

The cost of these mixed w-coordinates is $4\mathbf{M} + 1\mathbf{S} + 1\mathbf{D}$ for addition and $1\mathbf{M} + 3\mathbf{S} + 2\mathbf{D}$ for doubling and the total cost is $5\mathbf{M} + 4\mathbf{S} + 2\mathbf{D}$.

New Differential Addition. In this section we present two new differential addition formulas for *generalized Hessian* curve over binary field $\mathbb{F}_{2^m}$ with total cost of $5\mathbf{M} + 4\mathbf{S} + 1\mathbf{D}$ for both doubling and addition.

We modify the definition of the above rational function w, [5], and consider $w(x, y) = \frac{x^3 + y^3}{d^3}$. Using the differential addition formulas (14), by a straightforward calculations, we obtain the following formulas in affine coordinates.

$$w_3 + w_0 = \frac{w_1 w_2}{w_1^2 + w_2^2}, \tag{15}$$

$$w_3 w_0 = \frac{w_1^2 w_2^2 + (c^4 + c^3 d^3)/(d^{12})}{w_1^2 + w_2^2}. \tag{16}$$

Also, from the doubling formula (13), the following doubling formula is obtained.

$$w_4 = \frac{w_1^4 + (c^4 + c^3 d^3)/(d^{12})}{w_1^2}. \tag{17}$$

Cost of Projective w-Coordinates. To obtain the projective formulas, assume that w_i are given by the fractions W_i/Z_i for $i = 0, 1, 2, 3, 4$. From Eq. (15) the following explicit formulas give W_3/Z_3 by

$$\frac{W_3}{Z_3} = \frac{W_0(W_1 Z_2 + W_2 Z_1)^2 + Z_0(W_1 Z_2 W_2 Z_1)}{Z_0(W_1 Z_2 + W_2 Z_1)^2}. \tag{18}$$

Also, from Eq. (17), the doubling is given by

$$\frac{W_4}{Z_4} = \frac{W_1^4 + (c^4 + c^3 d^3)/(d^{12})\ Z_1^4}{W_1^2 Z_1^2}. \tag{19}$$

The cost of projective w-coordinates differential addition and doubling is $7\mathbf{M}+4\mathbf{S}+1\mathbf{D}$; see Eqs. (18) and (19). If we set $Z_0 = 1$ then we have the following mixed projective coordinates formulas with the total cost $5\mathbf{M} + 4\mathbf{S} + 1\mathbf{D}$.

$$A = W_1 Z_1, \quad B = W_1 Z_2, \quad C = W_2 Z_1$$
$$W_4 = (W_1 + \sqrt[4]{(c^4 + c^3 d^3)/d^{12}}\ Z_1)^4, \quad Z_4 = A^2,$$
$$Z_3 = (B + C)^2, \quad W_3 = BC + w_0 Z_3.$$

Here, the differential addition formulas use $4\mathbf{M} + 1\mathbf{S}$ and doubling formulas use $1\mathbf{M}+3\mathbf{S}+1\mathbf{D}$ and the total cost is $5\mathbf{M}+4\mathbf{S}+1\mathbf{D}$. So the computation of $1\mathbf{D}$ is saved. Notice, the projective w-coordinate differential addition and doubling formulas (18) and (19) are almost complete; the exceptional points are 3 torsion points P_0 where $w(P_0) = w(\mathcal{O}) = (1:0)$.

5 Binary Huff Curves

Huff model at first introduced by Huff [7] in 1948 to study a diophantine problem. Huff model are extended over fields of odd characteristic. Joye et al. [9], extended the Huff model and also introduced the binary partner for Huff curve. In 2011 Devigen and Joye [4] described the addition law for *Binary Huff* curve and compute formulas for addition, doubling and differential addition which the cost of their differential addition and doubling is $5\mathbf{M}+5\mathbf{S}+1\mathbf{D}$. Here, we improve their results to the cost of $5\mathbf{M} + 4\mathbf{S} + 1\mathbf{D}$.

The *binary Huff* curve is given by the equation

$$HF_{a,b} \ : \ ax(y^2 + y + 1) = by(x^2 + x + 1), \tag{20}$$

where a, b are in $\mathbb{F}_{2^m}$ such that $a, b \neq 0$ and $a \neq b$. This curve have three points at infinity, namely $(a : b : 0)$, $(1 : 0 : 0)$ and $(0 : 1 : 0)$. Binary Huff curve is birationally equivalent to the Weierstraß elliptic curve

$$v^2 + uv = u^3 + (a^2 + b^2)u^2 + a^2b^2u$$

via the map $(x, y) \longmapsto (u, v)$ defined by

$$u = \frac{ab}{xy}, \qquad v = \frac{ab(axy + b)}{x^2y}$$

with the inverse map

$$x = \frac{b(u + a^2)}{v}, \qquad y = \frac{a(u + b^2)}{v + (a + b)u}.$$

The neutral element of binary Huff curve is the point $(0, 0)$. The negation of the point (x, y) is $(\tilde{x}, \tilde{y})$ where

$$\tilde{x} = \frac{y(b + axy)}{a + bxy}, \qquad \tilde{y} = \frac{x(a + bxy)}{b + axy}.$$

Affine Addition. The sum of two points (x_1, y_1) and (x_2, y_2) on $HF_{a,b}$ is the point (x_3, y_3) defined as follows:

$$x_3 = \frac{(x_1y_1 + x_2y_2)(1 + y_1y_2)}{(y_1 + y_2)(1 + x_1x_2y_1y_2)}, \qquad y_3 = \frac{(x_1y_1 + x_2y_2)(1 + x_1x_2)}{(x_1 + x_2)(1 + x_1x_2y_1y_2)}. \tag{21}$$

Affine Doubling. The doubling of point (x_1, y_1) is the point (x_4, y_4) defined as follows:

$$x_4 = \frac{(a + b)x_1^2(1 + y_1^2)}{b(1 + x_1^2)(1 + x_1^2y_1^2)}, \qquad y_4 = \frac{(a + b)y_1^2(1 + x_1^2)}{a(1 + y_1^2)(1 + x_1^2y_1^2)}. \tag{22}$$

As $b \neq 0$ we can divide the Eq. (20) by b and for simplicity we can assume $b = 1$. So, we consider the binary Huff curve with the equation

$$ax(y^2 + y + 1) = y(x^2 + x + 1)$$

where $a \neq 0, 1$.

Differential Addition. Devigen and Joye, [4], proposed the rational function $w(x, y) = xy$ for the binary Huff curves. They obtained the following affine w-coordinates formulas

$$w_4 = \frac{(a^2 + 1)/aw_1^2}{1 + w_1^4}, \qquad w_3 = \frac{(w_1 + w_2)^2}{w_0(1 + w_1w_2)^2}.$$

The projective coordinates of the formulas are

$$W_4 = (a^2 + 1)/a(W_1Z_1)^2, \quad Z_4 = (W_1 + Z_1)^4,$$
$$W_3 = w_0(W_1Z_2 + W_2Z_1)^2, \quad Z_3 = (W_1W_2 + Z_1Z_2)^2.$$

The cost of this w-coordinates in one step of the Montgomery ladder is $5\mathbf{M} + 5\mathbf{S} + 1\mathbf{D}$.

New Differential Addition. Here, we modify the rational function $w(x,y) = xy$ on binary Huff curve by scaling to $w(x,y) = \frac{(a^2+1)}{a}xy$. This new rational function reduces the cost of differential addition by $1\mathbf{S}$. As before, we use the same notation for differential addition and doubling. From addition formulas (21), we obtain the following formulas in affine coordinates.

$$w_3 + w_0 = \frac{w_1 w_2}{(a/(a^2+1))^4 w_1^2 w_2^2 + 1}, \tag{23}$$

$$w_3 w_0 = \frac{w_1^2 + w_2^2}{(a/(a^2+1))^4 w_1^2 w_2^2 + 1}. \tag{24}$$

The doubling formula (22) provides the following affine doubling formula.

$$w_4 = \frac{w_1^2}{(a/(a^2+1))^4 w_1^4 + 1}. \tag{25}$$

Then, by Eqs. (23) and (24) we have

$$\frac{1}{w_3} + \frac{1}{w_0} = \frac{w_1 w_2}{(w_1 + w_2)^2}. \tag{26}$$

Cost of Projective w-Coordinates. Assume that w_i are given by the fractions W_i/Z_i for $i = 0,1,2,3,4$. By Eq. (26) the following explicit formulas give the output W_3/Z_3 by

$$\frac{Z_3}{W_3} = \frac{Z_0(W_1Z_2 + W_2Z_1)^2 + W_0(W_1Z_2W_2Z_1)}{W_0(W_1Z_2 + W_2Z_1)^2}. \tag{27}$$

Also, from Eq. (25) the explicit doubling formulas is obtained.

$$\frac{W_4}{Z_4} = \frac{W_1^2 Z_1^2}{W_1^4 + (a/(a^2+1))^4 \ Z_1^4}. \tag{28}$$

So, the cost of projective w-coordinates differential addition and doubling is $7\mathbf{M}+4\mathbf{S}+1\mathbf{D}$; see Eqs. (27) and (28). Let assume $W_0 = 1$. Then using the mixed projective coordinates, we have the following formulas for differential addition:

$$A = W_1Z_1, \quad B = W_1Z_2, \quad C = W_2Z_1,$$
$$Z_4 = (W_1 + (a/(a^2+1))Z_1)^4, \quad W_4 = A^2,$$
$$W_3 = (B + C)^2, \qquad Z_3 = BC + z_0 Z_3.$$

Here, the addition formulas use $4\mathbf{M} + 1\mathbf{S}$ and doubling formulas use $1\mathbf{M} + 3\mathbf{S} + 1\mathbf{D}$. The total cost is $5\mathbf{M} + 4\mathbf{S} + 1\mathbf{D}$ and one S is saved. Moreover the projective w-coordinates differential addition and doubling formulas (27) and (28) are almost complete.

6 Comparison with Previous Works

In Table 1, we compare our new differential addition formulas with other models of binary elliptic curves. The addition formulas for all binary elliptic are complete or almost complete which makes the Montgomery ladder work perfectly in cryptographic applications. The cost of almost complete formulas is $5\mathbf{M} + 4\mathbf{S} + 1\mathbf{D}$ that is the best known record. We believe this record may be obtained for any form of binary elliptic curve by a suitable rational function. The proposed formulas for general binary Edwards are improved in terms of efficiency and speed. The complete formulas for binary Edwards curves are the only known complete formulas for binary elliptic curves with the cost of $5\mathbf{M} + 4\mathbf{S} + 2\mathbf{D}$.

Table 1. Cost of differential addition and doubling for families of binary elliptic curves

Model	Projective differential	Mixed differential	Completeness
Short Weierstraß [2]	$7\mathbf{M} + 4\mathbf{S} + 1\mathbf{D}$	$5\mathbf{M} + 4\mathbf{S} + 1\mathbf{D}$	Almost
Binary Edwards (*general*) [3]	$8\mathbf{M} + 4\mathbf{S} + 4\mathbf{D}$	$6\mathbf{M} + 4\mathbf{S} + 4\mathbf{D}$	Yes
$(d_1 = d_2)$ [3]	$7\mathbf{M} + 4\mathbf{S} + 2\mathbf{D}$	$5\mathbf{M} + 4\mathbf{S} + 2\mathbf{D}$	Yes
$(d_1 = d_2)$ [10]	$7\mathbf{M} + 4\mathbf{S} + 2\mathbf{D}$	$5\mathbf{M} + 4\mathbf{S} + 1\mathbf{D}$	Almost
(general) this work	$7\mathbf{M} + 4\mathbf{S} + 2\mathbf{D}$	$5\mathbf{M} + 4\mathbf{S} + 2\mathbf{D}$	Yes
(general) this work	$7\mathbf{M} + 4\mathbf{S} + 1\mathbf{D}$	$5\mathbf{M} + 4\mathbf{S} + 1\mathbf{D}$	Almost
Binary Hessian [5]	$7\mathbf{M} + 4\mathbf{S} + 2\mathbf{D}$	$5\mathbf{M} + 4\mathbf{S} + 2\mathbf{D}$	Almost
This work	$7\mathbf{M} + 4\mathbf{S} + 1\mathbf{D}$	$5\mathbf{M} + 4\mathbf{S} + 1\mathbf{D}$	Almost
Binary Huff [4]	$6\mathbf{M} + 4\mathbf{S} + 2\mathbf{D}$	$5\mathbf{M} + 5\mathbf{S} + 1\mathbf{D}$	Almost
This work	$7\mathbf{M} + 4\mathbf{S} + 1\mathbf{D}$	$5\mathbf{M} + 4\mathbf{S} + 1\mathbf{D}$	Almost

Acknowledgment. The authors would like to thank anonymous reviewers for their useful comments. This research was in part supported by a grant from IPM (No. 93050416).

References

1. Bernstein, D.J., Chuengsatiansup, C., Kohel, D., Lange, T.: Twisted Hessian curves. In: Lauter, K., Rodríguez-Henríquez, F. (eds.) LATINCRYPT 2015. LNCS, vol. 9230, pp. 269–294. Springer, Heidelberg (2015). doi:10.1007/978-3-319-22174-8_15
2. Bernstein, D., Lange, T.: Explicit-formulas database. http://www.hyperelliptic.org/EFD/
3. Bernstein, D.J., Lange, T., Rezaeian Farashahi, R.: Binary Edwards curves. In: Oswald, E., Rohatgi, P. (eds.) CHES 2008. LNCS, vol. 5154, pp. 244–265. Springer, Heidelberg (2008). doi:10.1007/978-3-540-85053-3_16

4. Devigne, J., Joye, M.: Binary Huff curves. In: Kiayias, A. (ed.) CT-RSA 2011. LNCS, vol. 6558, pp. 340–355. Springer, Heidelberg (2011). doi:10.1007/978-3-642-19074-2_22
5. Farashahi, R.R., Joye, M.: Efficient arithmetic on Hessian curves. In: Nguyen, P.Q., Pointcheval, D. (eds.) PKC 2010. LNCS, vol. 6056, pp. 243–260. Springer, Heidelberg (2010). doi:10.1007/978-3-642-13013-7_15
6. Gaudry, P., Lubicz, D.: The arithmetic of characteristic 2 Kummer surface. Finite Fields Appl. 246–260 (2009)
7. Huff, G.B.: Diophantine problems in geometryand elliptic ternary forms. Duke Math. J. **15**, 246–260 (1948)
8. Joye, M., Quisquater, J.-J.: Hessian elliptic curves and side-channel attacks. In: Koç, Ç.K., Naccache, D., Paar, C. (eds.) CHES 2001. LNCS, vol. 2162, pp. 402–410. Springer, Heidelberg (2001). doi:10.1007/3-540-44709-1_33
9. Joye, M., Tibouchi, M., Vergnaud, D.: Huff's model for elliptic curves. In: Hanrot, G., Morain, F., Thomé, E. (eds.) ANTS 2010. LNCS, vol. 6197, pp. 234–250. Springer, Heidelberg (2010). doi:10.1007/978-3-642-14518-6_20
10. Kim, K.H., Lee, C.O., Negre, C.: Binary Edwards curves revisited. In: Meier, W., Mukhopadhyay, D. (eds.) INDOCRYPT 2014. LNCS, vol. 8885, pp. 393–408. Springer, Heidelberg (2014). doi:10.1007/978-3-319-13039-2_23
11. Koblitz, N.: Elliptic curves cryptosystem. Math. Comput. **48**, 203–209 (1987)
12. Kohel, D.: Efficient arithmetic on elliptic curves in characteristic 2. In: Galbraith, S., Nandi, M. (eds.) INDOCRYPT 2012. LNCS, vol. 7668, pp. 378–398. Springer, Heidelberg (2012). doi:10.1007/978-3-642-34931-7_22
13. Lopez, J., Dahab, R.: Improved algorithms for elliptic curve arithmetic in $GF(2^n)$ without precomputation. CHES, 220–254 (1999)
14. Miller, V.S.: Use of elliptic curves in cryptography. In: Williams, H.C. (ed.) CRYPTO 1985. LNCS, vol. 218, pp. 417–426. Springer, Heidelberg (1986). doi:10.1007/3-540-39799-X_31
15. Montgomery, P.L.: Speeding the polard and elliptic curves methods of factorization. Math. Comput. **48**, 243–264 (1987)
16. Montgomery, P.L.: Modular multiplication without trial division. Math. Comput. **48**, 243–264 (1987)
17. Smart, N.P.: The Hessian form of an elliptic curve. In: Koç, Ç.K., Naccache, D., Paar, C. (eds.) CHES 2001. LNCS, vol. 2162, pp. 118–125. Springer, Heidelberg (2001). doi:10.1007/3-540-44709-1_11
18. Washington, L.C.: Elliptic Curves Number Theory and Cryptography. CRC Press, Boca Raton (2008)

Adequate Elliptic Curves for Computing the Product of n Pairings

Loubna Ghammam[1,2(✉)] and Emmanuel Fouotsa[3,4]

[1] IRMAR, UMR CNRS 6625, Université Rennes 1,
Campus de Beaulieu, 35042 Rennes Cedex, France
ghammam.loubna@yahoo.fr

[2] Laboratoire d'électronique et de microélectronique,
FSM Monastir Université de Monastir, Monastir, Tunisia

[3] LMNO, UMR CNRS 5139 Université de Caen, Campus 2,
14032 Caen Cedex, France
emmanuelfouotsa@yahoo.fr

[4] Higher Teacher Training College, University of Bamenda,
P.O. Box 39, Bambili, Cameroon

Abstract. Many pairing-based protocols require the computation of the product and/or of a quotient of n pairings where $n > 1$ is a natural integer. Zhang et *al.* [1] recently showed that the Kachisa-Schafer and Scott family of elliptic curves with embedding degree 16 denoted KSS16 at the 192-bit security level is suitable for such protocols comparatively to the Baretto-Lynn and Scott family of elliptic curves of embedding degree 12 (BLS12). In this work, we provide important corrections and improvements to their work based on the computation of the optimal Ate pairing. We focus on the computation of the final exponentiation which represent an important part of the overall computation of this pairing. Our results improve by 864 multiplications in $\mathbb{F}_p$ the computations of Zhang et *al.* [1]. We prove that for computing the product or the quotient of 2 pairings, BLS12 curves are the best solution. In other cases, especially when $n > 2$ as mentioned in [1], KSS16 curves are recommended for computing product of n pairings. Furthermore, we prove that the curve presented by Zhang et *al.* [1] is not resistant against small subgroup attacks. We provide an example of KSS16 curve protected against such attacks.

Keywords: BN curves · KSS16 curves · BLS curves · Optimal Ate pairing · Product of n pairings · Subgroup attacks

1 Introduction

Pairing-based cryptography is another way of building cryptographic protocols. Thanks to the various and steady improvements for the computation of pairings

This work was supported in part by French ANR projects PEACE and ANR-12-INSE-0014 SIMPATIC, LIRIMA MACISA project and centre Henri Lebesgue, The Simons Foundations through Pole of Research in Mathematics with applications to Information Security, Subsaharan Africa.

S. Duquesne and S. Petkova-Nikova (Eds.): WAIFI 2016, LNCS 10064, pp. 36–53, 2016.
DOI: 10.1007/978-3-319-55227-9_3

on elliptic curves together with their implementation, several protocols have been published [2–6]. The BN [7] family of elliptic curves are the most suitable for implementing pairing-based cryptography at the 128-bit security level. At the high security level, the BLS12 [8] curves are recommended for computing the optimal Ate pairing according to the results presented in [9,10].

Many pairing-based protocols require the computation of products or quotients of pairings. Some of them require the computation of two pairings [11], others require three pairings [12] and even more than three pairings as in [13,14]. The few works that studied an efficient computation of products of pairings are those of Granger and Smart [1,15]. In particular, Zhang et *al.* [1] have recently shown that the KSS16 [16] elliptic curves are more suitable when computing products or quotients of optimal Ate pairings at the 192-bit security level. In their work they gave explicit formulas and cost evaluation for the Miller loop and developed interesting ways of computing the hard part of the final exponentiation. Unfortunately their results contain several forgotten operations costing 1332 multiplications in the base field $\mathbb{F}_p$. In this work we study the computation of the optimal Ate pairing on KSS16 curves. We present also a new multiple of the hard part of the final exponentiation of the optimal Ate pairing. This new multiple enabled us to improve the cost of the computation of the hard part of the final exponentiation with respect to the work of Zhang et *al.* [1]. We also compare the efficiency of KSS16 curves when computing product of pairings with respect to other common curves at the same security level. We also analyzed the resistance of the KSS16 curves to the small subgroup attack following the approach described in [17]. More precisely, the contribution of this work is as follows:

1. We first pointed out ignored operations in the computation of the optimal Ate pairing (final exponentiation) on KSS16 curves by Zhang et *al.* [1] and give detailed cost of operations with a magma code to verify the formulas [18]. Despite the improvement we obtained for the computation of the final exponentiation in this case and based on the fastest known result to date to our knowledge, we show that BLS12 curves are suitable for the computation of products of two pairings at the high security level and not KSS16 curves as recommended in [1]. We also proved that for computing n pairings where $n > 2$ then KSS16 curves are the best solution.
2. In [17], Barreto et *al.* recently studied the resistance of BN, BLS and KSS18 curves to small subgroup attacks. We extend the same analysis to KSS16 curves. In particular we show that the parameters used in [1] do not ensure protection of these curves to such attacks and we provide an example of KSS16 curve resistant to this attack.

The rest of this work is organized as follows: Sect. 2 recalls results from [1] on optimal Ate pairing on KSS16 curves. We point out the forgotten operations and bring corrections and improvements in the computation of the final exponentiation. In Sect. 3, we present our new multiple of the hard part of the final exponentiation d'. We prove that by using the new vector we saved 864M with respect to the corrected work of Zhang et *al.* in the computation of the optimal

Ate pairing over KSS16 curves. Section 4 defines products of pairings and their efficient computation. Detailed costs of the calculation and comparison are then done with commonly pairing-friendly curves at the high security level. The Sect. 5 concerns the resistance of the KSS16 curves against small subgroup attacks. We show that the curve used in [1] is not protected against small subgroup attack and provide an adequate example. We conclude our work in Sect. 6.

Notations: In this paper we denote by:

- M_k a multiplication in $\mathbb{F}_{p^k}$.
- S_k a squaring in $\mathbb{F}_{p^k}$.
- F_k a Frobenius map in $\mathbb{F}_{p^k}$.
- I_k an inversion in $\mathbb{F}_{p^k}$.
- S_c a cyclotomic squaring in $\mathbb{F}_{p^{16}}$.
- C_c a cyclotomic cube in $\mathbb{F}_{p^{16}}$.

A multiplication, a square and an inversion in $\mathbb{F}_p$ are denoted respectively by M, S and I.

2 Pairings at High Security Level

The 192-bit security level is one of the highest security level recommended when implementing cryptographic protocols based on pairings. Aranha et *al.* [9] recommended the implementation of optimal Ate pairing at this security level over BLS12 curves. Their results on BLS12 curves have been improved by Ghammam and Fouotsa in [10] and still confirm that BLS12 curves are a better solution for implementation at the 192-bit security level. Recently, Zhang et *al.* [1] considered the computation of the optimal Ate pairing over KSS16 curves at the same security level. They proved in particular that this family of curves is suitable for computing products or quotients of pairings generally involved in many pairing-based protocols. In this section we review their computation of the optimal Ate pairing and in particular we bring corrections to shortcomings in their work and give improvements in the computation of the hard part of the final exponentiation. The previous data on costs of computing optimal Ate pairing from the literature at the 192-security level are given in Table 1.

Table 1. Latest best costs of optimal Ate pairing at the 192-bit security level.

Elliptic curves	Size of p (bit)	Complexity of Miller loop	Complexity of the final exponentiation
BLS12 Curves [10]	640	10785M	8116M+6I
BLS24 Curves [10]	480	14574M	23864M+10I
BN Curves [9]	640	16553M	7218M+4I
KSS18 Curves [9]	480	13168M	23821M+8I

Remark 1. *Recently, Kim presented in [19] improvements in discrete logarithm computation in finite fields of the form $\mathbb{F}_{p^{12}}$. Then Jeong and Kim generalized it in [20]. They proved the same result for any composite extension degree n when the prime p is of a special form which is the case of BN, BLS and KSS curves which we studied in this paper. Therefore, these curves no longer provide a 192-bit security level. However, they still present a high security level since it is more than the 128-bit security level.*

2.1 The KSS16 Family of Elliptic Curves and Optimal Ate Pairing

Kachisa et al. proposed in [16] a family of pairing-friendly elliptic curves of embedding degree $k \in \{16, 18, 32, 36, 40\}$. The main idea of their construction of these families of curves is to use the minimal polynomial of the elements of the cyclotomic field rather than the cyclotomic polynomial $\phi_k(x)$ to define the cyclotomic field.

The family of curve with $k = 16$ which is called KSS16 curves is parameterised as follows:

$$\begin{cases} t = 1/35\,(2u^5 + 41u + 35) \\ r = u^8 + 48u^4 + 625 \\ p = \dfrac{1}{980}(u^{10} + 2u^9 + 5u^8 + 48u^6 + 152u^5 + 240u^4 + 625u^2 + \\ \quad 2398u + 3125) \end{cases} \tag{1}$$

and the equation of the elliptic curve defined over $\mathbb{F}_p$ is of the form

$$y^2 = x^3 + ax$$

where t is the trace of the Frobenius endomorphism on E, p is the field size and r presents the order the pairing-friendly subgroup. Let $G_1 = E(\mathbb{F}_p)[r]$ be the r-torsion subgroup of $E(\mathbb{F}_p)$ and $G_2 = E'(\mathbb{F}_{p^4})[r] \cap \mathrm{Ker}(\pi_p - [p])$ where E' is the quartic twist of E. The subgroup of $\mathbb{F}^{\star}_{p^{16}}$ consisting of r-th roots of unity is denoted by $G_3 = \mu_r$. Consider the function $f_{u,Q}$ with divisor $\mathrm{Div}(f_{u,Q}) = u(Q) - ([u]Q) - (u-1)(\mathcal{O})$ and $\ell_{R,S}$ the straight line passing through the points R and S of the elliptic curve.

Proposition 2. [1] *The optimal Ate pairing on the KSS16 curves is the bilinear and non degenerated map:*

$$e_{opt} : G_1 \times G_2 \longrightarrow G_3$$

$$(P, Q) \longmapsto \left((f_{u,Q}(P) l_{[u]Q,[p]Q}(P))^{p^3} l_{Q,Q}(P) \right)^{\frac{p^{16}-1}{r}}$$

The parameter u proposed by Zhang et *al.* [1] is

$$u = 2^{49} + 2^{26} + 2^{15} - 2^7 - 1$$

which is a 49-bit integer of Hamming weight equal to 5 so that r has a prime factor of 377 bits and p is a prime integer of 481 bits. The computation of pairing involves two main steps: the Miller loop and the final exponentiation.

2.2 The Miller Loop

In our case, to compute the optimal Ate pairing in Proposition 2, the Miller loop consists of the computation of $(f_{u,Q}(P) \cdot l_{[u]Q,[p]Q}(P))^{p^3} \cdot l_{Q,Q}(P)$. Let $u = u_n 2^n + \cdots + u_1 2 + u_0$ with $u_i \in \{-1, 0, 1\}$. The computation of the function $f_{u,Q}(P)$ is done thanks to the algorithm in Table 2 known as the Miller algorithm [21]. The Miller loop consists of computing $f_{u,Q}(P)$, $l_{[u]Q,[p]Q}(P)$, $l_{Q,Q}(P)$ and two sparse multiplications in $\mathbb{F}_{p^{16}}$ to multiply terms together and one p^3-Frobenius.

Table 2. Miller algorithm.

Miller algorithm: **Input:** $u = (u_n, u_{n-1}, \ldots, u_0)$,$P$,$Q$,
Output:$(f_{u,Q}(P) \cdot l_{[u]Q,[p]Q}(P))^{p^3} \cdot l_{Q,Q}(P)$

Set $f_1 \leftarrow 1$ and $R \leftarrow Q$
For $i = n-1$ **down to** 0 **do**
  $f_1 \leftarrow f_1^2 \cdot \ell_{R,R}(P)$, $R \leftarrow 2R$ Doubling step
  if $u_i = 1$ **then**
    $f_1 \leftarrow f_1 \cdot \ell_{R,Q}(P)$ $R \leftarrow R + Q$, **end if** Addition step
  if $u_i = -1$ **then**
    $f_1 \leftarrow f_1 \cdot \ell_{R,-Q}(P)$ $R \leftarrow R - Q$, **end if** Addition step
end For
return $f_1 = f_{u,Q}(P)$

The computation of $f_{u,Q}(P)$ costs 49 doubling steps with associated line evaluation, 4 addition steps with line evaluations, 48 squarings in $\mathbb{F}_{p^{16}}$ and 52 sparse multiplications in $\mathbb{F}_{p^{16}}$. We then need an extra $2p$-Frobenius maps for computing $[p]Q$ and $[u]Q$ is obtained through the computation of $f_{u,Q}(P)$. Thus we have to perform 8 multiplications in $\mathbb{F}_p$, a multiplication in $\mathbb{F}_{p^4}$ and one squaring in $\mathbb{F}_{p^4}$ plus $2p$-Frobenius to compute $l_{[u]Q,[p]Q}(P)$. We need also 8 multiplications in $\mathbb{F}_p$, 4 multiplications in $\mathbb{F}_{p^4}$, and one squaring in $\mathbb{F}_{p^4}$ to compute $l_{Q,Q}(P)$ (see [1] for formulas and complete details on the costs).

Therefore, the overall cost of the computation of the Miller loop, as mentioned in [1], is 49 doubling steps with associated line evaluations, 4 addition steps with line evaluations, 48 squarings in $\mathbb{F}_{p^{16}}$, 54 sparse multiplications in $\mathbb{F}_{p^{16}}$, $2p$, p^3 Frobenius maps in $\mathbb{F}_{p^{16}}$, 16 multiplications in $\mathbb{F}_p$, 5 multiplications in $\mathbb{F}_{p^4}$ and one squaring in $\mathbb{F}_{p^4}$. From Table 4 of [1], the Miller loop of the optimal Ate pairing on KSS16 curve costs about 10208 multiplications in $\mathbb{F}_p$.

2.3 The Final Exponentiation

The second step in computing the optimal Ate pairing is the final exponentiation which consists of raising the result f_1 of the Miller loop to the power $\frac{p^{16}-1}{r}$. Thanks to the cyclotomic polynomial, this expression is simplified and presented as follows:

$$f_1^{\frac{p^{16}-1}{r}} = (f_1^{p^8-1})^{\frac{p^8+1}{r}}.$$

First we have to compute $f = f_1^{p^8-1}$ which is called the simple part of the final exponentiation. This costs one p^8–Frobenius, an inversion and a multiplication in $\mathbb{F}_{p^{16}}$. Raising f to the power $\frac{p^8+1}{r}$ is called the hard part of the final exponentiation. In [1], Zhang et *al.* considered a multiple of the second part of the final exponentiation. So instead of computing f^d they computed $f^{857500d}$ where $d = \frac{p^8+1}{r}$. This choice enables them to only have integer coefficients in the representation of $d_1 = 857500d$ in base p which is a simple way for computing this hard part of the final exponentiation.

$$\frac{p^8+1}{r} = \sum_{i=0}^{\phi(16)-1} c_i p^i = c_0 + c_1 p + c_2 p^2 + \cdots + c_7 p^7$$

Where:

$$\begin{cases} c_0 = -11u^9 - 22u^8 - 55u^7 - 278u^5 - 1172u^4 - 1390u^3 + 1372 \\ c_1 = 15u^8 + 30u^7 + 75u^6 + 220u^4 + 1280u^3 + 1100u^2 \\ c_2 = 25u^7 + 50u^6 + 125u^5 + 950u^3 + 3300u^2 + 4750u \\ c_3 = -125u^6 - 250u^5 - 625u^4 - 3000u^2 - 13000u - 15000 \\ c_4 = -2u^9 - 4u^8 - 10u^7 + 29u^5 - 54u^4 + 154u^3 + 4704 \\ c_5 = -20u^8 - 40u^7 - 100u^6 - 585u^4 - 2290u^3 - 2925u^2 \\ c_6 = 50u^7 + 100u^6 + 250u^5 + 1025u^3 + 4850u^2 + 5125u \\ c_7 = 875u^2 + 1750u + 4375 \end{cases} \tag{2}$$

Then Zhang et *al.* presented a very nice decomposition of c_i where $i \in \{0, 1, 2, 3, 4, 5, 6, 7\}$. This representation enabled them to quickly compute the hard part of the final exponentiation. Let

$$A = u^3.B + 56 \text{ and } B = (u+1)^2 + 4, \text{ then}$$

$$\begin{cases} c_0 = -11(u^4A + 27u^3B + 28) + 19A; & c_4 = -(2u^4A + 55u^3B) + 84A \\ c_1 = 5(3u^3A + 44u^2B) = 5c_1'; & c_5 = -5(4u^3A + 117u^2B) = -5c_5' \\ c_2 = 25(u^2A + 38uB) = 25c_2'; & c_6 = 25(2u^2A + 41uB) = 25c_6' \\ c_3 = -125(uA + 24B) = -125c_3'; & c_7 = 125.7B = 125c_7' \end{cases}$$

The problem with this representation is that when we recomputed these expressions we discovered that there is a missing term in the expression of c_0. In fact

$$\begin{cases} c_0 = -11u^9 - 22u^8 - 55u^7 - 278u^5 - 1172u^4 - 1390u^3 + 1372 \\ \quad = -11(u^4A + 27u^3B + 28) + 19A + \mathbf{616} \end{cases} \tag{3}$$

We verified also the algorithm presented in Appendix A of [1] where the term f^{616} is missing in the computation of the final exponentiation. Fortunately, the expression of c_0 do not influence the rest of the expressions c_i with $0 < i < 8$. Therefore, we have to add this term to the final result of the hard part of the final exponentiation of the optimal Ate pairing. Using the square-and-multiply algorithm, the additional step f^{616} costs 8 squarings and 3 multiplications in $\mathbb{F}_{p^{16}}$ but we will not add this cost because they are terms precomputed in the algorithm of Zhang et *al.* We will add to their algorithm these operations after the first term of the original algorithm:

$$\begin{cases} A0 \leftarrow T3^8 \\ A1 \leftarrow A0 \cdot T3 \\ A2 \leftarrow A1 \cdot T2 \\ A3 \leftarrow T1^2 \\ A2 \leftarrow A3 \cdot A2 \end{cases} \tag{4}$$

By adding these operations we got in $A2$ the missing term f^{616}. At the end of the algorithm presented by Zhang et *al.* we have to add this term to the final result costing an extra multiplication. So the additional cost is 4 multiplications and 4 squarings in $\mathbb{F}_{p^{16}}$.

Other shortcomings with their algorithm that computed the hard part of the final exponentiation concern the computation of c_5', c_0' and c_4'. In fact, in the expression of c_0', the output of their algorithm is $-11(u^4A+55u^3B+28)+35A$ instead of the result $-11(u^4A+55u^3B+28)+19A$. Also, the expression of c_4' computed in their algorithm is $-(2u^4A+55u^3B)+148A$ not as mentioned in the development which is $-(2u^4A+55u^3B)+84A$.

The expression of c_5' is deduced by multiplying the term stocked in the temporary variable T_{11} by the term stocked in F_{14} and not by the one recorded in F_{25}. Also in the computation of c_7' we must perform the operation $\overline{F_5}.T_4$ instead of $\overline{F_5}.T_6$.

Therefore we must perform some modifications in the original algorithm to have the coherent result at the end. We presented the corrected algorithm in Appendix A, Table 9, and a magma code for the verification of formulas is available in [18]. The additional corrections cost 4 multiplications and 3 squarings in $\mathbb{F}_{p^{16}}$ instead of 3 multiplications and 4 squarings which is the cost of the operations before our modifications. Furthermore Zhang et *al.* claimed that in the final algorithm they used only 16 squarings, but it is not the case because by a simple count we found that one is forced to perform 38 squarings in $\mathbb{F}_{p^{16}}$.

As a consequence to compute the final exponentiation we have to perform 7 exponentiations by u, 2 exponentiations by $(u+1)$, one inversion, 44 cyclotomic squarings in $G_{\phi_2(p^8)}$, 38 multiplications in $\mathbb{F}_{p^{16}}$, 2 cyclotomic cubings in $\mathbb{F}_{p^{16}}$ and p, p^2, p^3, p^4, p^5, p^6, p^6, p^7, p^8-Frobenius maps.

In Table 3 we present the new cost of the final exponentiation of the optimal Ate pairing after our correction of the result of the work in [1]. Hence, by adding some modifications to the original result the overall cost of the optimal Ate pairing on KSS16 curve is 33870M+I. So we have extra 1332 multiplications in $\mathbb{F}_p$ than the cost presented in [1].

Table 3. Complexity of the optimal Ate pairing.

The method	Complexity of Miller loop	Complexity of the final exponentiation
Method of [1]	10208 M	22330M+I
Our correction	10208 M	23662M+I

3 A New Multiple of the Hard Part of the Final Exponentiation

An efficient method to compute the hard part is described by Scott *et al.* [22]. They suggested to write $d = \frac{\phi_k(p)}{r}$ in base p as $d = d_0 + d_1 p + \cdots + d_{\phi(k)-1} p^{\phi(k)-1}$ and find a short vector addition chain to compute f^d much more efficiently than the naive method. In [23], based on the fact that a fixed power of a pairing is still a pairing, Fuentes *et al.* [23] suggested to apply Scott *et al.*'s method with a power of any multiple d' of d with r not dividing d'. This could lead to a more efficient exponentiation than a direct computation of f^d. Their idea of finding the polynomial $d'(x)$ is to apply the *LLL*-algorithm to the matrix formed by $\mathbb{Q}$-linear combinations of the elements $d(x), xd(x), \ldots, x^{degr-1}d(x)$. In this paper we tried to find a new multiple of $d_1 = 857500 \cdot d$ (with r not dividing d). We use a lattice-based method to find d' such that $f^{d'}$ can be computed in a more efficient way than computing $f^{857500 \cdot d}$.

Thanks to the LLL algorithm [24], the best vector that we found is given by:

$$d'(u) = m_0 + m_1 p + m_2 p^2 + m_3 p^3 + m_4 p^4 + m_5 p^5 + m_6 p^6 + m_7 p^7 = s(u) d_1$$

$$\text{where} \begin{cases} s(u) = u^3/125 \\ m_0 = 2u^8 + 4u^7 + 10u^6 + 55u^4 + 222u^3 + 275u^2 \\ m_1 = -4u^7 - 8u^6 - 20u^5 - 75u^3 - 374u^2 - 375u \\ m_2 = -2u^6 - 4u^5 - 10u^4 - 125u^2 - 362u - 625 \\ m_3 = -u^9 - 2u^8 - 5u^7 - 24u^5 - 104u^4 - 120u^3 + 196 \\ m_4 = u^8 + 2u^7 + 5u^6 + 10u^4 + 76u^3 + 50u^2 \\ m_5 = 3u^7 + 6u^6 + 15u^5 + 100u^3 + 368u^2 + 500u \\ m_6 = -11u^6 - 22u^5 - 55u^4 - 250u^2 - 1116u - 1250 \\ m_7 = 7u^5 + 14u^4 + 35u^3 + 392 \end{cases} \quad (5)$$

Our aim in this section by presenting the new vector d' is to reduce the complexity of computing the hard part of the final exponentiation for the optimal Ate pairing in KSS16 curves and then the complexity of computing the product of n pairings. Let

$$\begin{cases} A = u^3 B + 56 \\ B = (u+1)^2 + 4 \end{cases}$$

then we can write the expressions of m_i where $0 < i < 8$ more simply as follows:

$$\begin{cases} m_0 = 2u^3A + 55u^2B; & m_4 = u^3A + 10u^2B \\ m_1 = -4u^2A - 75uB; & m_5 = 3u^2A + 100uB \\ m_2 = -2uA - 125B; & m_6 = -11uA - 250B \\ m_3 = -u^4A - 24u^3B + 196; & m_7 = 7A \end{cases}$$

These new expressions enabled us to be faster than Zhang et *al.* in the computation of the hard part of the final exponentiation. We detailed the computation of the final exponentiation in the algorithm presented in Appendix A, Table 8, and a magma code for the verification of formulas is available in [18]. The overall cost of this algorithm is then 7 exponentiations by u, 2 exponentiations by $(u+1)$, 34 cyclotomic squarings in $G_{\phi_2(p^8)}$, 32 multiplications in $\mathbb{F}_{p^{16}}$, 3 cyclotomic cubings in $\mathbb{F}_{p^{16}}$ and p, p^2, p^3, p^4, p^5, p^6, p^6, p^7, p^8-Frobenius maps.

Table 4. Comparison between Zhang et *al.* and our new development.

Method	Algorithm	Complexity			
		S_c	M_{16}	F_{16}	C_c
Zhang et al.	1	44	37	8	1
Our development	2	**34**	**32**	**8**	**3**

Our result of computing the hard par of the final exponentiation is compared with the corrected result presented in Sect. 2.3 in Table 4. For a full comparison, we consider the example presented in [1]. The extension tower is built as follows:

- $\mathbb{F}_{p^4} = \mathbb{F}_p[v]/(v^4 + 3))$
- $\mathbb{F}_{p^8} = \mathbb{F}_{p^4}[w]/(w^2 - v)$
- $\mathbb{F}_{p^{16}} = \mathbb{F}_{p^8}[z]/(z^2 - w)$

The cost of operations for computing the optimal Ate pairing on KSS16 curve are presented in Table 4 of [1].

Table 5. Comparison between the two vectors d and d'.

The result	Complexity of algorithm	Complexity of the hard part the final exponentiation
Corrected result of [1]	See cost in Table 8	23537M
Our new algorithm	See cost in Table 9	**22673M**

In Table 5 we compared the complexity in $\mathbb{F}_p$ of our result using a new multiple of the hard part of the final exponentiation and the corrected one of Zhang et *al.* In this table we remark that our computations are faster than those presented in [1] for computing the hard part of the final exponentiation. We saved about 864 multiplications in $\mathbb{F}_p$ which is an interesting result if one is interested in hardware or software implementations of the optimal Ate pairing at the 192-security level.

4 On Computing Products of n Pairings

In some protocols, for example in the BBG HIBE scheme [25], the BLS short group signature scheme [5], ABE scheme due to Waters [14], the non interactive proof systems proposed by Groth and Sahai [26] and others [11,13], it is necessary to compute the product or the quotient of two or more pairings. Scott in [27] and Granger et *al.* in [15] investigated the computation of the product of n pairings.

Let

$$e : G_1 \times G_2 \rightarrow G_3$$

a bilinear non-degenerated map from two additive groups G_1 and G_2 to G_3 a multiplicative group. The evaluation of a product of n pairings is of the form

$$e_n = \prod_{i=1}^{n} e(P_i, Q_i)$$

In this section we are interested by the computation of n pairings. We give a comparison of this computation for different category of curves at the 192-bit security level. For this security level it is recommended by Aranha et *al.* in [9] to use the BLS12 curves to compute the optimal Ate pairing. In this section and in the case where one computes the product of n optimal Ate pairings, we will prove that this category of curves are not a solution for all n specially where $n > 2$. We prove also that the KSS16 curves, proposed as the best solution for computing the product of n pairings by Zhang et *al.* in [1] are not the best for $n = 2$. We First recall in Table 6 the different formulas for the optimal Ate pairing over common families of pairing-friendly curves such as KSS16, KSS18, BN, BLS12 and BLS24 curves. For computing the optimal Ate pairing we have two steps: The Miller loop and the final exponentiation. The computation of the product of n pairings consists only of the computation of the product of n Miller loops followed by the evaluation of the result of the final exponentiation. Recall that in the Miller loop (see the algorithm in Table 2) we have to compute the following step:

$$f \leftarrow f^2 l(Q) \tag{6}$$

Table 6. Optimal Ate pairing on elliptic curves.

Curve	Optimal Ate pairing: $(P, Q) \rightarrow$
KSS16 [1]	$\left((f_{u,Q}(P) l_{[u]Q,[p]Q}(P))^{p^3} l_{Q,Q}(P)\right)^{\frac{p^{16}-1}{r}}$
KSS18 [9]	$(f_{u,Q}(P) f_{3,Q}^{p} l_{[u]Q,[3p]Q}(P))^{\frac{p^{18}-1}{r}}$
BN [9]	$((f_{6u+2,Q}(P) l_{[6u+2]Q,[p]Q}(P) l_{[6u+2]Q,[-p^2]Q}(P)))^{\frac{p^{12}-1}{r}}$
BLS12 [9]	$(f_{u,Q}(P))^{\frac{p^{12}-1}{r}}$
BLS24 [9]	$(f_{u,Q}(P))^{\frac{p^{24}-1}{r}}$

where l is the tangent to the curve at a point depending on Q and depending on the loop iteration in Miller's algorithm. To compute the product of Eq. (6), each doubling function-evaluation step becomes

$$f \leftarrow f^2 \prod_{i=1}^{n} l_i(Q_i) \tag{7}$$

Therefore one needs only to calculate a single squaring in the extension field per doubling rather than n squarings using the naive method of the computation of the product of n pairings.

So to evaluate the cost of the computation of the product of n optimal Ate pairings we have to compute at first:

- **Cost1**: Full squarings in the Miller loop (squarings in Eq. 7).
- **Cost2**: Other operations in the Miller loop (point operations and line evaluation).
- **Cost3**: Final exponentiation.

Then we have to sum **Cost1**, n**Cost2** and **Cost3** to find the overall cost of the product of n pairings.

Table 7. Costs comparison of product of n pairings at the 192-bit security levels.

Costs	KSS16 Zhang	KSS16	BLS12 [10]	BN [9]	KSS18 [9]
Full squarings for DBL	2592M	2592M	5892M	8837M	4158M
Others in Miller loop	7616M	7616M	10760M	16720M	9544M
Final exponentiation	23662M+I	22888M +I	12574M+6I	11145M+6I	23821M+8I
Total cost for $n = 1$	33870M+I	33096M+I	29226M+6I	36702M+6I	37523M+8I
Total cost for $n = 2$	41486M+I	40712M+I	39986M+6I	53422M+6I	47067M+8I
Total cost for $n = 3$	49102M+I	48328M+I	50746M+6I	64567M+6I	56611M+8I
Total cost for $n = 7$	79656M+I	78792M+I	93786M+6I	109147M+6I	94784M +8I

In Table 7, we present the costs for computing the product of n pairings considering common curves in Table 6. From Table 7, we can deduce that for $n = 2$, meaning when we would like to compute the product of two parings, it is better to use BLS12 curves. In the case of $n > 2$ as mentioned in [1] KSS16 curves can give the fastest computations of products or quotients of n pairings.

Security of Cryptographic protocols is important in practice. That's why, when we compute optimal Ate pairing on KSS16 curves we have to verify the security of the parameters of the elliptic curve. In the next section we will present a detailed study of the security of the computation of the optimal Ate pairing and more precisely the resistance against the subgroup attacks.

5 Subgroup Security for KSS16 Pairing-Friendly Curves

A detailed study on subgroup security for pairing-friendly curves was recently studied by Baretto et *al.* [17]. They focus on common families of elliptic curves having twists of order six such as BN, BL12, BLS24 and KSS18 curves. In particular they provided parameters that enable the aforementioned curves to be resistant against subgroups attacks. In this section, we extend the same analysis to the KSS family of elliptic curves having quartic twists and of embedding degree 16. We first recall the definition of *subgroup secure curves* concept from [17] The subgroup security concept explicitly described on pairing-friendly curves by Barreto et *al.* [17], is a property that strengthens the resistance of pairing-friendly curves against subgroup attacks. Let E be an elliptic curve of embedding degree k and parameterised by $p(u)$, $t(u)$, $r(u) \in \mathbb{Q}[u]$. Let d be the degree of the twist of the elliptic curve E and let $E'(\mathbb{F}_{p^{k/d}})$ its twists. Let $h_1(u) = \frac{|E(\mathbb{F}_p)(u)|}{r(u)}$, $h_2(u) = \frac{|E'(\mathbb{F}_{p^{k/d}})(u)|}{r(u)}$ and $h_T = \frac{|G_{\phi_k}(p(u))|}{r(u)}$ be the indices of the three groups on which a pairing is defined.

Definition 3. [17] *The curve E is* subgroup secure *if all $\mathbb{Q}[u]$-irreducible factors of $h_1(u)$, $h_2(u)$, $h_T(u)$ that represent primes and that have degree at least the degree of $r(u)$, contain no prime factor smaller than $r(u_0) \in \mathbb{Z}$ when evaluated at $u = u_0$.*

In the case of KSS16, the indices are given in the following proposition:

Proposition 4. *Let $p(u)$, $t(u)$, $r(u) \in \mathbb{Q}[u]$ be the parameters of the KSS16 pairing-friendly elliptic curve. The indice $h_T = \frac{p(u)^8+1}{r(u)}$ is a polynomial in u of degree 72. Also $h_1(u) = (125/2)(u^2 + 2u + 5)$ and the order of the quartic twist $E'(\mathbb{F}_{p^4})$ is $|E'(\mathbb{F}_{p^4})| = h_2(u) \cdot r(u)$ where $h_2(u) = (1/15059072)(u^{32} + 8u^{31} + 44u^{30} + 152u^29 + 550u^{28} + 2136u^{27} + 8780u^{26} + 28936u^{25} + 83108u^{24} + 236072u^{23} + 754020u^{22} + 2287480u^{21} + 5986066u^{20} + 14139064u^{19} + 35932740u^{18} + 97017000u^{17} + 237924870u^{16} + 498534968u^{15} + 1023955620u^{14} + 2353482920u^{13} + 5383092978u^{12} + 10357467880u^{11} + 17391227652u^{10} + 31819075896u^{9} + 65442538660u^{8} + 117077934360u^{7} + 162104974700u^{6} + 208762740168u^{5} + 338870825094u^{4} + 552745197960u^{3} + 632358687500u^{2} + 414961135000u + 126854087873)$.*

Proof. The order of the group $E(\mathbb{F}_{p^4})$ is $|E(\mathbb{F}_{p^4})| = p^4 + 1 - t_4$ where $t_4 = t^4 - 4pt^2 + 2p^2$ (see [28, Theorem 4.12]). The order of the correct quartic twist $E'(\mathbb{F}_{p^4})$ is given by $|E'(\mathbb{F}_{p^4})| = p^4 + 1 + v_4$ where $v_4^2 = 4p^4 - t_4^2$ (see [29, Proposition 2]). A direct calculation gives the cofactor as $h_2(u) = \frac{p^4+1+v_4}{r(u)}$.

Remark 5. *The value used in [1] for the computation of optimal pairing on $KSS16$ curves is $u_0 = 2^{49} + 2^{26} + 2^{15} - 2^7 - 1$. With this value we see that $h_2(u_0)$ has the factorisation $2 \cdot 1249 \cdot 366593 \cdot c_{1515}$ where c_{1515} is still a composite integer of 1515 bits. This means that the corresponding curve fails to satisfy the small subgroup attack property. In the following section we search for a parameter u to avoid subgroup attack on this curve.*

For the 192-bit security level, the u_0 which gives corresponding sizes of r and p must be an integer of bit size at least 49. Also, the good u_0 must be such that $p(u_0), r(u_0), h_2(u_0)$ and $h_T(u_0)$ are simultaneously prime. Since $u \equiv \pm 25 \mod 70$ (for p to represent integers) one can easily see that $h_2(u) \equiv 0 \mod 2$ and $h_T(u) \equiv 0 \mod 2$. We will therefore search for u_0 such that $p(u_0), r(u_0)$, $h_2(u_0)/2$ and $h_T(u_0)/2$ are simultaneously prime. One can have a chance to obtain such a u_0 if and only if those polynomials satisfy the Bunyakovsky's property. A quick verification enables to see that the prime number 17 divides these polynomials when evaluated at $n \in \mathbb{N}$. Therefore it is enough to search for prime numbers with 2 and/or 17 as factors. The Batemann-Horn conjecture then ensures that they are approximately 24500 values of $u_0 \in [2^{49}, 2^{53}]$ with $p(u_0), r'(u_0), h_2'(u_0)$ and $h_T'(u_0)$ simultaneously prime, where $r(u) = 17^{n_1} \cdot r'(u)$, $h_2(u) = 2 \cdot 17^{n_2} \cdot h_2'(u)$ and $h_T = 2 \cdot 17^{n_3} \cdot h_T'(u)$ for some positive or zero integers n_1, n_2 and n_3. A careful search enabled us, after several long tries starting with x_0 of Hamming weight 5, to obtain the following value

$$u_0 = 2^{50} + 2^{47} - 2^{38} + 2^{32} + 2^{25} - 2^{15} - 2^5 - 1$$

which gives a prime p of 492 bits, $r(u_0) = r'(u_0)$ prime of 386 bits, $h_2(u_0) = 2 \cdot 17 \cdot h_2'(u_0)$ and $h_T = 2 \cdot 17 \cdot h_T'(u_0)$ where $h_2'(u_0)$ and $h_T'(u_0)$ are prime numbers of 3544 bits and 1577 bits respectively. For the value of p obtained the extension field $\mathbb{F}_{p^{16}}$ is built using the following tower of extensions:

- $\mathbb{F}_{p^2} = \mathbb{F}_p[\alpha]/(\alpha^2 - 11)$
- $\mathbb{F}_{p^4} = \mathbb{F}_{p^2}[\beta]/(\beta^2 - \alpha))$
- $\mathbb{F}_{p^8} = \mathbb{F}_{p^4}[\gamma]/(\gamma^2 - \beta)$
- $\mathbb{F}_{p^{16}} = \mathbb{F}_{p^8}[\theta]/(\theta^2 - \gamma)$

An example of elliptic curve E over $\mathbb{F}_p$ that satisfies $|E(\mathbb{F}_p)| = p + 1 - t$ has the equation $E: \; y^2 = x^3 + 17x$. The corresponding quartic twist E' over $\mathbb{F}_{p^4}$ with order $|E'(F_{p^4})| = 2 \cdot 17 \cdot h_2'(u_0) \cdot r(u_0)$ is the curve $E': \; y^2 = x^3 + 17/\beta x$.

6 Conclusion

In many pairing-based protocols the evaluation of the product or the quotient of many pairings is required. In this paper we were interested in the computation of the product of n optimal Ate pairings at the high security level.

This problem was first considered by Zhang et *al.* [1]. They suggested the KSS16 curves as a best choice for computing n pairings. We checked their results on the computation of the hard part of the final exponentiation of the optimal Ate pairing. We found that they missed 1332 multiplications in $\mathbb{F}_p$ in their complexity calculation. We corrected their algorithm and we presented a new algorithm for the computation of the final exponentiation based on a new multiple of the hard part of the final exponentiation. With this new vector we saved about 864 multiplications in the basic field which is an important result if one thinks about hardware or software implementations. We implemented our new algorithms in Magma to verify their correctness [18]. We computed also the product of n pairings. We proved that for $n = 2$ it is better to use BLS12 curves and for $n > 2$ KSS16 curves are the best solution. Finally we proposed a new parameter u for the KSS16 curves to ensure the resistance against the small subgroup attacks.

A Algorithms

In these tables and to have the same expressions as Zhang et *al.* we denote by f the result of Miller loop and by M the result of the first part of the final exponentiation.

Table 8. Final exponentiation with a new exponent. See [18] for the magma code for the verification.

Operations	Terms computed	Cost
$E1 = f^{p^8} E2 = E1 \cdot f^{-1}$	$M = f^{p^8-1}$	
$T0 = M^2; T1 = T0^2$	$M^2; M^4$	$2S_{16}$
$T2 = M^{u+1}; T3 = T2^{u+1}$	$M^{u+1}; M^{(u+1)^2}$	$2F_u$
$T4 = T3 \cdot T1$	$M^{(u+1)^2+4} = M^B$	$1M_{16}$
$T5 = T4^u; T6 = T4^5$	$M^{uB}; M^{5B}$	$1E_u + 1M_{16} + 2S_{16}$
$T7 = T1^8; T8 = T7^2$	$M^{32}; M^{64}$	$4S_{16}$
$T9 = T7 \cdot T1^{-1}; T10 = T9^2$	$M^{28}; M^{56}$	$1M_{16} + 1S_{16}$
$T11 = T5^u; T12 = T11^u$	$M^{u^2B}; M^{u^3B}$	$2E_u$
$T01 = T12 \cdot T10$	$M^{u^3B+56} = M^A$	$1M_{16}$
$T14 = T01^u; T13 = T14^{-2}$	$M^{uA}; M^{-2uA}$	$1E_u + 1S_{16}$
$T00 = T6^5; T15 = T00^5$	$M^{25B}; M^{125B}$	$2M_{16} + 4S_{16}$
$T0 = T13 \cdot T15^{-1}$	$M^{-2uA-125B} = M^{c_2}$	$1M_{16}$
$T16 = T0^2; T17 = T13^4$	$M^{2c_2}; M^{-8uA}$	$3S_{16}$
$T18 = T17 \cdot T14$	M^{-7uA}	$1M_{16}$
$T2 = T16 \cdot T18$	$M^{2c_2-7uA} = M^{c_6}$	$1M_{16}$
$T19 = T14^u; T20 = T19^u$	$M^{u^2A}; M^{u^3A}$	$2E_u$
$T21 = T20^u; T22 = T19^2$	$M^{u^4}; M^{2u^7A}$	$1E_u + 1S_{16}$
$T23 = T5^5; T24 = T23^5$	$M^{5uB}; M^{25uB}$	$2M_{16} + 4S_{16}$

Table 8. (*continued*)

Operations	Terms computed	Cost
$T25 = T24^3; T26 = T24 \cdot T25$	$M^{75uB}; M^{100uB}$	$1C_{16} + 1M_{16}$
$T27 = T22^2$	M^{4u^2A}	$1S_{16}$
$T37 = (T27 \cdot T25)^{-1}$	$M^{-4u^2A-75uB} = M^{c_1}$	$1M_{16}$
$T28 = T27 \cdot T19^{-1}$	M^{3u^2A}	$1M_{16}$
$T3 = T28 \cdot T26$	$M^{3u^2A+100xB} = M^{c_5}$	$1M_{16}$
$T29 = T11^5; T30 = T29^2$	$M^{5u^2B}; M^{10u^2B}$	$1M_{16} + 3S_{16}$
$T4 = T20 \cdot T30$	$M^{u^3A+10u^2B} = M^{c_4}$	$1M_{16}$
$S0 = T20^2; S1 = T30^5$	$M^{2u^3A}; M^{50u^2B}$	$1M_{16} + 3S_{16}$
$S2 = S1 \cdot T29; S3 = S0 \cdot S2$	$M^{55u^2B}; M^{2u^3A-55u^2B} = M^{c_0}$	$2M_{16}$
$T31 = T12^{24}$	M^{24u^3B}	$1C_{16} + 3S_{16}$
$T5 = T21^{-1} \cdot T31^{-1}$	$M^{-u^4A-24u^3B}$	$1M_{16}$
$T6 = T8^3 \cdot T1$	M^{196}	$1M_{16} + 1C_{16}$
$T7 = T5 \cdot T6$	$M^{-u^4A-24u^3B+196} = M^{c_3}$	$1M_{16}$
$T8 = T1^7$	$M^{7A} = M^{c_7}$	$2M_{16} + 2S_{16}$
$T32 = T37^p \cdot T7^{p^3} \cdot T3^{p^5} \cdot T8^{p^7}$	$M^{c_1p+c_3p^3+c_5p^5+c_7p^7}$	$3M_{16} + 4(15M)$
$T33 = T0^{p^2} \cdot T2^{p^6}$	$M^{c_2p^2+c_6p^6}$	$1M_{16} + 2(12M)$
$T = S3 \cdot T32 \cdot T33 \cdot T4^{p^4}$	$M^{\frac{p^8+1}{r}}$	$3M_{16} + 1(8M)$

Table 9. Corrected version of the final exponentiation in [1]. See [18] for the magma code for the verification.

Operations	Terms computed	Cost
$E1 = f^{p^8}$ $E2 = E1 \cdot f^{-1}$	$M = f^{p^8-1}$	
$T1 = E2^4; T2 = T1^8; T3 = T2^2$		$6S_{16}$
$A0 = T3^8; A1 = A0 \cdot T3$		$1M_{16} + 3S_{16}$
$A2 = A1 \cdot T2; A3 = T1^2$		$1M_{16} + 1S_{16}$
$A2 = A3 \cdot A2$		$1M_{16}$
$F1 = T2 \cdot T1^{-1}; F2 = F1^2$		$1M_{16} + 1S_{16}$
$F3 = E2^{u+1}; F4 = F3^{u+1}$		$2E_{u+1}$
$F5 = F4 \cdot T1; T4 = F5^8$	$F5 = M^B$	$1M_{16} + 3S_{16}$
$F6 = F5^u; F7 = F5^{-1} \cdot T4$	$F7 = M^{c_7'}$	$1E_u + 1M_{16}$
$F8 = T4^3; T5 = F6^8$		$1C_{16} + 3S_{16}$
$F9 = F6^u; F10 = T5 \cdot F6^{-1}$		$1E_u + 1M_{16}$
$F11 = F10^2; T6 = F9^8$		$4S_{16}$
$F12 = F9^u; F13 = T6 \cdot F9^{-1}$		$1E_u + 1M_{16}$
$F14 = F13^2; F15 = F12 \cdot F2$	$F15 = M^A$	$1S_{16} + 1M_{16}$

Table 9. (*continued*)

Operations	Terms computed	Cost
$T7 = F15^2; T8 = T7^4$		$3S_{16}$
$S1 = T8^2; S2 = T7^2$		$2S_{16}$
$S3 = S2 \cdot S1; S4 = S3 \cdot F15^{-1}$		$2M_{16}$
$T9 = S1^4; S5 = S3 \cdot T9$		$1M_{16} + 2S_{16}$
$S6 = F14^2; F16 = F15^u$		$1E_u + 1S_{16}$
$F22 = F16 \cdot F8$	$F22 = M^{c'_3}$	$1M_{16}$
$F23 = F22^u; F24 = F23 \cdot F11$	$F24 = M^{c'_2}$	$1E_u + 1M_{16}$
$T10 = F23^2; F25 = F23^u$		$1E_u + 1S_{16}$
$F26 = T10 \cdot F10^{-1}; T11 = F25^4$	$F26 = M^{c'_6}$	$1M_{16} + 2S_{16}$
$F27 = F25^u; F28 = T11 \cdot F25^{-1}$		$1E_u + 1M_{16}$
$F29 = F13 \cdot F14; F30 = T11 \cdot F29$	$F30 = M^{c'_5}$	$2M_{16}$
$F31 = F28 \cdot S6^{-1}; F32 = F12^2$		$1M_{16} + 1S_{16}$
$F33 = F32 \cdot F12; F34 = F27 \cdot F33$		$2M_{16}$
$F35 = F34^2; F36 = F35 \cdot F12$		$1M_{16} + 1S_{16}$
$F37 = F36^{-1} \cdot S5; F38 = F34 \cdot F1$	$F37 = M^{c'_4}$	$2M_{16}$
$F39 = F38^2; F40 = F39^2$		$2S_{16}$
$F41 = F40^2; F42 = F39 \cdot F38$		$1M_{16} + 1S_{16}$
$F43 = F41 \cdot F42; F44 = F43^{-1} \cdot S4$		$2M_{16}$
$H1 = F7^{p^7}; H2 = F22^{p^3}$		$2(14M)$
$H3 = F24^{p^2}; H4 = F26^{p^6}$		$2(12M)$
$H5 = F30^{p^5}; H6 = F31^p$		$2(14M)$
$H7 = F37^{p^4}; H8 = H1 \cdot H2^{-1}$		$1M_{16} + 1(8M)$
$H9 = H8^2; H10 = H9^2$		$2S_{16}$
$H11 = H10 \cdot H8; H12 = H11 \cdot H3$		$2M_{16}$
$H13 = H12 \cdot H4; H14 = H13^2$		$1M_{16} + 1S_{16}$
$H15 = H14^2; H16 = H15 \cdot H13$		$1M_{16} + 1S_{16}$
$H17 = H16 \cdot H6; H18 = H17 \cdot H5^{-1}$		$2M_{16}$
$H19 = H18^2; H20 = H19^2$		$2S_{16}$
$H21 = H20 \cdot H18; H22 = H21 \cdot H7$		$2M_{16}$
$H23 = H22 \cdot F44$	$H23 = M^{d'}$	$1M_{16}$

References

1. Zhang, X., Lin, D.: Analysis of optimum pairing products at high security levels. In: Galbraith, S., Nandi, M. (eds.) INDOCRYPT 2012. LNCS, vol. 7668, pp. 412–430. Springer, Heidelberg (2012). doi:10.1007/978-3-642-34931-7_24
2. Boneh, D., Franklin, M.: Identity-based encryption from the Weil pairing. In: Kilian, J. (ed.) CRYPTO 2001. LNCS, vol. 2139, pp. 213–229. Springer, Heidelberg (2001). doi:10.1007/3-540-44647-8_13

3. Cocks, C.: An identity based encryption scheme based on quadratic residues. In: Honary, B. (ed.) Cryptography and Coding 2001. LNCS, vol. 2260, pp. 360–363. Springer, Heidelberg (2001). doi:10.1007/3-540-45325-3_32
4. Libert, B., Quisquater, J.-J.: Identity based undeniable signatures. In: Okamoto, T. (ed.) CT-RSA 2004. LNCS, vol. 2964, pp. 112–125. Springer, Heidelberg (2004). doi:10.1007/978-3-540-24660-2_9
5. Boneh, D., Lynn, B., Shacham, H.: Short signatures from the Weil pairing. J. Cryptol. **17**(4), 297–319 (2004)
6. Goyal, V., Pandey, O., Sahai, A., Waters, B.: Attribute-based encryption for fine-grained access control of encrypted data. In: Linawati, Mahendra, M.S., Neuhold, E.J., Tjoa, A.M., You, I. (eds.) ICT-EurAsia 2014. LNCS, vol. 8407, pp. 89–98. Springer, Heidelberg (2006). doi:10.1007/978-3-642-55032-4_60
7. Barreto, P.S.L.M., Naehrig, M.: Pairing-friendly elliptic curves of prime order. In: Preneel, B., Tavares, S. (eds.) SAC 2005. LNCS, vol. 3897, pp. 319–331. Springer, Heidelberg (2006). doi:10.1007/11693383_22
8. Barreto, P.S.L.M., Lynn, B., Scott, M.: Constructing elliptic curves with prescribed embedding degrees. In: Cimato, S., Persiano, G., Galdi, C. (eds.) SCN 2002. LNCS, vol. 2576, pp. 257–267. Springer, Heidelberg (2003). doi:10.1007/3-540-36413-7_19
9. Aranha, D.F., Fuentes-Castañeda, L., Knapp, E., Menezes, A., Rodríguez-Henríquez, F.: Implementing pairings at the 192-bit security level. In: Abdalla, M., Lange, T. (eds.) Pairing 2012. LNCS, vol. 7708, pp. 177–195. Springer, Heidelberg (2013). doi:10.1007/978-3-642-36334-4_11
10. Ghammam, L., Fouotsa, E.: On the computation of the optimal ate pairing at the 192-bit security level. IACR Cryptology ePrint Archive, 2016:130 (2016)
11. Chen, L., Cheng, Z., Smart, N.P.: A built-in decisional function and security proof of id-based key agreement protocols from pairings. IACR Cryptology ePrint Archive, 2006:160 (2006)
12. Boneh, D., Boyen, X., Shacham, H.: Short group signatures. In: Franklin, M. (ed.) CRYPTO 2004. LNCS, vol. 3152, pp. 41–55. Springer, Heidelberg (2004). doi:10.1007/978-3-540-28628-8_3
13. Abdalla, M., Catalano, D., Dent, A.W., Malone-Lee, J., Neven, G., Smart, N.P.: Identity-based encryption gone wild. In: Bugliesi, M., Preneel, B., Sassone, V., Wegener, I. (eds.) ICALP 2006. LNCS, vol. 4052, pp. 300–311. Springer, Heidelberg (2006). doi:10.1007/11787006_26
14. Waters, B.: Efficient identity-based encryption without random oracles. In: Cramer, R. (ed.) EUROCRYPT 2005. LNCS, vol. 3494, pp. 114–127. Springer, Heidelberg (2005). doi:10.1007/11426639_7
15. Granger, R., Smart, N.P.: On computing products of pairings. IACR Cryptology ePrint Archive, 2006:172 (2006)
16. Kachisa, E.J., Schaefer, E.F., Scott, M.: Constructing Brezing-Weng pairing friendly elliptic curves using elements in the cyclotomic field. IACR Cryptology ePrint Archive 2007:452 (2007)
17. Barreto, P.S.L.M., Costello, C., Misoczki, R., Naehrig, M., Pereira, G.C.C.F., Zanon, G.: Subgroup security in pairing-based cryptography. In: Lauter, K., Rodríguez-Henríquez, F. (eds.) LATINCRYPT 2015. LNCS, vol. 9230, pp. 245–265. Springer, Cham (2015). doi:10.1007/978-3-319-22174-8_14
18. Fouotsa, E., Ghammam, L.: http://www.camercrypt.org/KSS16-finalexponentiation
19. Kim, T.: Extended tower number field sieve: a new complexity for medium prime case. IACR Cryptology ePrint Archive, 2015:1027 (2015)

20. Jeong, J., Kim, T.: Extended tower number field sieve with application to finite fields of arbitrary composite extension degree. IACR Cryptology ePrint Archive, 2016:526 (2016)
21. Miller, V.S.: The Weil pairing, and its efficient calculation. J. Cryptol. **17**(4), 235–261 (2004)
22. Scott, M., Benger, N., Charlemagne, M., Dominguez Perez, L.J., Kachisa, E.J.: On the final exponentiation for calculating pairings on ordinary elliptic curves. In: Shacham, H., Waters, B. (eds.) Pairing 2009. LNCS, vol. 5671, pp. 78–88. Springer, Heidelberg (2009). doi:10.1007/978-3-642-03298-1_6
23. Fuentes-Castañeda, L., Knapp, E., Rodríguez-Henríquez, F.: Faster hashing to $\mathbb{G}_2$. In: Miri, A., Vaudenay, S. (eds.) SAC 2011. LNCS, vol. 7118, pp. 412–430. Springer, Heidelberg (2012). doi:10.1007/978-3-642-28496-0_25
24. Smeets, I., Lenstra, A.K., Lenstra, H., Lovász, L., van Emde Boas, P.: The history of the LLL-algorithm. In: Nguyen, P.Q., Vallée, B. (eds.) The LLL Algorithm - Survey and Applications, pp. 1–17. Springer, Heidelberg (2010)
25. Boneh, D., Boyen, X., Goh, E.-J.: Hierarchical identity based encryption with constant size ciphertext. In: Cramer, R. (ed.) EUROCRYPT 2005. LNCS, vol. 3494, pp. 440–456. Springer, Heidelberg (2005). doi:10.1007/11426639_26
26. Groth, J., Sahai, A.: Efficient non-interactive proof systems for bilinear groups. In: Smart, N. (ed.) EUROCRYPT 2008. LNCS, vol. 4965, pp. 415–432. Springer, Heidelberg (2008). doi:10.1007/978-3-540-78967-3_24
27. Scott, M.: Computing the tate pairing. In: Menezes, A. (ed.) CT-RSA 2005. LNCS, vol. 3376, pp. 293–304. Springer, Heidelberg (2005). doi:10.1007/978-3-540-30574-3_20
28. Washington, L.C.: Elliptic Curves Number Theory and Cryptography. Discrete Mathematics and Its Applications. Chapman and Hall, London (2008)
29. Hesse, F., Smart, N.P., Vercauteren, F.: The eta pairing revisited. IEEE Trans. Inf. Theory **52**(10), 4595–4602 (2006)

On Pseudorandom Properties of Certain Sequences of Points on Elliptic Curve

László Mérai(✉)

Johann Radon Institute for Computational and Applied Mathematics,
Austrian Academy of Sciences, Altenbergerstr. 69, 4040 Linz, Austria
laszlo.merai@oeaw.ac.at

Abstract. In this paper we study the pseudorandom properties of sequences of points on elliptic curves. These sequences are constructed by taking linear combinations with small coefficients (e.g. $-1, 0, +1$) of the orbit elements of a point with respect to a given endomorphism of the curve. We investigate the linear complexity and the distribution of these sequences. The result on the linear complexity answers a question of Igor Shparlinski.

1 Introduction

For a prime power $q = p^n$ we denote by $\mathbb{F}_q$ the field of q elements. Let $\mathbf{E}$ be a non-singular elliptic curve over $\mathbb{F}_q$ defined by

$$\mathbf{E}: Y^2 + (a_1X + a_3)Y = X^3 + a_2X^2 + a_4X + a_6,$$

with some $a_1, \ldots, a_6 \in \mathbb{F}_q$ (see [12]). We recall that the $\mathbb{F}_q$-rational points $\mathbf{E}(\mathbb{F}_q)$ of the curve $\mathbf{E}$ with the usual addition $\oplus$ form an Abelian group with the point at infinity $\mathcal{O}$ as a neutral element. We write every point $P \neq \mathcal{O}$ on $\mathbf{E}$ as $P = (x(P), y(P))$.

Every integer $m \neq 0$ has a unique *non-adjacent binary expansion*, also called binary NAF, for some length k:

$$m = \sum_{j=0}^{k-1} \mu_j 2^j, \quad \text{with } (\mu_0, \ldots, \mu_{k-1}) \in \mathcal{M}_k, \tag{1}$$

where $\mathcal{M}_k$ is the set of k-tuples with components $0, \pm 1$ such that there are no two consecutive non-zero components:

$$\mathcal{M}_k = \{(\mu_0, \ldots, \mu_{k-1}) \in \{0, \pm 1\}^k \mid \mu_j \mu_{j+1} = 0\}.$$

This expansion provides a faster scalar multiplication compared to the double and add algorithm as the number of additions is reduced and the number of doublings is kept constant, as additions and subtractions of points cost about the same. We remark that the Hamming weight of this representation of m is

S. Duquesne and S. Petkova-Nikova (Eds.): WAIFI 2016, LNCS 10064, pp. 54–63, 2016.
DOI: 10.1007/978-3-319-55227-9_4

minimal among all signed-digit representations of m. It is not hard to show (see [2]) that for $k \geq 2$

$$\#\mathcal{M}_k = \frac{4}{3}2^k + O(1).$$

We also remark that given an integer m (in binary form) one can efficiently computes its NAF representation.

This concept is generalized from the endomorphism doubling δ to arbitrary endomorphisms by Lange and Shparlinski [8]. Given an $\mathbb{F}_q$-rational point $P \in \mathbf{E}(\mathbb{F}_q)$ and an endomorphism σ on $\mathbf{E}$ we consider the set of points

$$P_{\sigma,\mathbf{m}} = \sum_{j=0}^{k-1} \mu_j \sigma^j(P), \quad \mathbf{m} = (\mu_0, \ldots, \mu_{k-1}) \in \mathcal{M}_k.$$

The endomorphism doubling $\delta(P) = 2P$ is defined for any elliptic curve over any finite field. In this case we have

$$P_{\delta,\mathbf{m}} = mP,$$

where $\mathbf{m}$ is the NAF representation (1) of m.

Two further examples of endomorphisms were also considered in [8]. For $a \in \mathbb{F}_2$ we define the *Koblitz curve* $\mathbf{E}_a$ over $\mathbb{F}_2$ by the Weierstrass equation

$$\mathbf{E}_a : Y^2 + XY = X^3 + aX^2 + 1$$

introduced in [5]. We define the *Frobenius endomorphism* φ which acts on an $\mathbb{F}_{2^n}$-rational point $P = (x, y)$ as

$$\varphi(P) = (x^2, y^2).$$

Finally, we also consider one of the GLV curves introduced by Gallant, Lambert and Vanstone [4]. Let $p > 3$ be a prime number such that -7 is a quadratic residue modulo p (that is, $p \equiv 1, 2, 4 \pmod 7$). Define the elliptic curve $\mathbf{E}_{\mathrm{GLV}}$ over $\mathbb{F}_p$ as

$$\mathbf{E}_{\mathrm{GLV}} : Y^2 = X^3 - \frac{3}{4}X^2 - 2X - 1.$$

If $b = (1 + \sqrt{-7})/2$ and $c = (b - 3)/4$, then the map ψ defined as

$$\psi(P) = \left(\frac{x^2 - b}{b^2(x - c)}, \frac{y(x^2 - 2cx + b)}{b^3(x - c)^2} \right)$$

for $P = (x, y) \in \mathbf{E}_{\mathrm{GLV}}$ is an endomorphism of $\mathbf{E}_{\mathrm{GLV}}$.

The behavior of the point set $P_{\sigma,\mathbf{m}}$ ($\mathbf{m} \in \mathcal{M}_k$) was studied by Lange and Shaprlinski [7,8]. First they gave an upper bound on the number of collisions of these points.

Proposition 1. *Let $P \in \mathbf{E}(\mathbb{F}_q)$ be of prime order ℓ and let $N_k(Q)$ be the number of representations*

$$P_{\sigma,\mathbf{m}} = Q, \quad \mathbf{m} \in \mathcal{M}_k.$$

If σ is one of the following endomorphisms:

- δ *for an arbitrary curve* $\mathbf{E}$,
- φ *for a Koblitz curve* $\mathbf{E} = \mathbf{E}_a$, $a = 0, 1$,
- ψ *for the GLV curve* $\mathbf{E} = \mathbf{E}_{\text{GLV}}$,

then for any integer r *with* $1 \le r \le k$ *and* $2^r \le \ell/8$ *we have* $N_k(Q) \le \#\mathcal{M}_{k-r}$.

In particular, the bound of Proposition 1 implies that if $2^k < \ell/8$, then the points $P_{\sigma,\mathbf{m}}$ are all distinct. For larger k choosing $r = \lfloor \log_2 \ell \rfloor - 3$ we obtain $N_k(Q) = O(2^k \ell^{-1})$.

Next, Lange and Shaprlinski [8] also studied the distribution of the points $P_{\sigma,\mathbf{m}}$ ($\mathbf{m} \in \mathcal{M}_k$). For a non-trivial additive character χ, they proved that

$$\sum_{\mathbf{m} \in \mathcal{M}_k} \chi(x(P_{\sigma,\mathbf{m}})) \ll \#\mathcal{M}_k \left(q^{1/4\nu} \ell^{-1/2\nu} + 2^{-k/2\nu} q^{(\nu+1)/4\nu^2} \right) \qquad (2)$$

holds with any fixed integer $\nu \ge \log q / 2k$ where σ is one of the endomorphisms in Proposition 1. From the bound (2) they proved that the set of points $P_{\sigma,\mathbf{m}}$ ($\mathbf{m} \in \mathcal{M}_k$) has good uniformity of distribution properties.

In this paper we study the *sequence* of points $P_{\sigma,\mathbf{m}}$ arranged in a sequence by ordering the vectors $\mathbf{m} \in \mathcal{M}_k$ lexicographically. First, we give a lower bound to the linear complexity of the coordinate-sequence $x(P_{\sigma,\mathbf{m}})$ in Sect. 2. This result gives an answer to a question of Igor Shparlinki (Question 31 in [11]).

Next, in Sect. 3 we extend the result (2). For a vector $\mathbf{m} \in \mathcal{M}_k$, let $\tau(\mathbf{m}) \in \mathcal{M}_k$ be the successor of $\mathbf{m}$ with respect to the lexicographic ordering. Then we study the distribution of vectors

$$\left(x\left(P_{\sigma,\mathbf{m}}\right), x\left(P_{\sigma,\tau(\mathbf{m})}\right), \ldots, x\left(P_{\sigma,\tau^{s-1}(\mathbf{m})}\right)\right), \quad \mathbf{m} \in \mathcal{M}_k : \ \tau^{s-1}(\mathbf{m}) \in \mathcal{M}_k. \ (3)$$

First we give an upper bound to the character sum

$$S_{\sigma,k,s}(\chi) = \max_{(a_0,\ldots,a_{s-1}) \neq (0,\ldots,0)} \sum_{\mathbf{n} \in \mathcal{M}_k^*} \chi\left(\sum_{i=0}^{s-1} a_i x\left(P_{\sigma,\tau^i(\mathbf{n})}\right) \right),$$

where in the sum we exclude the last $s-2$-many elements of $\mathcal{M}_k$ and χ is a non-trivial additive character. We apply this result to show that the vectors (3) have good uniformity of distribution properties.

2 Linear Complexity

The *linear complexity* of a sequence (s_n) of length M over a ring R is the length L of a shortest linear recurrence relation

$$s_{n+L} = c_{L-1} s_{n+L-1} + \cdots + c_1 s_{n+1} + c_0 s_n, \quad n = 0, \ldots, M - L - 1$$

for some $c_0, \ldots, c_{L-1} \in R$, that (s_n) satisfies.

The linear complexity measures the unpredictability of a sequence and thus its suitability in cryptography. For more details see [9, 10, 13].

Theorem 1. *Let $P \in \mathbf{E}(\mathbb{F}_q)$ be of prime order ℓ. Then for any $k \geq 1$ the linear complexity of $(x(P_{\sigma,\mathbf{m}}))_{\mathbf{m}\in\mathcal{M}_k}$ satisfies*

$$L(x(P_{\sigma,\mathbf{m}})) \gg \min\left\{\#\mathcal{M}_{\lceil k/2\rceil}, \ell\right\}.$$

The following lemma is a crucial step in the proof.

Lemma 1. *If P has prime order ℓ and ℓ does not divide the constant term of the characteristic polynomial of σ, then $\sigma(P) \neq \mathcal{O}$. It happens if $\ell > 2$ and σ is one of the selected endomorphism.*

Proof. Let $\chi_\sigma = aT^2 + bT + c \in \mathbb{F}_q[T]$ be the characteristic polynomial of σ. If $\sigma(P) = \mathcal{O}$, then $\mathcal{O} = -a\sigma^2(P) - b\sigma(P) = cP$, thus the order ℓ of P divides c. Finally, we remark, that if σ is one of the endomorphism in Proposition 1, then the constant term of χ_σ is 2 (see [8]). □

Proof (Theorem 1). For $k = 1, 2$ the result is trivial, so we may assume that $k \geq 3$. Put $r = \lfloor k/2 \rfloor$ and consider the set of vectors

$$\mathbf{n} = (\nu_0, \ldots, \nu_{k-r-2}, 0, \nu_{k-r}, \ldots, \nu_{k-1}) \in \mathcal{M}_k.$$

Clearly, $\mathbf{n}$ can be written as $\mathbf{n} = (\mathbf{u}, 0, \mathbf{v})$ with $\mathbf{u} \in \mathcal{M}_{k-r-1}$, $\mathbf{v} \in \mathcal{M}_r$.

Let L be the linear complexity of the sequence $x(P_{\sigma,\mathbf{m}})$ and N be the number of distinct points $P_{\sigma,\mathbf{v}}$, $\mathbf{v} \in \mathcal{M}_r$. By Proposition 1, we have $N \geq \#\mathcal{M}_r / \max_Q N_r(Q) \geq \#\mathcal{M}_r / \max\{1, \#\mathcal{M}_{r-\lfloor \log_2 \ell \rfloor + 3}\} \gg \min\{\#\mathcal{M}_r, \ell\}$. If $N \leq L$, then the theorem holds, otherwise there are vectors $\mathbf{d}_0, \ldots, \mathbf{d}_L \in \mathcal{M}_r$, such that the points $P_{\sigma,\mathbf{d}_0}, \ldots, P_{\sigma,\mathbf{d}_L}$ are all distinct. Fix these vectors and for each $j = 0, \ldots, L$ define the sequence

$$x_j(\mathbf{s}) = x\left(P_{\sigma,(\mathbf{s},0,\mathbf{d}_j)}\right), \quad \mathbf{s} \in \mathcal{M}_{k-r-1},$$

where again the elements $x_j(\mathbf{s})$ arranged in a sequence by ordering the vectors $\mathbf{s}$ lexicographically. The sequences $x_j(\mathbf{s})$ are subsequences of $x(P_{\sigma,\mathbf{m}})$, thus they satisfy the same linear recurrence relation of order L. Then these sequences are in a vector space of dimension at most L so they are linearly dependent, i.e. there are constants $c_0, \ldots, c_L \in \mathbb{F}_q$, not all of them are zero, such that

$$c_0 x\left(P_{\sigma,(\mathbf{s},0,\mathbf{d}_0)}\right) + \cdots + c_L x\left(P_{\sigma,(\mathbf{s},0,\mathbf{d}_L)}\right) = 0, \quad \mathbf{s} \in \mathcal{M}_{k-r-1},$$

i.e.,

$$c_0 x\left(P_{\sigma,\mathbf{s}} + \sigma^{k-r}\left(P_{\sigma,\mathbf{d}_0}\right)\right) + \cdots + c_L x\left(P_{\sigma,\mathbf{s}} + \sigma^{k-r}\left(P_{\sigma,\mathbf{d}_L}\right)\right) = 0, \quad \mathbf{s} \in \mathcal{M}_{k-r-1}.$$

By Lemma 1 the points $\sigma^{k-r}\left(P_{\sigma,\mathbf{d}_0}\right), \ldots, \sigma^{k-r}\left(P_{\sigma,\mathbf{d}_L}\right)$ are pairwise distinct so the function

$$F(Q) = c_0 x\left(Q + \sigma^{k-r}\left(P_{\sigma,\mathbf{d}_0}\right)\right) + \cdots + c_L x\left(Q + \sigma^{k-r}\left(P_{\sigma,\mathbf{d}_L}\right)\right)$$

has $L+1$ poles and the points $P_{\sigma,\mathbf{s}}$, $\mathbf{s} \in \mathcal{M}_{k-r-1}$ are all zeros of it.

Since there are at least

$$\#\mathcal{M}_{k-r-1}/\max_{Q} N_{k-r-1}(Q) \geq \#\mathcal{M}_{k-r-1}/\max\{1, \#\mathcal{M}_{k-r-1-\lfloor\log_2 \ell\rfloor+3}\}$$
$$\gg \min\{\#\mathcal{M}_r, \ell\}$$

distinct points among them, we get the result comparing the number of poles and zeros of F. □

3 Distribution

In this section we first prove a bound on a character sum. This bound immediately implies results about pseudorandomness of the points $P_{\sigma,\mathbf{m}}$.

Theorem 2. *Let $P \in \mathbf{E}(\mathbb{F}_q)$ be of prime order ℓ and χ be a non-trivial additive character of $\mathbb{F}_q$. Then for any $k \geq 1$ and $s \geq 2$ we have that*

$$S_{\sigma,k,s}(\chi) \ll \#\mathcal{M}_k\, s \left(q^{1/4\nu}\ell^{-1/2\nu} + 2^{-k/2\nu}q^{(\nu+1)/4\nu^2}\right)$$

holds with any fixed integer

$$\nu > \frac{\log q}{\min\{2k, 2(\log \ell - 3)\}}$$

if σ is one of the endomorphisms in Proposition 1.

The proof of the theorem is based on the following character sum estimate [1]. Let $\mathbb{F}_q(\mathbf{E})$ be the function field of the curve $\mathbf{E}$. If $f \in \mathbb{F}_q(\mathbf{E})$ and χ is an additive character, write $\chi(f(Q)) = 0$ whenever Q is a pole of f.

Lemma 2. *Let $\mathbf{E}$ be an elliptic curve defined over $\mathbb{F}_q$. Let $f \in \mathbb{F}_q(\mathbf{E})$ and suppose that $f \neq z^p - z$ for all $z \in \overline{\mathbb{F}_q}(\mathbf{E})$. Let χ be a non-trivial additive character of $\mathbb{F}_q$. Then the bound*

$$\left|\sum_{Q\in\mathbf{E}(\mathbb{F}_q)} \chi\left(f(Q)\right)\right| \leq 2 \deg f q^{1/2}$$

holds.

Proof (Theorem 2). Let us fix some positive integer $r < \min\{k, \log \ell - 3\}$.

For $j = 0, 1$ we define $\mathcal{U}_j$ to be the subset of vectors $\mathbf{u} = (u_0, \ldots, u_{r-1}) \in \mathcal{M}_r$ with $u_{r-1} = \pm j$. To form a vector in $\mathcal{M}_k$, a vector from $\mathcal{U}_0$ can be appended by any vector from $\mathcal{V}_0 = \mathcal{M}_{k-r}$, while a vector from $\mathcal{U}_1$ requires the following digit to be zero. Hence, we put

$$\mathcal{V}_1 = \{(0, \mathbf{w}) : \ \mathbf{w} \in \mathcal{M}_{k-r-1}\}.$$

We have

$$\sum_{\mathbf{m}\in\mathcal{M}_k^*} \chi\left(\sum_{i=0}^{s-1} a_i x\left(P_{\sigma,\tau^i(\mathbf{m})}\right)\right) = R_{\sigma,0} + R_{\sigma,1},$$

where

$$R_{\sigma,j} = \sum_{\substack{\mathbf{u}\in\mathcal{U}_j\\(\mathbf{u},\mathbf{v})\in\mathcal{M}_k^*}} \sum_{\mathbf{v}\in\mathcal{V}_j} \chi\left(\sum_{i=0}^{s-1} a_i x\left(P_{\sigma,\tau^i((\mathbf{u},\mathbf{v}))}\right)\right)$$

$$= \sum_{\mathbf{u}\in\mathcal{U}_j^*} \sum_{\mathbf{v}\in\mathcal{V}_j} \chi\left(\sum_{i=0}^{s-1} a_i x\left(P_{\sigma,\tau^i(\mathbf{u})} + \sigma^r\left(P_{\sigma,\mathbf{v}}\right)\right)\right) + O\left(s\cdot\#\mathcal{V}_j\right),$$

where the asterisks indicates that we exclude the last $s-1$ elements on $\mathcal{U}_j$.

For $j=0$ we have by the Hölder inequality that

$$\left(\sum_{\mathbf{v}\in\mathcal{V}_0}\left|\sum_{\mathbf{u}\in\mathcal{U}_0^*} \chi\left(\sum_{i=0}^{s-1} a_i x\left(P_{\sigma,\tau^i(\mathbf{u})} + \sigma^r\left(P_{\sigma,\mathbf{v}}\right)\right)\right)\right|\right)^{2\nu}$$

$$\le \#\mathcal{V}_0^{2\nu-1} \sum_{\mathbf{v}\in\mathcal{V}_0}\left|\sum_{\mathbf{u}\in\mathcal{U}_0^*} \chi\left(\sum_{i=0}^{s-1} a_i x\left(P_{\sigma,\tau^i(\mathbf{u})} + \sigma^r\left(P_{\sigma,\mathbf{v}}\right)\right)\right)\right|^{2\nu}. \tag{4}$$

Now $\#\mathcal{V}_0 = \#\mathcal{M}_{k-r} = O(2^{k-r})$. Then by Proposition 1 and Lemma 1 we have that every point Q is represented by $\sigma^r(P_{\sigma,\mathbf{v}})$ ($\mathbf{v}\in\mathcal{V}_0$) in at most $O(2^{k-r}\ell^{-1}+1)$ times. So (4) is

$$\ll 2^{(k-r)(2\nu-1)}(2^{k-r}\ell^{-1}+1) \sum_{Q\in\mathbf{E}(\mathbb{F}_p)}\left|\sum_{\mathbf{u}\in\mathcal{U}_0^*} \chi\left(\sum_{i=0}^{s-1} a_i x\left(P_{\sigma,\tau^i(\mathbf{u})} + Q\right)\right)\right|^{2\nu}.$$

As for every complex number z we have $|z|^2 = z\overline{z}$ and for any $u\in\mathbb{F}_q$ we have $\overline{\chi(u)} = \chi(-u)$, we get

$$\sum_{Q\in\mathbf{E}(\mathbb{F}_p)}\left|\sum_{\mathbf{u}\in\mathcal{U}_0^*} \chi\left(\sum_{i=0}^{s-1} a_i x\left(P_{\sigma,\tau^i(\mathbf{u})} + Q\right)\right)\right|^{2\nu}$$

$$= \sum_{\mathbf{u}_1,\dots,\mathbf{u}_{2\nu}\in\mathcal{U}_0^*} \sum_{Q\in\mathbf{E}(\mathbb{F}_p)} \chi\left(\sum_{h=1}^{\nu}\sum_{i=0}^{s-1} a_i \left(x\left(P_{\sigma,\tau^i(\mathbf{u}_h)} + Q\right) - x\left(P_{\sigma,\tau^i(\mathbf{u}_{h+\nu})} + Q\right)\right)\right). \tag{5}$$

If all the values of $\mathbf{u}_1,\dots,\mathbf{u}_{2\nu}$ occur more than once in $(\mathbf{u}_1,\dots,\mathbf{u}_{2\nu})$, we estimate the sum trivially by $\#\mathbf{E}(\mathbb{F}_q)\ll q$. It happens at most $(\nu\cdot\#\mathcal{U}_0^*)^\nu$ times. If there is at least one unique value in $(\mathbf{u}_1,\dots,\mathbf{u}_{2\nu})$, we show that the function

$$F(Q) = \sum_{h=1}^{\nu} \sum_{i=0}^{s-1} a_i \left(x\left(P_{\sigma,\tau^i(\mathbf{u}_h)} + Q\right) - x\left(P_{\sigma,\tau^i(\mathbf{u}_{h+\nu})} + Q\right)\right)$$

cannot have the form $z^p - z$ with $z \in \overline{\mathbb{F}_q(\mathbf{E})}$.

Let $\mathbf{u}$ be the least value in $(\mathbf{u}_1, \dots, \mathbf{u}_{2\nu})$ with respect to the lexicographic ordering, such that

$$\sum_{\substack{1 \le h \le \nu \\ \mathbf{u}_h = \mathbf{u}}} x\left(P_{\sigma,\mathbf{u}_h} + Q\right) \neq \sum_{\substack{\nu+1 \le h \le 2\nu \\ \mathbf{u}_h = \mathbf{u}}} x\left(P_{\sigma,\mathbf{u}_h} + Q\right). \tag{6}$$

Since there is at least one unique value in $(\mathbf{u}_1, \dots, \mathbf{u}_{2\nu})$, there exists such a $\mathbf{u}$.

Since $2^r < \ell/8$, we get from Proposition 1, that $P_{\sigma,\mathbf{u}'} \neq P_{\sigma,\mathbf{u}''}$ for any $\mathbf{u}' \neq \mathbf{u}''$, $\mathbf{u}', \mathbf{u}'' \in \mathcal{U}_0^* \subset \mathcal{M}_r$. Let $<_l$ denote the lexicographic ordering in $\mathcal{M}_r$. As $\tau^i(\mathbf{u}) <_l \tau^j(\mathbf{u}')$ if $i < j$ or $\mathbf{u}' \neq \mathbf{u}$ is a vector satisfying (6), we have $P_{\sigma,\tau^i(\mathbf{u})} \neq P_{\sigma,\tau^j(\mathbf{u}')}$ with the same conditions on i, j and $\mathbf{u}, \mathbf{u}'$. Thus if i is the least index such that $a_i \neq 0$, then the term $x\left(P_{\sigma,\tau^i(\mathbf{u})} + Q\right)$ does not vanish in F by the choice of $\mathbf{u}$, and $-P_{\sigma,\tau^i(\mathbf{u})}$ is not a pole of any other term of F, thus F cannot have the form $z^p - z$. Then from Lemma 2 we get that (5) is

$$\sum_{\mathbf{u}_1, \dots, \mathbf{u}_{2\nu} \in \mathcal{U}_0^*} \sum_{Q \in \mathbf{E}(\mathbb{F}_q)} \chi(F(Q)) \ll \#\mathcal{U}_0^{2\nu} s q^{1/2} + \#\mathcal{U}_0^{\nu} q.$$

Since $\#\mathcal{U}_0 \le \#\mathcal{M}_r = O(2^r)$, we have that

$$R_{\sigma,0}^{2\nu} \ll 2^{(k-r)(2\nu-1)}(2^{k-r}\ell^{-1} + 1)(2^{2r\nu} s q^{1/2} + 2^{r\nu} q) + s^{2\nu} 2^{(k-r)2\nu}.$$

Using the same argument one can also obtain

$$R_{\sigma,1}^{2\nu} \ll 2^{(k-r-1)(2\nu-1)}(2^{k-r-1}\ell^{-1} + 1)(2^{2r\nu} s q^{1/2} + 2^{r\nu} q) + s^{2\nu} 2^{(k-r-1)2\nu}.$$

Thus

$$|S_{\sigma,k,s}(\chi)|^{2\nu} \ll 2^{(k-r)(2\nu-1)}(2^{k-r}\ell^{-1} + 1)(2^{2r\nu} s q^{1/2} + 2^{r\nu} q) + s^{2\nu} 2^{(k-r)2\nu}.$$

Choosing

$$r = \left\lfloor \frac{\log q}{2\nu} \right\rfloor < k$$

we get that the terms $2^{2r\nu} q^{1/2}$ and $2^{r\nu} q$ are $O(q^{3/2})$. Hence

$$\begin{aligned} |S_{\sigma,k,s}(\chi)|^{2\nu} &\ll s 2^{(k-r)2\nu} \ell^{-1} q^{3/2} + 2^{(k-r)(2\nu-1)} q^{3/2} + s^{2\nu} 2^{(k-r)2\nu} \\ &\ll s 2^{2\nu k + \log q/2} \ell^{-1} + 2^{2\nu k - k + (\nu+1)\log q/2\nu} + s^{2\nu} 2^{2\nu k - \log q} \end{aligned}$$

which proves the result. □

The most interesting case is when the $\mathbf{E}(\mathbb{F}_q)$ has a cyclic group structure with prime order ℓ. Then by the Hasse-Weil theorem, we have that $\ell = q^{1+o(1)}$. If $k = (1 + o(1)) \log q$, then choosing $\nu = 2$, Theorem 2 implies

$$S_{\sigma,k,s}(\chi) \ll \#\mathcal{M}_k \, s q^{-1/16}.$$

As an application of Theorem 2, we get results about pseudorandomness of the vectors (3).

Let $\beta_1, \ldots, \beta_n$ be a fixed basis of $\mathbb{F}_{p^n}$ over $\mathbb{F}_p$. For a fixed subset $J \subset \{1, \ldots, n\}$, and fixed elements $c_{i,j} \in \mathbb{F}_q$, $i \in \{0, 1, \ldots, s-1\}$, $j \in I$, put

$$N\big(J, (c_{i,j})\big) = \# \left\{ \mathbf{m} \in \mathcal{M}_k : \ x\left(P_{\sigma, \tau^i(\mathbf{m})}\right)_j = c_{i,j}, i \in \{0, 1 \ldots, s-1\}, j \in J \right\}.$$

Using the standard techniques (see e.g. [8]) to express the deviation of $N\big(J, (c_{i,j})\big)$ from its expected value $\mathcal{M}_k / p^{\# J \cdot s}$ by character sums we get from Theorem 2.

Corollary 1. *Let p be a prime number and let $P \in \mathbf{E}(\mathbb{F}_{p^n})$ be of prime order ℓ. Then for any integers $k, s \geq 1$ the bound*

$$\max_{J, (c_{i,j})} \left| N\big(J, (c_{i,j})\big) - \frac{\# \mathcal{M}_k}{p^{\# J \cdot s}} \right| \ll \# \mathcal{M}_k \, s \left(q^{1/4\nu} \ell^{-1/2\nu} + 2^{-k/2\nu} q^{(\nu+1)/4\nu^2} \right)$$

holds with any fixed integer

$$\nu > \frac{\log q}{\min\{2k, 2(\log \ell - 3)\}},$$

where σ is one of the following endomorphisms:

- *δ for an arbitrary curve $\mathbf{E}$,*
- *φ for a Koblitz curve $\mathbf{E} = \mathbf{E}_a$, $a = 0, 1$.*

If again $\ell = q^{1+o(1)}$ and $k = \lfloor \log q \rfloor$, then choosing $\nu = 2$, Corollary 1 implies that for any $\alpha < 1/16$ the components of (3) on any $\# \mathcal{J} \leq \alpha n$ positions are uniformity distributed.

If the point P is $\mathbb{F}_p$-rational, then the sequence $x(P_{\sigma, \mathbf{m}})$ is a sequence in a prime field $\mathbb{F}_p$. In this case we can study the *discrepancy* of this sequence. More precisely, let $D_{\psi, k, s}$ be the discrepancy of the vectors

$$\left(\frac{x\left(P_{\psi, \mathbf{m}}\right)}{p}, \frac{x\left(P_{\psi, \tau(\mathbf{m})}\right)}{p}, \ldots, \frac{x\left(P_{\psi, \tau^{s-1}(\mathbf{m})}\right)}{p} \right), \quad \mathbf{m} \in \mathcal{M}_k, \tag{7}$$

in the s-dimensional unite cube. That is

$$D_{\psi, k, s} = \sup_{I \subset [0,1)^s} \left| \frac{T(I)}{\# \mathcal{M}_k} - |I| \right|,$$

where $T(I)$ is the number of points (7) which hit the s-dimensional interval $I = [\alpha_1, \beta_1) \times \cdots \times [\alpha_s, \beta_s)$ of size $|I| = (\beta_1 \alpha_1) \cdots (\beta_s - \alpha_s)$.

Using the *Erdős-Turán inequality*, see [3,6], relating the discrepancy and exponential sums we immediately derive:

Corollary 2. *Let p be a prime number and let $P \in \mathbf{E}(\mathbb{F}_p)$ be of prime order ℓ. Then for any integers $k, s \geq 1$ the bound*

$$D_{\psi,k,s} \ll \#\mathcal{M}_k \, s \left(q^{1/4\nu} \ell^{-1/2\nu} + 2^{-k/2\nu} q^{(\nu+1)/4\nu^2} \right)$$

holds with any fixed integer

$$\nu > \frac{\log q}{\min\{2k, 2(\log \ell - 3)\}}.$$

where σ is one of the following endomorphisms:

– *δ for an arbitrary curve $\mathbf{E}$,*
– *ψ for the GLV curve $\mathbf{E} = \mathbf{E}_{\mathrm{GLV}}$.*

Acknowledgements. The author is partially supported by the Austrian Science Fund FWF Project F5511-N26 which is part of the Special Research Program "Quasi-Monte Carlo Methods: Theory and Applications" and by Hungarian National Foundation for Scientific Research, Grant No. K100291.

References

1. Beelen, P.H.T., Doumen, J.M.: Pseudorandom sequences from elliptic curves. In: Mullen, G.L., Stichtenoth, H., Tapia-Recillas, H. (eds.) Finite Fields with Applications to Coding Theory, Cryptography and Related Areas (Oaxaca, 2001), pp. 37–52. Springer, Berlin (2002). doi:10.1007/978-3-642-59435-9_3
2. Bosma, W.: Signed bits and fast exponentiation. J. Théor. Nombres Bordx. **13**(1), 27–41 (2001). 21st Journées Arithmétiques (Rome, 2001)
3. Drmota, M., Tichy, R.: Sequences, Discrepancies and Applications. Springer, Berlin (1997)
4. Gallant, R.P., Lambert, R.J., Vanstone, S.A.: Faster point multiplication on elliptic curves with efficient endomorphisms. In: Kilian, J. (ed.) CRYPTO 2001. LNCS, vol. 2139, pp. 190–200. Springer, Heidelberg (2001). doi:10.1007/3-540-44647-8_11
5. Koblitz, N.: CM-curves with good cryptographic properties. In: Feigenbaum, J. (ed.) CRYPTO 1991. LNCS, vol. 576, pp. 279–287. Springer, Heidelberg (1992). doi:10.1007/3-540-46766-1_22
6. Kuipers, L., Niederreiter, H.: Uniform Distribution of Sequences. John Wiley, New York (1974)
7. Lange, T., Shparlinski, I.E.: Collisions in fast generation of ideal classes and points on hyperelliptic and elliptic curves. Appl. Algebra Eng. Commun. Comput. **15**, 329–337 (2005)
8. Lange, T., Shparlinski, I.E.: Distribution of some sequences of points on elliptic curves. J. Math. Cryptol. **1**(1), 1–11 (2007)
9. Meidl, W., Winterhof, A.: Linear complexity of sequences and multisequences. In: Mullen, G., Panario, D. (eds.) Handbook of Finite Fields, pp. 324–336. Chapman & Hall, London (2013)
10. Niederreiter, H.: Linear complexity and related complexity measures for sequences. In: Johansson, T., Maitra, S. (eds.) INDOCRYPT 2003. LNCS, vol. 2904, pp. 1–17. Springer, Heidelberg (2003). doi:10.1007/978-3-540-24582-7_1

11. Shparlinski, I.E.: Pseudorandom number generators from elliptic curves. Recent Trends Cryptogr. Contemp. Math. **477**, 121–141 (2009). American Mathematical Society, Providence, RI
12. Silverman, J.H.: The Arithmetic of Elliptic Curves. Springer, Berlin (1995)
13. Winterhof, A.: Linear complexity and related complexity measures. In: Selected Topics in Information and Coding Theory, pp. 3–40. World Scientific, Singapore (2010)

Applications

Linear Complexity and Expansion Complexity of Some Number Theoretic Sequences

Richard Hofer(✉) and Arne Winterhof

Johann Radon Institute for Computational and Applied Mathematics, Austrian Academy of Sciences, Altenberger Str. 69, 4040 Linz, Austria
{richard.hofer,arne.winterhof}@oeaw.ac.at

Abstract. We study the predictability of some number theoretic sequences over finite fields and thus their suitability in cryptography. First we analyze the non-periodic binary sequence $\mathcal{T} = (t_n)_{n\geq 0}$ with $t_n = 1$ whenever n is the sum of three integer squares. We show that it has a large Nth linear complexity, which is necessary but not sufficient for unpredictability. However, it also has a very small expansion complexity and thus is rather predictable.

Next we prove that some linear combinations of p-periodic sequences of binomial coefficients modulo a prime p have a very small expansion complexity and are predictable despite of a high linear complexity.

Finally, we consider the Legendre sequence and verify that it does not belong to this class of predictable sequences.

Keywords: Expansion complexity · Linear complexity · Generating function · Three-square theorem · Automatic sequence · Linear recurrence sequence · Binomial coefficients · Legendre sequence

1 Introduction

The expansion complexity for sequences over finite fields $\mathbb{F}_q$ was introduced by Diem in [7]. For a sequence $\mathcal{S} = (s_n)_{n\geq 0}$ over $\mathbb{F}_q$ the *generating function* $G(x)$ of $\mathcal{S}$ is

$$G(x) = \sum_{n=0}^{\infty} s_n x^n.$$

For a positive integer N, the *Nth expansion complexity* $E_N(\mathcal{S})$ of $\mathcal{S}$ is defined by $E_N(\mathcal{S}) = 0$ if $s_n = 0$ for $0 \leq n \leq N-1$ and otherwise as the least total degree of a nonzero polynomial $h(x,y) \in \mathbb{F}_q[x,y]$ with

$$h(x, G(x)) \equiv 0 \bmod x^N.$$

By a famous result of Christol [5,6] the *expansion complexity*

$$E(\mathcal{S}) = \sup_{N\geq 1} E_N(\mathcal{S})$$

S. Duquesne and S. Petkova-Nikova (Eds.): WAIFI 2016, LNCS 10064, pp. 67–74, 2016.
DOI: 10.1007/978-3-319-55227-9_5

is finite, that is $G(x)$ is algebraic, if and only if $\mathcal{S}$ can be generated by a finite automaton and $\mathcal{S}$ is called *automatic*. For more details on automatic sequences we refer to the monograph of Allouche and Shallit [1].

The *Nth linear complexity* $L_N(\mathcal{S})$ of $\mathcal{S}$ over $\mathbb{F}_q$ is defined by $L_N(\mathcal{S}) = 0$ if $s_n = 0$ for $0 \le n \le N-1$, $L_N(\mathcal{S}) = N$ if $s_n = 0 \ne s_{N-1}$ for $0 \le n \le N-2$ and otherwise as the length of a shortest linear recurrence

$$s_{n+L_N(\mathcal{S})} + \sum_{\ell=0}^{L_N(\mathcal{S})-1} c_\ell s_{n+\ell} = 0, \quad 0 \le n \le N - L_N(\mathcal{S}) - 1,$$

for some $c_\ell \in \mathbb{F}_q$, which is satisfied by the first N terms of the sequence.

The *linear complexity* $L(\mathcal{S})$ is

$$L(\mathcal{S}) = \sup_{N \ge 1} L_N(\mathcal{S}).$$

It is well-known, see for example [9, Chap. 8], that $L(\mathcal{S})$ is finite if and only if $\mathcal{S}$ is ultimately periodic, that is, $\mathcal{S}$ is a *linear recurring sequence*. See also the surveys about linear complexity and related measures [10,13–15].

Expansion complexity and linear complexity are both measures for the unpredictability of a sequence.

First we study the non-periodic automatic sequence $\mathcal{T} = (t_n)_{n\ge0}$ over $\mathbb{F}_2$ defined by

$$t_n = \begin{cases} 1 \text{ if } n = u^2 + v^2 + w^2 \text{ for some integers } u, v, w, \\ 0 \text{ otherwise.} \end{cases}$$

By the Three-Square Theorem this is equivalent to

$$t_n = \begin{cases} 0 & \text{if there exist non-negative integers } a, k \\ & \text{such that } n = 4^a(8k+7), \\ 1 & \text{otherwise.} \end{cases} \tag{1}$$

We show in Sect. 2 that

$$E(\mathcal{T}) \le 12 \tag{2}$$

and

$$L_N(\mathcal{T}) \ge (N-7)/4 \quad \text{for } N \ge 1. \tag{3}$$

Note that lower bounds on the Nth linear complexity of many other automatic sequences including the Thue-Morse sequence, the Rudin-Shapiro sequence, and the regular paper-folding sequence were obtained in [11]. Roughly speaking, for the class of non-periodic automatic sequences the linear complexity is a much weaker measure for the unpredictability of a sequence than the expansion complexity.

Next we study periodic sequences. Mérai, Niederreiter and the second author [12] recently proved that

$$E_N(\mathcal{S}) \le E(\mathcal{S}) = L(\mathcal{S}) + 1, \quad N \ge 1,$$

for any purely periodic sequence $\mathcal{S}$ over $\mathbb{F}_q$. We will provide examples of p-periodic sequences over $\mathbb{F}_p$ with large linear complexity but small $E_p(\mathcal{S})$. More precisely, for the sequences $\mathcal{A}_{u,v} = (a_n)_{n\geq 0}$ of the form

$$a_n = \sum_{k=u}^{v} \lambda_k \binom{n+k}{k} \bmod p, \quad n \geq 0, \tag{4}$$

with $\lambda_u\lambda_v \neq 0$, $\lambda_k \in \mathbb{F}_p$ and $0 \leq u < v \leq p-1$ we prove in Sect. 3

$$E_p(\mathcal{A}_{u,v}) \leq \min\left\{(u+1)\left\{\frac{p}{v+1}\right\} + (v-u)\frac{p}{v+1}, v+2\right\} =: B(\mathcal{A}_{u,v}), \tag{5}$$

where $\{x\} = x - \lfloor x \rfloor$ is the fractional part of x. On the one hand the bound can be very small if v is large with respect to p and $v-u$ is small. For the case $u = v$ see [12]. On the other hand we have $L(\mathcal{A}_{u,v}) = v+1$ by [3, Theorem 8]. Hence, there are many p-periodic sequences over $\mathbb{F}_p$ of large linear complexity but small pth expansion complexity and we have the following hierarchy of complexity measures for p-periodic sequences

$$E_p(\mathcal{A}_{u,v}) \leq B(\mathcal{A}_{u,v}) \leq L(A_{u,v}) + 1 = v+2,$$

where B is defined in (5).

Note that any p-periodic sequence can be written in the form (4). More precisely, any p-periodic sequence $\mathcal{S} = (s_n)_{n\geq 0}$ over $\mathbb{F}_p$ can be defined by $s_n = f(n)$, $n \geq 0$, with a unique polynomial $f(x)$ over $\mathbb{F}_p$ of degree at most $p-1$. Now the polynomials

$$f_k(x) = (k!)^{-1}(x+k)(x+k-1)\cdots(x+1) = \binom{x+k}{k}, \quad k = 0, \ldots, p-1,$$

of degree k are a basis of the linear space of polynomials over $\mathbb{F}_p$ of degree at most $p-1$. Hence, the sequences $\left(\binom{n+k}{k}\right)_{n\geq 0}$, $k = 0, \ldots, p-1$, are a basis of the linear space of p-periodic sequences over $\mathbb{F}_p$ and any p-periodic sequence is a linear combination of these basis sequences.

Further, we consider the Legendre sequence $\mathcal{L} = (\ell_n)_{n\geq 0}$ over $\mathbb{F}_p$ of period p defined by

$$\ell_n = \begin{cases} 1 \text{ if } n \text{ is a quadratic residue mod } p, \\ 0 \text{ otherwise,} \end{cases}$$

or equivalently

$$\ell_n = \frac{n^{p-1} + n^{\frac{p-1}{2}}}{2} \bmod p.$$

The Legendre sequence has many desirable features of pseudorandomness, see for example [8,14]. Unfortunately, a lower bound on $E_p(\mathcal{L})$ seems to be out of reach. However, we prove

$$B(\mathcal{L}) = p + O\left(p^{\frac{1}{4\sqrt{e}}+\varepsilon}\right) \quad \text{for any } \varepsilon > 0 \tag{6}$$

in Sect. 4.

2 The Characteristic Sequence of the Set of Sums of Three Squares

In this section we prove (2) and (3) for the sequence $\mathcal{T}$ over $\mathbb{F}_2$ defined by (1). The generating function $G(x)$ of $\mathcal{T}$ is given by

$$G(x) = \sum_{n=0}^{\infty} t_n x^n = \sum_{n=0}^{\infty} x^n + \sum_{a=0}^{\infty}\sum_{k=0}^{\infty} x^{4^a(8k+7)}$$
$$= \frac{1}{x+1} + \sum_{a=0}^{\infty}\left(x^7 \sum_{k=0}^{\infty} x^{8k}\right)^{4^a} = \frac{1}{x+1} + \sum_{a=0}^{\infty}\left(\frac{x^7}{(x+1)^8}\right)^{4^a}.$$

It holds

$$G(x) + G(x)^4 = \frac{1}{x+1} + \frac{1}{(x+1)^4} + \frac{x^7}{(x+1)^8}$$

and therefore we get the equation

$$(x+1)^8(G(x)+G(x)^4) + x^6 + x^5 + x^3 + x^2 + x = 0. \tag{7}$$

Thus we have found a nonzero polynomial $h(x,y) \in \mathbb{F}_2[x,y]$, namely $h(x,y) = (x+1)^8(y+y^4) + x^6 + x^5 + x^3 + x^2 + x$ such that $h(x, G(x)) = 0$. Hence, $E(\mathcal{T}) \le \deg(h) = 12$.

Assume $G(x)$ is a rational function, that is

$$G(x) = \frac{f(x)}{g(x)}, \quad f, g \in \mathbb{F}_2[x],\ g \neq 0,$$

with $\gcd(f,g) = 1$. Then from (7) we get

$$(x+1)^8(fg^3 + f^4) + (x^6 + x^5 + x^3 + x^2 + x)g^4 = 0.$$

Hence $(x+1)^8 \mid g^4$, that is $(x+1)^2 \mid g$. Also $g^3 \mid (x+1)^8$ since $\gcd(f,g) = 1$. This is only possible if $g(x) = x^2+1$. Now $(x^2+1)G(x) = f(x)$ implies $t_{n+2} = t_n$ for $n \ge \deg(f)$. However, if $n \equiv 7 \bmod 8$ and thus $n+2 \equiv 1 \bmod 8$, we have $1 = t_{n+2} \neq t_n = 0$. Consequently, $G(x)$ is not rational. Moreover, the four zeros of $h(x,y)$ are obviously $y = G(x) + \alpha$ with $\alpha \in \mathbb{F}_4$ and none of them is rational.

Put $L_N = L_N(\mathcal{T})$ and let

$$\sum_{\ell=0}^{L_N} c_\ell t_{n+\ell} = 0, \quad 0 \le n \le N - L_N - 1, \quad c_{L_N} = 1,$$

be a shortest linear recurrence satisfied by the first N elements of $\mathcal{T}$. Then with

$$g(x) = \sum_{\ell=0}^{L_N} c_{L_N-\ell}x^\ell$$

we get

$$g(x)G(x) \equiv f(x) \bmod x^N$$

for some polynomial $f(x)$ of degree at most $L_N - 1$. Then

$$(x+1)^8(fg^3+f^4)+(x^6+x^5+x^3+x^2+x)g^4 = K(x)x^N$$

with $K(x) \neq 0$ since $h(x,y)$ has no rational zero. Comparing the degrees of both sides gives $4L_N + 7 \geq N$, that is $L_N \geq (N-7)/4$.

3 Expansion Complexity of p-periodic Sequences over $\mathbb{F}_p$

In this section we prove the bound (5) on $E_p(\mathcal{A}_{u,v})$. Note that any p-periodic sequence can be uniquely written as some $\mathcal{A}_{u,v}$.

Lemma 1. *The generating function of the sequence $\mathcal{A}_{u,v} = (a_n)_{n\geq 0}$ defined by* (4) *is*

$$G(x) = \sum_{k=u}^{v} \frac{\lambda_k}{(1-x)^{k+1}}.$$

Proof. For $u = v = k$ and $\lambda_k = 1$ we have (cf. proof of [12, Lemma 2])

$$(1-x)^p G(x) = (1-x^p)G(x) = \sum_{n=0}^{p-1} a_n x^n = \sum_{n=0}^{p-1-k} \binom{n+k}{k} x^n$$
$$= \sum_{n=0}^{p-1-k} \binom{p-1-k}{n} (-x)^n = (1-x)^{p-1-k}$$

since

$$\binom{p-1-k}{n}(-1)^n \equiv \prod_{j=1}^{n} \frac{k+j}{j} \equiv \binom{n+k}{n} \equiv \binom{n+k}{k} \bmod p.$$

The generating function of $\mathcal{A}_{u,v}$ is the linear combination of the generating functions of $\mathcal{A}_{k,k}$ with $k = u, \ldots, v$. □

Theorem 1. *The p-th expansion complexity of $\mathcal{A}_{u,v} = (a_n)_{n\geq 0}$, $u < v$, of the form* (4) *can be bounded by*

$$\min\left\{\left\lceil \frac{p}{v+2} \right\rceil, v+2\right\} \leq E_p(\mathcal{A}_{u,v})$$
$$\leq \min\left\{(u+1)\left\{\frac{p}{v+1}\right\} + (v-u)\frac{p}{v+1}, v+2\right\}.$$

Proof. The bound

$$\min\left\{\left\lceil \frac{p}{v+2}\right\rceil, v+2\right\} \le E_p(\mathcal{A}_{u,v}) \le v+2$$

follows from [12, Theorem 1] and [3, Theorem 8].

By Lemma 1 we have

$$G(x) = \sum_{k=u}^{v} \frac{\lambda_k}{(1-x)^{k+1}} = \frac{1}{(1-x)^{v+1}} \sum_{k=u}^{v} \lambda_k (1-x)^{v-k}.$$

Put

$$d = \left\lfloor \frac{p}{v+1} \right\rfloor$$

and take

$$h(x,y) = y^d - \left(\sum_{k=u}^{v} \lambda_k (1-x)^{v-k}\right)^d (1-x)^{p-d(v+1)}.$$

Then

$$\begin{aligned} h(x,G(x)) &= \frac{\left(\sum_{k=u}^{v} \lambda_k (1-x)^{v-k}\right)^d - \left(\sum_{k=u}^{v} \lambda_k (1-x)^{v-k}\right)^d (1-x)^p}{(1-x)^{d(v+1)}} \\ &= \frac{\left(\sum_{k=u}^{v} \lambda_k (1-x)^{v-k}\right)^d x^p}{(1-x)^{d(v+1)}} \equiv 0 \bmod x^p \end{aligned}$$

since $\gcd(x,(1-x)) = 1$. Hence

$$\begin{aligned} E_p(\mathcal{A}_{u,v}) &\le \deg(h) = \max\{d, d(v-u) + p - d(v+1)\} \\ &= \max\{d, p - d(u+1)\} = p - d(u+1) \end{aligned}$$

and the result follows.

4 The Legendre Sequence

In this section we prove (6).

Assume $\mathcal{L} = \mathcal{A}_{u,v}$. Then we need $v = p-1$.

If $p \equiv 1 \bmod 4$, then $p-1$ is a quadratic residue mod p, thus $\ell_{p-1} = 1$, and hence we must have $u = 0$ since otherwise $\ell_{p-1} = a_{p-1} = 0$.

If $p \equiv 3 \bmod 4$, then $p-1$ is not a quadratic residue mod p and also $p - u, \ldots, p-1$ must be quadratic non-residues mod p since $\ell_n = a_n = 0$ for $n =$

$p-u,\ldots,p-1$. This simply means that $1,\ldots,u$ are quadratic residues mod p since the product of two quadratic non-residues is a quadratic residue. Thus

$$u = O\left(p^{\frac{1}{4\sqrt{e}}+\varepsilon}\right) \quad \text{for any } \varepsilon > 0,$$

by the Burgess bound [4, Theorem 2]. Assuming the Extended Riemann Hypothesis we get the better result $u = O((\log p)^2)$ by Ankeny's theorem [2].

Hence, the Legendre sequence does not belong to the class of sequences for which the bound (5) is small.

5 Final Remarks and Open Problems

- Find lower bounds on $E_N(\mathcal{L})$ for the Legendre sequence both considered over $\mathbb{F}_p$ and over $\mathbb{F}_2$.
- Find lower bounds on E_N for other interesting sequences of high linear complexity profile such as inversive pseudorandom number generators or Sidelnikov sequences, see [10,14,15].
- Find classes of t-periodic sequences $\mathcal{S}$ over $\mathbb{F}_{p^r}$ with $\gcd(t,p) = 1$ and $E_t(\mathcal{S})$ of smaller order of magnitude than $L(\mathcal{S})$.
- A relation between Nth expansion complexity and Nth linear complexity is stated in [12, Theorem 3].

Acknowledgements. The authors are partially supported by the Austrian Science Fund FWF Project 5511-N26 which is part of the Special Research Program "Quasi-Monte Carlo Methods: Theory and Applications".

References

1. Allouche, J.P., Shallit, J.: Automatic Sequences: Theory, Applications, Generalizations. Cambridge University Press, Cambridge (2003)
2. Ankeny, N.C.: The least quadratic non residue. Ann. Math. **2**(55), 65–72 (1952)
3. Blackburn, S.R., Etzion, T., Paterson, K.G.: Permutation polynomials, de Bruijn sequences, and linear complexity. J. Comb. Theory Ser. A **76**(1), 55–82 (1996)
4. Burgess, D.A.: The distribution of quadratic residues and non-residues. Mathematika **4**, 106–112 (1957)
5. Christol, G.: Ensembles presque periodiques k-reconnaissables. Theor. Comput. Sci. **9**(1), 141–145 (1979)
6. Christol, G., Kamae, T., Mendés-France, M., Rauzy, G.: Suites algébriques, automates et substitutions. Bull. Soc. Math. Fr. **108**(4), 401–419 (1980)
7. Diem, C.: On the use of expansion series for stream ciphers. LMS J. Comput. Math. **15**, 326–340 (2012)
8. Hofer, R., Winterhof, A.: On the arithmetic autocorrelation of the Legendre sequence. Adv. Math. Commun. **11**(1), 237–244 (2017)
9. Lidl, R., Niederreiter, H.: Finite fields. In: Encyclopedia of Mathematics and Its Applications, vol. 20. Addison-Wesley Publishing Company, Advanced Book Program, Reading (1983)

10. Meidl, W., Winterhof, A.: Linear complexity of sequences and multisequences. In: Handbook of Finite Fields. Discrete Mathematics and its Applications, pp. 324–336. CRC Press, Boca Raton (2013)
11. Mérai, L., Winterhof, A.: On the Nth linear complexity of p-automatic sequences over $\mathbb{F}_p$ (2016, preprint)
12. Mérai, L., Niederreiter, H., Winterhof, A.: Expansion complexity and linear complexity of sequences over finite fields. Cryptogr. Commun. 1–9 (2016)
13. Niederreiter, H.: Linear complexity and related complexity measures for sequences. In: Johansson, T., Maitra, S. (eds.) INDOCRYPT 2003. LNCS, vol. 2904, pp. 1–17. Springer, Heidelberg (2003). doi:10.1007/978-3-540-24582-7_1
14. Topuzoğlu, A., Winterhof, A.: Pseudorandom sequences. In: Garcia, A., Stichtenoth, H. (eds.) Topics in Seometry Coding Theory and Cryptography. Algebra and Applications, vol. 6, pp. 135–166. Springer, Dordrecht (2007)
15. Winterhof, A.: Linear complexity and related complexity measures. Selected topics in information and coding theory. Coding Theory and Cryptology, vol. 7, pp. 3–40. World Scientific Publishing, Hackensack (2010)

Irreducible Polynomials

On Sets of Irreducible Polynomials Closed by Composition

Andrea Ferraguti[1], Giacomo Micheli[2](✉), and Reto Schnyder[3]

[1] DPMMS, Centre for Mathematical Sciences, University of Cambridge, Wilbeforce Rd, Cambridge CB3 0WB, UK
af612@cam.ac.uk

[2] Mathematical Institute, University of Oxford, Woodstock Rd, Oxford OX2 6GG, UK
giacomo.micheli@maths.ox.ac.uk

[3] Institute of Mathematics, University of Zurich, Winterthurerstrasse 190, Zurich 8057, Switzerland
reto.schnyder@math.uzh.ch

Abstract. Let $\mathcal{S}$ be a set of monic degree 2 polynomials over a finite field and let C be the compositional semigroup generated by $\mathcal{S}$. In this paper we establish a necessary and sufficient condition for C to be consisting entirely of irreducible polynomials. The condition we deduce depends on the finite data encoded in a certain graph uniquely determined by the generating set $\mathcal{S}$. Using this machinery we are able both to show examples of semigroups of irreducible polynomials generated by two degree 2 polynomials and to give some non-existence results for some of these sets in infinitely many prime fields satisfying certain arithmetic conditions.

Keywords: Finite fields · Irreducible polynomials · Semigroups · Graphs

1 Introduction

Since irreducible polynomials play a fundamental role in applications and in the whole theory of finite fields (see for example [2,4,13–16]), related questions have a long history (see for example [3,8,9,11,12,17,18]). In this paper we specialize on irreducibility questions regarding compositional semigroups of polynomials. This kind of question has been addressed in the specific case of semigroups generated by a single quadratic polynomial, see for example in [1,2,10–12,15], for analogous results related to additive polynomials, see [5,6]. It is worth mentioning that one of these results [12, Lemma 2.5] has been recently used in [7] by the first and the second author of the present paper to prove [3, Conjecture 1.2].

Throughout the paper, q will be an odd prime power, $\mathbb{F}_q[x]$ the univariate polynomial ring over the finite field $\mathbb{F}_q$ and $\mathrm{Irr}(\mathbb{F}_q[x])$ the set of irreducible polynomials in $\mathbb{F}_q[x]$. Let us give an example which motivates this paper. For a prime q congruent to 1 modulo 4, we can fix in $\mathbb{F}_q[x]$ two quadratic polynomials

S. Duquesne and S. Petkova-Nikova (Eds.): WAIFI 2016, LNCS 10064, pp. 77–83, 2016.
DOI: 10.1007/978-3-319-55227-9_6

$f = (x-a)^2 + a$ and $g = (x-a-1)^2 + a$ such that both a and $a+1$ are non-squares in $\mathbb{F}_q$. One can experimentally check that any possible composition of a sequence of f's and g's is irreducible (for a concrete example, take $q = 13$, $(x-5)^2+5$ and $g = (x-6)^2+5$). Let us denote the set of such compositions by C. A couple of observations are now necessary:

- In principle, it is unclear whether a finite number of irreducibility checks will ensure that C is a subset of $\mathrm{Irr}(\mathbb{F}_q[x])$.
- The fact that $C \subseteq \mathrm{Irr}(\mathbb{F}_q[x])$ is indeed pretty unlikely to happen by chance, as the density of degree 2^n monic irreducible polynomials over $\mathbb{F}_q$ is roughly $1/2^n$. Thus, if C satisfies this property, one reasonably expects that there must be an algebraic reason for that.

We address these issues by giving a necessary and sufficient condition for the semigroup $C \subset \mathbb{F}_q[x]$ to be contained in $\mathrm{Irr}(\mathbb{F}_q[x])$. In addition, this condition is algebraic and can be checked by performing only a finite amount of computation over $\mathbb{F}_q$, answering both points above.

In Sect. 2 we describe the criterion (Theorem 2.4 and Corollary 2.5) and provide a non-trivial example (Example 2.7) of a compositional semigroup in $\mathbb{F}_q[x]$ contained in $\mathrm{Irr}(\mathbb{F}_q[x])$ and generated by two polynomials.

In Sect. 3 we show the non-existence of such C whenever q is a prime congruent to 3 modulo 4 and the generating polynomials are of a certain form (Proposition 3.2). Example 3.3 shows that these conditions are indeed sharp.

2 A General Criterion

In order to state our main result, we first need the following definition, which describes how to build a finite graph encoding only the useful (to our purposes) information contained in the generating set of the semigroup.

Definition 2.1. *Let q be an odd prime power, $\mathbb{F}_q$ the finite field of order q and S a subset of $\mathbb{F}_q[x]$. We denote by G_S the directed multigraph defined as follows:*

- *the set of nodes of G_S is $\mathbb{F}_q$;*
- *for any node $a \in \mathbb{F}_q$ and any polynomial $f \in S$, there is a directed edge $a \to f(a)$. We label that edge with f.*

Before stating the next definition, we recall that for any monic polynomial f of degree 2 there exists a unique pair $(a_f, b_f) \in \mathbb{F}_q^2$ such that $f = (x-a_f)^2 - b_f$.

Definition 2.2. *Let S be a subset of $\mathbb{F}_q[x]$ consisting of monic polynomials of degree 2. We call the set $D_S = \{-b_f \mid f \in S\} \subseteq \mathbb{F}_q$, the S-*distinguished set *of $\mathbb{F}_q$.*

The following result is just an inductive extension of the classical Capelli's Lemma.

Lemma 2.3 (Recursive Capelli's Lemma). *Let K be a field and $f_1, \dots, f_\ell$ be a set of irreducible polynomials in $K[X]$. The polynomial $f_1(f_2(\cdots(f_\ell)\cdots))$ is irreducible if and only if the following conditions are satisfied:*

- *f_1 is irreducible over $K[X]$,*
- *$f_2 - \alpha_1$ is irreducible over $K(\alpha_1)[X]$ for a root α_1 of f_1,*
- *$f_3 - \alpha_2$ is irreducible over $K(\alpha_1, \alpha_2)[X]$ for a root α_2 of $f_2 - \alpha_1$,*

...

- *$f_\ell - \alpha_{\ell-1}$ is irreducible over $K(\alpha_1, \dots, \alpha_{\ell-1})[X]$ for a root $\alpha_{\ell-1}$ of $f_{\ell-1} - \alpha_{\ell-2}$.*

Proof. Given Capelli's Lemma [12, Lemma 2.4], the proof is straightforward by induction. □

We are now ready to state and prove the main theorem.

Theorem 2.4. *Let $\mathcal{S}$ be a set of generators for a compositional semigroup $C \subseteq \mathbb{F}_q[x]$. Suppose that $\mathcal{S}$ consists of polynomials of degree 2. Then we have that $C \subseteq \mathrm{Irr}(\mathbb{F}_q[x])$ if and only if no element of $-D_\mathcal{S} = \{b_f \mid f \in \mathcal{S}\} \subseteq \mathbb{F}_q$ is a square and in $G_\mathcal{S}$ there is no path of positive length from a node of $D_\mathcal{S}$ to a square of $\mathbb{F}_q$.*

Proof. It is clear that C contains a reducible polynomial of degree 2 if and only if one element of $-D_\mathcal{S}$ is a square. Thus we can assume that $\mathcal{S}$ consists only of irreducible polynomials.

We now show that in $G_\mathcal{S}$ there is a path of positive length from a node of $D_\mathcal{S}$ to a square if and only if C contains a reducible polynomial of degree greater or equal than 4.

First, suppose that the composition $f_1 f_2 \cdots f_{\ell+1}$ is a reducible polynomial of minimal degree, with $f_i \in \mathcal{S}$ and $f_i = (x - a_i)^2 - b_i$, for $i \in \{1, \dots, \ell+1\}$ and $\ell \geq 1$. Whenever β is not a square in $\mathbb{F}_q$, we denote by $\sqrt{\beta}$ a root of the polynomial $T^2 - \beta$ in the algebraic closure of $\mathbb{F}_q$. By Capelli's Lemma applied to the composition of $f_1 \cdots f_\ell$ and by the minimality of the degree of $f_1 f_2 \cdots f_{\ell+1}$, we have that the following elements are not squares in their field of definition:

$$
\begin{aligned}
\beta_0 &= b_1 \in \mathbb{F}_q \\
\beta_1 &= b_2 + a_1 + \sqrt{\beta_0} \in \mathbb{F}_{q^2} \\
\beta_2 &= b_3 + a_2 + \sqrt{\beta_1} \in \mathbb{F}_{q^{2^2}} \qquad (1)\\
&\cdots \\
\beta_{\ell-1} &= b_\ell + a_{\ell-1} + \sqrt{\beta_{\ell-2}} \in \mathbb{F}_{q^{2^{\ell-1}}}.
\end{aligned}
$$

To see this, note that f_1 has the root $\alpha_1 = a_1 + \sqrt{\beta_0}$, so $f_2 - \alpha_1 = (x - a_2)^2 - \beta_1$ is irreducible if and only if β_1 is nonsquare; and so on. On the other hand, $\beta_\ell = b_{\ell+1} + a_\ell + \sqrt{\beta_{\ell-1}} \in \mathbb{F}_{q^{2^\ell}}$ is necessarily a square. For $j < i$, let us denote by $N_i^j : \mathbb{F}_{q^{2^i}} \to \mathbb{F}_{q^{2^j}}$ the usual norm map. We claim that the $\mathbb{F}_q$-norm $N_\ell^0 : \mathbb{F}_{q^{2^\ell}} \to \mathbb{F}_q$ maps β_ℓ to $f_1(\cdots f_\ell(-b_{\ell+1})\cdots)$, and this defines a path in $G_\mathcal{S}$ from $-b_\ell$ to a square. This can be easily seen by first decomposing N_ℓ^1:

$$
N_\ell^1 = N_2^1 \circ N_3^2 \circ \dots N_\ell^{\ell-1} \qquad (2)
$$

and then by directly computing $N_2^1 \circ N_3^2 \circ \ldots N_\ell^{\ell-1}(\beta_\ell)$. It is important indeed that $\beta_0, \beta_1, \ldots, \beta_{\ell-1}$ are not squares, as the computation above only gives the desired result when $(\sqrt{\beta_i})^{q^{2^i}} = -\sqrt{\beta_i}$.

Conversely, suppose that in $G_\mathcal{S}$ there is a path to a square s. Choose such a path of minimal length, starting at some $-b_f$ in the distinguished set, for some $f \in \mathcal{S}$. Consider now the composition associated to this path: if

$$s = f_1 f_2 \cdots f_\ell(-b_f), \tag{3}$$

set $f_{\ell+1} = f$ and let $g = f_1 f_2 \cdots f_{\ell+1} \in \mathbb{F}_q[x]$. One can construct the β_i's as before, i.e. $\beta_0 = b_1$ and for $i \in \{1, \ldots, \ell\}$, $\beta_i = b_{i+1} + a_i + \sqrt{\beta_{i-1}}$. We can suppose that the β_i's for $i < \ell$ are all non-squares as otherwise, by taking the smallest d such that β_d is square, we find a composition $f_1 f_2 \cdots f_{d+1}$ that is reducible by Recursive Capelli's Lemma, and then we are done.

As all the β_i's, for $i < \ell$, can be supposed to be non-squares, we have as above that $N_\ell^0(\beta_\ell) = f_1 f_2 \cdots f_\ell(-b_{\ell+1}) = s$, which we have assumed to be a square. Now, recall that an element of a finite field is a square if and only if its norm is a square: this shows that g is reducible by Recursive Capelli's Lemma. □

The reader should observe that this theorem generalizes [12, Proposition 2.3], as the condition given by our graph is the same as the stability condition in [12, Proposition 2.3] whenever the semigroup we are considering has only one generator. It is useful to mention the following corollary, which is immediate.

Corollary 2.5. *Let $\mathcal{S}$ be a set of irreducible degree two polynomials and C defined as in Theorem 2.4. Then $C \subseteq \mathrm{Irr}(\mathbb{F}_q[x])$ if and only if there is no path of positive length from a node of $D_\mathcal{S}$ to a square of $\mathbb{F}_q$.*

Proof. It is enough to observe that whenever $\mathcal{S} \subseteq \mathrm{Irr}(\mathbb{F}_q[x])$ then $-D_\mathcal{S}$ consists of non-squares. □

Remark 2.6. Given that C is generated by degree 2 polynomials, it is easy to observe that the datum of $\mathcal{S}$ is equivalent to the datum of C.

The following example shows a way to find examples of semigroups contained in $\mathrm{Irr}(\mathbb{F}_q[x])$ when $q \equiv 1 \mod 4$.

Example 2.7. Let $q \equiv 1 \mod 4$ be a prime power, and let $a \in \mathbb{F}_q$ such that both a and $b = a + 1$ are non-squares. Define $f = (x - a)^2 + a$ and $g = (x - b)^2 + a$. In this situation, we have $D_\mathcal{S} = \{a\}$, and by assumption, $-a$, a and b are all non-squares. Since $f(a) = g(b) = a$ and $f(b) = g(a) = b$, all paths in $G_\mathcal{S}$ starting from a end in a non-square, and the conditions of Theorem 2.4 are satisfied. Figure 1 shows the relevant part of the graph $G_\mathcal{S}$. The reader should observe that this is indeed the example mentioned in the introduction.

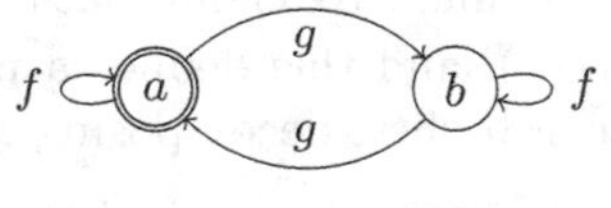

Fig. 1. The nodes of $G_\mathcal{S}$ reachable from $D_\mathcal{S}$.

3 The Case $p \equiv 3 \mod 4$

Whenever $q = p$ is a prime congruent to 3 modulo 4, we have the following non-existence results for polynomials without a linear term.

Proposition 3.1. *Let $p \equiv -1 \bmod 8$ be a prime, and let $f = x^2 - b$ be a polynomial in $\mathbb{F}_p[x]$. Let C be the semigroup generated by f. Then C contains a reducible polynomial.*

Proof. Assume for contradiction that $C \subset \mathrm{Irr}(\mathbb{F}_p[x])$. First note that if b is a square, then f is reducible, so we can assume that b is not a square, and thus $-b$ is a square. Consider the set of iterates $T = \{f(-b), f^2(-b), \ldots\} \subseteq \mathbb{F}_p$. By Corollary 2.5, C contains only irreducible polynomials if and only if T contains only nonsquares. So assume that this condition holds. Since T is finite, there exist $k < m \in \mathbb{N}_{>0}$ such that $f^m(-b) = f^k(-b)$. Choose k to be minimal. Now there are two cases: if $k > 1$, then there exist two distinct elements $u, v \in T$ such that $u^2 - b = v^2 - b$. Thus, $u = -v$, which implies that one between u and v is a square, a contradiction. If on the other hand $k = 1$, then we have $f^m(-b) = f(-b) = b^2 - b$, and so $f^{m-1}(-b)$ is either $-b$ or b. It can't be $-b$, since that is a square, so we must have $f^{m-1}(-b) = b \in T$. Setting $u = f^{m-2}(-b)$, we get that $u^2 - b = b$ and so $u^2 = 2b$, which is a contradiction because 2 is a square in $\mathbb{F}_p$ and consequently $2b$ is not. □

Proposition 3.2. *Let $p \equiv 3 \bmod 4$ be a prime. Let $f = x^2 - b_f$ and $g = x^2 - b_g$ be polynomials in $\mathbb{F}_p[x]$ with b_f, b_g distinct non-squares. Let $\mathcal{S} = \{f, g\}$ and let C be the semigroup generated by $\mathcal{S}$. Then C contains a reducible polynomial.*

Proof. Let $G_{\mathcal{S}}$ be the graph attached to $\mathcal{S}$ as in Definition 2.1. Let $G'_{\mathcal{S}}$ be the induced subgraph consisting of all nodes of $G_{\mathcal{S}}$ that are reachable by some path of positive length starting from $-b_f$ or $-b_g$. That is, the edges of $G'_{\mathcal{S}}$ are just the edges of $G_{\mathcal{S}}$ starting and ending at a node in $G'_{\mathcal{S}}$. From now on, when we speak of nodes and edges, we will always be referring to nodes and edges in $G'_{\mathcal{S}}$. We call an edge from u to v an *f-edge* if it comes from the relation $f(u) = v$, while we call it a *g-edge* if it comes from $g(u) = v$. Since b_f and b_g are assumed nonsquare, we have by Corollary 2.5 that C contains a reducible polynomial if and only if at least one of the nodes of $G'_{\mathcal{S}}$ is a square. In the following, we assume for contradiction that $G'_{\mathcal{S}}$ consists only of non-squares.

Let us observe the following: suppose that there exists a node v of $G'_{\mathcal{S}}$ which is the target of two f-edges. By definition, this means that there exist two distinct nodes $u, u' \in G'_{\mathcal{S}}$ such that $u^2 - b_f = u'^2 - b_f = v$. This implies that $u' = -u$, and thus one between u and u' is a square, since -1 is not a square in $\mathbb{F}_p$. This contradicts our assumption. By symmetry, the same applies to g-edges.

By the argument above, we see that every nodo is the target of at most one f-edge and one g-edge, and by counting edges that it is indeed exactly one of each.

Now, consider the sum

$$\sum_{v \in G'_S} (f(v) - g(v)). \tag{4}$$

On one hand, each node $u \in G'_S$ appears exactly once as $f(v)$ and once as $g(v')$ for some $v, v' \in G'_S$, so the sum is zero. On the other hand, it clearly holds that $f(v) - g(v) = b_g - b_f$ for all v. Letting n be the number of nodes in G'_S, we get the equation

$$0 = n(b_g - b_f) \text{ in } \mathbb{F}_p. \tag{5}$$

Since $b_f \neq b_g$ by hypothesis, we must have $p \mid n$. This is impossible however, since G'_S is not empty and consists only of nonsquares, so $1 \leq n \leq \frac{p-1}{2}$. □

The fact that the polynomials of Proposition 3.2 don't have a linear term is of crucial importance. Let us see why by giving an explicit example of a semigroup of irreducible polynomials in $\mathbb{F}_p[x]$ for which Proposition 3.2 does not apply (but $p \equiv 3 \mod 4$).

Example 3.3. Let us fix $p = 7$ and

$$\begin{aligned} f &= (x-1)^2 - 5 = x^2 + 5x + 3 \in \mathbb{F}_7[x] \\ g &= (x-4)^2 - 5 = x^2 + 6x + 4 \in \mathbb{F}_7[x]. \end{aligned} \tag{6}$$

The set $\mathcal{S} = \{f, g\}$ has distinguished set $D_{\mathcal{S}} = \{-5\}$ and graph as in Fig. 2.

Since 5 is not a square, and we only look at paths of positive length, the final claim follows by checking that 3 and -1 are not squares modulo 7.

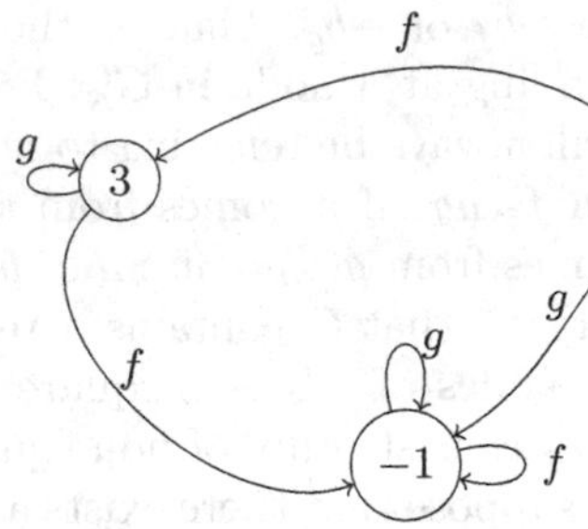

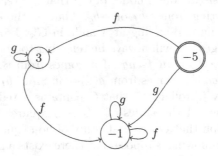

Fig. 2. The nodes of $G_{\mathcal{S}}$ reachable from -5.

Acknowledgements. The first author was supported by the Swiss National Science Foundation grant number 168459. The second author was supported by Swiss National Science Foundation grant number 161757. The third author was supported in part by Swiss National Science Foundation grant number 149716 and *Armasuisse*.

References

1. Ahmadi, O.: A note on stable quadratic polynomials over fields of characteristic two. arXiv preprint arXiv:0910.4556 (2009)
2. Ahmadi, O., Luca, F., Ostafe, A., Shparlinski, I.E.: On stable quadratic polynomials. Glasg. Math. J. **54**(02), 359–369 (2012)
3. Andrade, J., Miller, S.J., Pratt, K., Trinh, M.T.: Special sets of primes in function fields. arXiv preprint arXiv:1309.5597 (2013)
4. Barbulescu, R., Gaudry, P., Joux, A., Thomé, E.: A heuristic quasi-polynomial algorithm for discrete logarithm in finite fields of small characteristic. In: Nguyen, P.Q., Oswald, E. (eds.) EUROCRYPT 2014. LNCS, vol. 8441, pp. 1–16. Springer, Heidelberg (2014). doi:10.1007/978-3-642-55220-5_1
5. Batra, A., Morton, P.: Algebraic dynamics of polynomial maps on the algebraic closure of a finite field, I. Rocky Mt. J. Math. **24**(2), 453–481 (1994)
6. Batra, A., Morton, P.: Algebraic dynamics of polynomial maps on the algebraic closure of a finite field, II. Rocky Mt. J. Math. **24**(3), 905–932 (1994)
7. Ferraguti, A., Micheli, G.: On the existence of infinite, non-trivial F-sets. J. Number Theory (2016, to appear)
8. Gao, S., Howell, J., Panario, D.: Irreducible polynomials of given forms. Contemp. Math. **225**, 43–54 (1999)
9. Gao, S., Panario, D.: Tests and constructions of irreducible polynomials over finite fields. In: Cucker, F., Shub, M. (eds.) Foundations of Computational Mathematics, pp. 346–361. Springer, Heidelberg (1997)
10. Gomez, D., Nicolás, A.P.: An estimate on the number of stable quadratic polynomials. Finite Fields Appl. **16**(6), 401–405 (2010)
11. Jones, R.: An iterative construction of irreducible polynomials reducible modulo every prime. J. Algebra **369**, 114–128 (2012)
12. Jones, R., Boston, N.: Settled polynomials over finite fields. Proc. Am. Math. Soc. **140**(6), 1849–1863 (2012)
13. Lidl, R., Niederreiter, H.: Finite Fields, vol. 20. Cambridge University Press, Cambridge (1997)
14. Mullen, G.L., Panario, D.: Handbook of Finite Fields. CRC Press, Boca Raton (2013)
15. Ostafe, A., Shparlinski, I.E.: On the length of critical orbits of stable quadratic polynomials. Proc. Am. Math. Soc. **138**(8), 2653–2656 (2010)
16. Rabin, M.O., et al.: Fingerprinting by random polynomials. Center for Research in Computing Techn., Aiken Computation Laboratory, Univ. (1981)
17. Shoup, V.: New algorithms for finding irreducible polynomials over finite fields. Math. Comp. **54**(189), 435–447 (1990)
18. von ZurGathen, J.: Irreducible trinomials over finite fields. Math. Comp. **72**(244), 1987–2000 (2003)

A Note on the Brawley-Carlitz Theorem on Irreducibility of Composed Products of Polynomials over Finite Fields

Akihiro Munemasa[(✉)] and Hiroko Nakamura

Research Center for Pure and Applied Mathematics,
Graduate School of Information Sciences, Tohoku University, Sendai, Japan
munemasa@math.is.tohoku.ac.jp, nakamu@ims.is.tohoku.ac.jp

Abstract. We give a new proof of the Brawley-Carlitz theorem on irreducibility of the composed products of irreducible polynomials. Our proof shows that associativity of the binary operation for the composed product is not necessary. We then investigate binary operations defined by polynomial functions, and give a sufficient condition in terms of degrees for the requirement in the Brawley-Carlitz theorem.

Keywords: Finite field · Composed product · Irreducible polynomial

1 Introduction

For a prime power q, we denote by $\mathbb{F}_q$ a finite field with q elements. If m and n are relatively prime positive integers, then the composite field of $\mathbb{F}_{q^m}$ and $\mathbb{F}_{q^n}$ is $\mathbb{F}_{q^{mn}}$. In fact, if $\mathbb{F}_{q^m} = \mathbb{F}_q(\alpha)$ and $\mathbb{F}_{q^n} = \mathbb{F}_q(\beta)$, then $\mathbb{F}_{q^{mn}} = \mathbb{F}_q(\alpha+\beta) = \mathbb{F}_q(\alpha\beta)$. In other words, both $\alpha + \beta$ and $\alpha\beta$ have minimal polynomial of degree mn over $\mathbb{F}_q$. Brawley and Carlitz generalized this fact by introducing the method of composed products in order to construct irreducible polynomials of large degree from polynomials of lower degree. A basic material of their construction is a binary operation on a subset of $\overline{\mathbb{F}_q}$ having certain properties, where $\overline{\mathbb{F}_q}$ is the algebraic closure of $\mathbb{F}_q$. Let G be a non-empty subset of $\overline{\mathbb{F}_q}$, which is invariant under the Frobenius map $\alpha \mapsto \alpha^q$. A binary operation $\diamond : G \times G \to G$ is called a *diamond product* on G if

$$\sigma(\alpha \diamond \beta) = \sigma(\alpha) \diamond \sigma(\beta) \tag{1}$$

holds for all $\alpha, \beta \in G$. Let $M_G[q, x]$ denote the set of all monic polynomials f in $\mathbb{F}_q[x]$ such that $\deg f \geq 1$, and all of the roots of f lie in G. Let $f(x) = \prod_{i=1}^{m}(x - \alpha_i)$ and $g(x) = \prod_{i=1}^{n}(x - \beta_i)$ be in polynomials in $M_G[q, x]$, where $\alpha_1, \ldots, \alpha_m, \beta_1, \ldots, \beta_n \in G$. We define the *composed product* $f \diamond g$ as

$$(f \diamond g)(x) = \prod_{i=1}^{m}\prod_{j=1}^{n}(x - \alpha_i \diamond \beta_j).$$

A. Munemasa—The work of this author is partially supported by JSPS KAKENHI Grant Number 26400003.

S. Duquesne and S. Petkova-Nikova (Eds.): WAIFI 2016, LNCS 10064, pp. 84–92, 2016.
DOI: 10.1007/978-3-319-55227-9_7

Theorem 1 ([1, Theorem 2]). *Let $\diamond$ be a diamond product on a non-empty subset G of $\overline{\mathbb{F}_q}$. Suppose that $(G, \diamond)$ is a group and let f, g be polynomials in $M_G[q, x]$ with $\deg f = m$ and $\deg g = n$. Then the composed product $f \diamond g$ is irreducible if and only if f and g are both irreducible with $\gcd(m, n) = 1$.*

The purpose of this paper is to give a new proof of Theorem 1 with weaker hypotheses. In order to explain the weakened hypothesis, we need a definition. For a positive integer m, let

$$\mathcal{F}_m(q) = \{\alpha \in \mathbb{F}_{q^m} \mid \mathbb{F}_q(\alpha) = \mathbb{F}_{q^m}\}.$$

Clearly,

$$\mathcal{F}_{kl}(q) \subset \mathcal{F}_l(q^k) \tag{2}$$

for positive integers k, l.

Definition 2. Let $\diamond$ be a diamond product on a subset $G \subset \overline{\mathbb{F}_q}$ containing $\mathcal{F}_m(q) \cup \mathcal{F}_n(q)$. We say that $\diamond$ satisfies *weak cancellation* on $\mathcal{F}_m(q) \times \mathcal{F}_n(q)$, if

$$\alpha \diamond \beta = \alpha \diamond \beta' \Longrightarrow \beta = \beta', \tag{3}$$
$$\alpha \diamond \beta = \alpha' \diamond \beta \Longrightarrow \alpha = \alpha' \tag{4}$$

for all $\alpha, \alpha' \in \mathcal{F}_m(q)$ and $\beta, \beta' \in \mathcal{F}_n(q)$.

We will show in Sect. 2 that the conclusion of Theorem 1 holds if $\diamond$ satisfies weak cancellation on $\mathcal{F}_m(q) \times \mathcal{F}_n(q)$. In other words, associativity of the product $\diamond$ is unnecessary. In Sect. 3, we consider a diamond product defined by a polynomial function, and show that such a diamond product satisfies weak cancellation if the degree is small (see Theorem 9 for details). In Sect. 4, the optimality of the degree bound for weak cancellation is investigated. This leads us to a conjecture on the existence of irreducible polynomials all of whose coefficients except the constant term belong to the prime field.

2 The Brawley-Carlitz Theorem

Throughout this paper, we let q be a prime power, and $\sigma : \overline{\mathbb{F}_q} \to \overline{\mathbb{F}_q}$ denote the Frobenius map $\alpha \mapsto \alpha^q$. For positive integers k and r, we denote by $\mathrm{ord}_k(r)$ the multiplicative order of r modulo k. For a nonzero $\alpha \in \overline{\mathbb{F}_q}$, we denote by $|\alpha|$ the multiplicative order of α. Then (see, for example, [3, Corollary 2.15]), we have, for $m > 1$,

$$\begin{aligned}\mathcal{F}_m(q) &= \{\alpha \in \mathbb{F}_{q^m} \mid \alpha, \sigma(\alpha), \dots, \sigma^{m-1}(\alpha) : \text{pairwise distinct}\} \\ &= \{\alpha \in \mathbb{F}_{q^m} \mid \{l \in \mathbb{Z} \mid \sigma^l(\alpha) = \alpha\} = m\mathbb{Z}\} \\ &= \{\alpha \in \mathbb{F}_{q^m} \mid \alpha \neq 0,\ \mathrm{ord}_{|\alpha|}(q) = m\}.\end{aligned}$$

Our proof of the Brawley-Carlitz theorem relies on the following lemma in group theory.

Lemma 3. *Let Γ be a finite group of order mn having subgroups M and N of order m and n, respectively. Assume $\Gamma = M \times N$ and $(m,n) = 1$. If K is a subgroup of Γ, then $K = (K \cap M)(K \cap N)$.*

Proof. Since $(m,n) = 1$, there exist integers r, s such that $rm + sn = 1$. Let $z \in K$. Since $\Gamma = M \times N$, there exist $x \in M$ and $y \in N$ such that $z = xy$. Then $z = z^{sn}z^{rm}$ with $z^{sn} = x^{sn} \in K \cap M$, $z^{rm} = y^{rm} \in K \cap N$. Since $z \in K$ was arbitrary, we conclude $K \subset (K \cap M)(K \cap N)$. Since the reverse containment is obvious, we obtain the desired result.

Theorem 4. *Suppose G is a non-empty subset of $\overline{\mathbb{F}_q}$. Let $\diamond$ be a diamond product on G satisfying (3) and (4). Let $f, g \in M_G[q,x]$, $\deg f = m$ and $\deg g = n$. Then the following are equivalent:*

(i) $f \diamond g$ is irreducible in $\mathbb{F}_q[x]$,
(ii) f and g are irreducible in $\mathbb{F}_q[x]$, and $\gcd(m,n) = 1$.

Proof. (i) $\Longrightarrow$ (ii). Since $(f \diamond g)(x)$ is irreducible, clearly $f(x)$ and $g(x)$ are irreducible. Let α and β be roots of $f(x)$ and $g(x)$, respectively. Then $\alpha \diamond \beta$ is a root of $(f \diamond g)(x)$ which is an irreducible polynomial of degree mn. This implies $\mathrm{ord}_{|\alpha\diamond\beta|}(q) = mn$. Let ℓ be the least common multiple of m and n. Then $\sigma^\ell(\alpha \diamond \beta) = \sigma^\ell(\alpha) \diamond \sigma^\ell(\beta) = \alpha \diamond \beta$. Thus $\mathrm{ord}_{|\alpha\diamond\beta|}(q)$ divides ℓ, and hence, $\ell = mn$. This implies $\gcd(m,n) = 1$.

(ii) $\Longrightarrow$ (i) Let $\alpha \in \mathcal{F}_m(q)$ and $\beta \in \mathcal{F}_n(q)$ be roots of f and g, respectively. The Frobenius automorphism σ generates the group $F = \langle \sigma \rangle$ of order mn acting on $\mathbb{F}_{q^{mn}}$. Moreover, setting

$$M = \langle \sigma^n \rangle = \{h \in F \mid h(\beta) = \beta\}, \tag{5}$$
$$N = \langle \sigma^m \rangle = \{h \in F \mid h(\alpha) = \alpha\}, \tag{6}$$

we have $|M| = m$ and $|N| = n$, so $F = M \times N$. Let

$$K = \{h \in F \mid h(\alpha \diamond \beta) = \alpha \diamond \beta\}.$$

Then

$$|F \cdot (\alpha \diamond \beta)| = |F : K|. \tag{7}$$

We claim $K \cap M = K \cap N = 1$. Indeed, if $h \in K \cap M$, then

$$\begin{aligned} \alpha \diamond \beta &= h(\alpha \diamond \beta) \\ &= h(\alpha) \diamond h(\beta) \qquad \text{(by (1))} \\ &= h(\alpha) \diamond \beta, \end{aligned}$$

so $\alpha = h(\alpha)$ by (4). This implies $h \in N$. Since $h \in M$ and $M \cap N = 1$, we conclude $h = 1$. This proves $K \cap M = 1$. Similarly, we can prove $K \cap N = 1$ using (3).

Now, by Lemma 3, we obtain $K = 1$. This implies $|F \cdot (\alpha \diamond \beta)| = |F|$ by (7). Therefore, the degree of the minimal polynomial of $\alpha \diamond \beta$ over $\mathbb{F}_q$ is $|F| = mn$. Since $\deg(f \diamond g) = mn$ and $(f \diamond g)(\alpha \diamond \beta) = 0$, we conclude that $f \diamond g$ is the minimal polynomial of $\alpha \diamond \beta$ over $\mathbb{F}_q$, and hence irreducible in $\mathbb{F}_q[x]$.

3 Diamond Products Defined by Polynomial Functions

Stichtenoth [5] classified associative diamond products defined by a polynomial function, under a certain condition. As we have seen in the previous section, associativity is irrelevant for the Brawley-Carlitz theorem. This prompts us to classify diamond products satisfying weak cancellation instead. In this section, we consider diamond products defined by a polynomial function, and give a sufficient condition in terms of degrees in order that the associated diamond product satisfies weak cancellation. It turns out that, in general, a wider class of polynomials than those classified in [5] can be used as a diamond product.

Let m be a positive integer, and let $\psi : \mathcal{F}_m(q) \to \mathcal{F}_m(q)$ be a function. We say that ψ satisfies the *restricted injectivity* on $\mathcal{F}_m(q)$ if, for all $\alpha \in \mathcal{F}_m(q)$ and $k \in \mathbb{Z}$,

$$\psi(\alpha) = \psi(\sigma^k(\alpha)) \Longrightarrow \alpha = \sigma^k(\alpha). \tag{8}$$

If $\psi : \mathcal{F}_m(q) \to \overline{\mathbb{F}_q}$ is a function taking values in $\mathbb{F}_{q^m}$ such that ψ commutes with σ, then $\psi(\sigma^k(\alpha)) = \sigma^k(\psi(\alpha))$. Thus, (8) is equivalent to

$$\psi(\alpha) \in \mathcal{F}_m(q). \tag{9}$$

In particular, this equivalence holds when ψ is a polynomial function with coefficients in $\mathbb{F}_q$.

Lemma 5. *Let $\psi(x) \in \mathbb{F}_q[x]$ be a polynomial with $\deg \psi \geq 1$. Then for $\alpha \in \overline{\mathbb{F}_q}$,*

$$\deg \psi \geq [\mathbb{F}_q(\alpha) : \mathbb{F}_q(\psi(\alpha))].$$

Proof. Let $\psi_0(x) = \psi(x) - \psi(\alpha) \in \mathbb{F}_q(\psi(\alpha))[x]$. Then $\psi_0(\alpha) = 0$, so ψ_0 is divisible by the minimal polynomial of α over $\mathbb{F}_q(\psi(\alpha))$. This implies

$$\begin{aligned}[\mathbb{F}_q(\alpha) : \mathbb{F}_q(\psi(\alpha))] &\leq \deg \psi_0 \\ &= \deg \psi.\end{aligned}$$

Lemma 6. *Let $m > 1$ be an integer, and let m_1 be the smallest prime divisor of m. If $\psi(x) \in \mathbb{F}_q[x]$ is a monic polynomial with $0 < \deg \psi < m_1$, then the function defined by ψ satisfies the restricted injectivity on $\mathcal{F}_m(q)$.*

Proof. For $\alpha \in \mathcal{F}_m(q)$, we have

$$\begin{aligned} m_1 &> \deg \psi \\ &\geq [\mathbb{F}_q(\alpha) : \mathbb{F}_q(\psi(\alpha))] && \text{(by Lemma 5)} \\ &= [\mathbb{F}_{q^m} : \mathbb{F}_q(\psi(\alpha))]. \end{aligned}$$

Since $[\mathbb{F}_{q^m} : \mathbb{F}_q(\psi(\alpha))]$ is a divisor of m and m_1 is the smallest prime divisor of m, we conclude that $[\mathbb{F}_{q^m} : \mathbb{F}_q(\psi(\alpha))] = 1$, that is, $\mathbb{F}_q(\psi(\alpha)) = \mathbb{F}_{q^m}$. This implies (9).

Lemma 7. *Let m_1 be the smallest prime divisor of a positive integer $m > 1$, and let k be an integer not divisible by m. Then for $\alpha \in \mathcal{F}_m(q)$, $\alpha - \sigma^k(\alpha), \alpha^2 - \sigma^k(\alpha^2), \ldots, \alpha^{m_1-1} - \sigma^k(\alpha^{m_1-1})$ are linearly independent over $\mathbb{F}_q$.*

Proof. Suppose $\alpha - \sigma^k(\alpha), \alpha^2 - \sigma^k(\alpha^2), \ldots, \alpha^{m_1-1} - \sigma^k(\alpha^{m_1-1})$ are linearly dependent. Then there exist $a_1, \ldots, a_{m_1-1} \in \mathbb{F}_q$, $(a_1, \ldots, a_{m_1-1}) \neq (0, \ldots, 0)$, and

$$\sum_{i=1}^{m_1-1} a_i(\alpha^i - \sigma^k(\alpha^i)) = 0.$$

Let

$$a_0 = \sum_{i=1}^{m_1-1} a_i\alpha^i \in \mathbb{F}_{q^m},$$

$$f(x) = \sum_{i=1}^{m_1-1} a_i x^i - a_0 \in \mathbb{F}_q(a_0)[x].$$

Then $\sigma^k(a_0) = a_0$, so $f \in \mathbb{F}_{q^{\gcd(k,m)}}[x]$. Since $f(\alpha) = 0$, f is divisible by the minimal polynomial of α over $\mathbb{F}_{q^{\gcd(k,m)}}$. This implies

$$\begin{aligned} m_1 &> \deg f \\ &\geq [\mathbb{F}_{q^{\gcd(k,m)}}(\alpha) : \mathbb{F}_{q^{\gcd(k,m)}}] \\ &= [\mathbb{F}_{q^m} : \mathbb{F}_{q^{\gcd(k,m)}}] \\ &= \frac{m}{\gcd(k,m)}. \end{aligned}$$

Since m_1 is the smallest prime divisor of m, we obtain $\gcd(k, m) = m$, that is, $m \mid k$. This is a contradiction.

Lemma 8. *If m and n are relatively prime positive integers, then $\mathcal{F}_n(q) \subset \mathcal{F}_n(q^m)$. In particular, if $\alpha \in \mathcal{F}_m(q)$, $\beta \in \mathcal{F}_n(q)$, $k \in \mathbb{Z}$ and*

$$\varphi(x, y) = \sum_{i=0}^{n-1} \psi_i(x)y^i \in \mathbb{F}_q[x, y]$$

satisfy $\varphi(\sigma^k(\alpha), \beta) = \varphi(\alpha, \beta)$, then $\psi_i(\sigma^k(\alpha)) = \psi_i(\alpha)$ for $0 \leq i \leq n-1$.

Proof. The first part is immediate from [3, Corollary 3.47]. Since $\varphi(\sigma^k(\alpha), \beta) = \varphi(\alpha, \beta)$, we have

$$\sum_{i=0}^{n-1} (\psi_i(\sigma^k(\alpha)) - \psi_i(\alpha))\beta^i = 0.$$

Since $\beta \in \mathcal{F}_n(q) \subset \mathcal{F}_n(q^m)$ and $\psi_i(\sigma^k(\alpha)) - \psi_i(\alpha) \in \mathbb{F}_{q^m}$, linear independence of $1, \beta, \ldots, \beta^{n-1}$ over $\mathbb{F}_{q^m}$ shows $\psi_i(\sigma^k(\alpha)) = \psi_i(\alpha)$ for $0 \leq i \leq n-1$.

Theorem 9. *Let q be a prime power, and let $m, n > 1$ be relatively prime positive integers. Suppose m_1 is the smallest prime divisor of m, n_1 is the smallest prime divisor of n. Let $\varphi(x, y) \in \mathbb{F}_q[x, y]$ be a polynomial with $0 < \deg_x \varphi < m_1$ and $0 < \deg_y \varphi < n_1$. Then the diamond product on $\overline{\mathbb{F}_q}$ defined by φ satisfies weak cancellation on $\mathcal{F}_m(q) \times \mathcal{F}_n(q)$.*

Proof. We need to show

$$\varphi(\alpha, \beta) = \varphi(\sigma^k(\alpha), \beta) \Longrightarrow \alpha = \sigma^k(\alpha), \tag{10}$$

$$\varphi(\alpha, \beta) = \varphi(\alpha, \sigma^k(\beta)) \Longrightarrow \beta = \sigma^k(\beta). \tag{11}$$

It suffices to show only (10), as the proof of (11) is similar. Suppose $\alpha \in \mathcal{F}_m(q)$, $\beta \in \mathcal{F}_n(q)$, $k \in \mathbb{Z}$. Let

$$\varphi(x, y) = \sum_{i=0}^{n_1-1} \psi_i(x) y^i, \tag{12}$$

$$\psi_i(x) = \sum_{j=0}^{m_1-1} a_{ij} x^j \quad (0 \le i \le n_1 - 1). \tag{13}$$

If $\varphi(\alpha, \beta) = \varphi(\sigma^k(\alpha), \beta)$, then, by Lemma 8, $\psi_i(\alpha) - \psi_i(\sigma^k(\alpha)) = 0$ for $0 \le i \le n_1 - 1$. This implies

$$\sum_{j=1}^{m_1-1} a_{ij}(\alpha^j - \sigma^k(\alpha^j)) = 0 \quad (0 \le i \le n_1 - 1).$$

If $\alpha \neq \sigma^k(\alpha)$, then k is not divisible by m. Then by Lemma 7, we obtain $a_{ij} = 0$ for $0 \le i \le n_1 - 1$ and $1 \le j \le m_1 - 1$. This implies $\deg_x \varphi = 0$, which contradicts the assumption. Therefore, $\alpha = \sigma^k(\alpha)$.

4 Irreducible Polynomials All of Whose Coefficients Except the Constant Term Belong to the Prime Field

In this section, we show that the hypotheses $\deg_x \varphi < m_1$ and $\deg_y \varphi < n_1$ in Theorem 9 are necessary. We believe that these upper bounds cannot be relaxed for any prime power q and relatively prime positive integers m and n. This leads to a conjecture on the existence of irreducible polynomials all of whose coefficients except the constant term belong to the prime field.

Proposition 10. *Let m and n be relatively prime integers with $m, n > 1$. Let m_1 and n_1 be the smallest prime divisor of m and n, respectively. Then the following are equivalent:*

(i) there exists a polynomial $\varphi(x, y) \in \mathbb{F}_q[x, y]$ with $\deg_x \varphi = m_1$, $0 < \deg_y \varphi < n_1$, such that $\sigma^k(\alpha) \neq \alpha$ and $\varphi(\sigma^k(\alpha), \beta) = \varphi(\alpha, \beta)$ for some $\alpha \in \mathcal{F}_m(q)$ and $\beta \in \mathcal{F}_n(q)$,

(ii) there exists a polynomial $\psi(x) \in \mathbb{F}_q[x]$ *with* $\deg \psi = m_1$ *which fails to satisfy the restricted injectivity on* $\mathcal{F}_m(q)$,

(iii) $\mathcal{F}_{m/m_1}(q) \cap \{\alpha^{m_1} + \sum_{i=1}^{m_1-1} c_i\alpha^i \mid \alpha \in \mathcal{F}_m(q),\ c_1,\dots,c_{m_1-1} \in \mathbb{F}_q\} \neq \emptyset$,

(iv) there exists a monic irreducible polynomial $f(x) \in \mathbb{F}_{q^{m/m_1}}[x]$ *of degree* m_1 *such that* $f(x) - f(0) \in \mathbb{F}_q[x]$ *and* $f(0) \in \mathcal{F}_{m/m_1}(q)$.

Proof. (i) $\Longrightarrow$ (ii). Let $\varphi(x,y)$ be as in (12), where $\psi_i(x) \in \mathbb{F}_q[x]$ for $0 \le i \le n_1 - 1$. Then by Lemma 8, $\psi_i(\sigma^k(\alpha)) = \psi_i(\alpha)$ for $0 \le i \le n_1 - 1$. By the assumption, there exists $i \in \{0,1,\dots,n_1-1\}$ such that $\deg \psi_i = m_1$, and this ψ_i fails to satisfy the restricted injectivity on $\mathcal{F}_m(q)$.

(ii) $\Longrightarrow$ (iii). We may assume without loss of generality that ψ is monic. Replacing $\psi(x)$ by $\psi(x) - \psi(0)$, we may further assume that $\psi(0) = 0$. By the assumption, there exists $\alpha \in \mathcal{F}_m(q)$ and $k \in \mathbb{Z}$ such that $\sigma^k(\alpha) \neq \alpha$ and $\psi(\sigma^k(\alpha)) = \psi(\alpha)$. Since $\psi(x) \in \mathbb{F}_q[x]$, the latter implies $\sigma^k(\psi(\alpha)) = \psi(\alpha)$. Thus $\psi(\alpha) \in \mathbb{F}_{q^{\gcd(k,m)}}$. Since $\sigma^k(\alpha) \neq \alpha$, k is not a multiple of m. This implies that $\mathbb{F}_{q^{\gcd(k,m)}}$ is a proper subfield of $\mathbb{F}_{q^m}$. Therefore, there exists a divisor $d > 1$ of m such that $\psi(\alpha) \in \mathbb{F}_{q^{m/d}}$. By Lemma 5, we have

$$\begin{aligned} m_1 &\ge [\mathbb{F}_q(\alpha) : \mathbb{F}_q(\psi(\alpha))] \\ &\ge [\mathbb{F}_{q^m} : \mathbb{F}_{q^{m/d}}] \\ &= d. \end{aligned}$$

Since m_1 is the smallest prime divisor of m, we obtain $m_1 = d$. This forces $\mathbb{F}_q(\psi(\alpha)) = \mathbb{F}_{q^{m/m_1}}$, and hence $\psi(\alpha) \in \mathcal{F}_{m/m_1}(q)$.

(iii) $\Longrightarrow$ (iv). Suppose $\alpha \in \mathcal{F}_m(q)$, $c_1,\dots,c_{m_1-1} \in \mathbb{F}_q$, and

$$c_0 = \alpha^{m_1} + \sum_{i=1}^{m_1-1} c_i\alpha^i \in \mathcal{F}_{m/m_1}(q).$$

Define

$$f(x) = x^{m_1} + \sum_{i=1}^{m_1-1} c_i x^i - c_0 \in \mathbb{F}_{q^{m/m_1}}[x].$$

Then $f(x) - f(0) \in \mathbb{F}_q[x]$ and $f(0) = -c_0 \in \mathcal{F}_{m/m_1}(q)$. We claim $f(x)$ is irreducible in $\mathbb{F}_{q^{m/m_1}}[x]$. Indeed, since $f(\alpha) = 0$, $f(x)$ is divisible by the minimal polynomial of α over $\mathbb{F}_{q^{m/m_1}}$. On the other hand, since $\mathbb{F}_{q^{m/m_1}}(\alpha) \supset \mathbb{F}_q(\alpha) = \mathbb{F}_{q^m}$, the minimal polynomial of α over $\mathbb{F}_{q^{m/m_1}}$ has degree at least $[\mathbb{F}_{q^m} : \mathbb{F}_{q^{m/m_1}}] = m_1 = \deg f$. Therefore, $f(x)$ is the minimal polynomial of α over $\mathbb{F}_{q^{m/m_1}}$, and hence is irreducible $\mathbb{F}_{q^{m/m_1}}[x]$.

(iv) $\Longrightarrow$ (i). Define

$$k = \frac{m}{m_1},$$
$$\varphi(x,y) = (f(x) - f(0))y \in \mathbb{F}_q[x,y].$$

Then, $\deg_x \varphi = m_1$, $\deg_y \varphi = 1$. Let α be a root of $f(x)$. Since $f(0) = -(f(\alpha) - f(0)) \in \mathbb{F}_q(\alpha)$, we have $\mathbb{F}_q(\alpha) = \mathbb{F}_q(f(0), \alpha) = \mathbb{F}_{q^{m/m_1}}(\alpha) = \mathbb{F}_{q^m}$. Thus $\alpha \in \mathcal{F}_m(q)$. Moreover, for an arbitrary $\beta \in \mathcal{F}_n(q)$, we have

$$\begin{aligned}\varphi(\sigma^k(\alpha), \beta) &= \sigma^k(f(\alpha) - f(0))\beta \\ &= -\sigma^k(f(0))\beta \\ &= -f(0)\beta \\ &= (f(\alpha) - f(0))\beta \\ &= \varphi(\alpha, \beta).\end{aligned}$$

Proposition 10 shows that hypotheses $0 < \deg_x \varphi < m_1$ and $0 < \deg_y \varphi < n_1$ in Theorem 9 are best possible, provided that any of the four equivalent conditions are satisfied. We conjecture that this is always the case.

Conjecture 1. *Let q be a prime power, and let k, l be positive integers. Then there exists a monic irreducible polynomial $f(x) \in \mathbb{F}_{q^k}[x]$ of degree l such that $f(x) - f(0) \in \mathbb{F}_q[x]$ and $f(0) \in \mathcal{F}_k(q)$.*

Note that Conjecture 1 is slightly stronger than Proposition 10(iv) in the sense that l is not necessarily the smallest prime divisor of kl.

Conjecture 2. *Let p be a prime, and let k, l be positive integers. Then there exists a monic irreducible polynomial $f(x) \in \mathbb{F}_{p^k}[x]$ of degree l such that $f(x) - f(0) \in \mathbb{F}_p[x]$ and $f(0) \in \mathcal{F}_k(p)$.*

Clearly, validity of Conjecture 1 for all prime power q implies that of Conjecture 2. Conversely, suppose that Conjecture 2 is true. Let $q = p^r$, where p is a prime. Then there exists a monic irreducible polynomial $f(x) \in \mathbb{F}_{p^{rk}}[x]$ of degree l such that $f(x) - f(0) \in \mathbb{F}_p[x]$ and $f(0) \in \mathcal{F}_{rk}(p)$. In particular, $f(x)$ is a monic irreducible polynomial in $\mathbb{F}_{q^k}[x]$ of degree l such that $f(x) - f(0) \in \mathbb{F}_q[x]$ and $f(0) \in \mathcal{F}_k(q)$ by (2). Therefore, the two conjectures are equivalent.

The existence problem of a monic irreducible polynomial of two prescribed coefficients dates back to Carlitz [2]. See [4, Part II, Sect. 3.5] for more recent work. Conjecture 2 is a similar but different problem, in the sense that all coefficients except the constant term are required to be in the prime field.

Conjecture 2 is trivially true for $l = 1$ or $k = 1$. Moreover, it is true for the following special cases:

Proposition 11. *Conjecture 2 is true if $l = p$.*

Proof. It is known (see for example [2]) that there exists $a \in \mathcal{F}_k(p)$ such that $\mathrm{Tr}_{\mathbb{F}_{p^k}}(a) = 1$. Then, by [3, Corollary 3.79], $x^l - x - a$ is irreducible in $\mathbb{F}_{p^k}[x]$.

Proposition 12. *Let l be a positive integer each of whose prime factor divides $p^k - 1$. Assume further that, $p^k \equiv 1 \pmod 4$ if $l \equiv 0 \pmod 4$. Then Conjecture 2 is true.*

Proof. Let a be a primitive element of $\mathbb{F}_{p^k}$. Then $x^l - a$ is irreducible in $\mathbb{F}_{p^k}[x]$ by [3, Theorem 3.75].

By Propositions 11 and 12, Conjecture 2 is true for $l = 2$, or $l = 3$ and k even. We have verified Conjecture 2 for $p^{kl} \leq 10^{20}$ by computer.

References

1. Brawley, J.V., Carlitz, L.: Irreducibles and the composed product for polynomials over a finite field. Discret. Math. **65**, 115–139 (1987)
2. Carlitz, L.: A theorem of Dickson on irreducible polynomials. Proc. Am. Math. Soc. **3**, 693–700 (1952)
3. Lidl, R., Niederreiter, H.: Finite Fields. Cambridge University Press, Cambridge (1997)
4. Mullen, G.L., Panario, D.: Handbook of Finite Fields. CRC Press, Boca Raton (2013)
5. Stichtenoth, H.: A note on composed products of polynomials over finite fields. Des. Codes Cryptogr. **73**, 27–32 (2014)

Invited Talk II

On Arcs and Quadrics

Simeon Ball(✉)
Departament de Matemàtiques, Universitat Politècnica de Catalunya,
Jordi Girona 1-3, Mòdul C3, Campus Nord, 08034 Barcelona, Spain
simeon@ma4.upc.edu

Abstract. An arc is a set of points of the $(k-1)$-dimensional projective space over the finite field with q elements $\mathbb{F}_q$, in which every k-subset spans the space. In this article, we firstly review Glynn's construction of large arcs which are contained in the intersection of quadrics. Then, for q odd, we construct a series of matrices F_n, where n is a non-negative integer and $n \leqslant |G| - k - 1$, which depend on a small arc G. We prove that if G can be extended to a large arc S of size $q + 2k - |G| + n - 2$ then, for each vector v of weight three in the column space of F_n, there is a quadric ψ_v containing $S \setminus G$. This theorem is then used to deduce conditions for the existence of quadrics containing all the vectors of S.

2010 Mathematics Subject Classification: 51E21 · 15A03 · 94B05 · 05B35

1 Introduction

Let $\mathrm{PG}_{k-1}(\mathbb{F}_q)$ denote the $(k-1)$-dimensional projective space over $\mathbb{F}_q$, the finite field with q elements. For any homogeneous polynomials $f_1, \ldots, f_r$ in k variables, let $V(f_1, \ldots, f_r)$ denote the points of $\mathrm{PG}_{k-1}(\mathbb{F}_q)$ which are zeros of all the polynomials $f_1, \ldots, f_r$.

An *arc* of $\mathrm{PG}_{k-1}(\mathbb{F}_q)$ is a set S of points of $\mathrm{PG}_{k-1}(\mathbb{F}_q)$ with the property that every k-subset spans $\mathrm{PG}_{k-1}(\mathbb{F}_q)$. The set of columns of a generator matrix of a k-dimensional linear maximum distance separable (MDS) code over $\mathbb{F}_q$ (when viewed as points in the corresponding projective space) is an arc of $\mathrm{PG}_{k-1}(\mathbb{F}_q)$ and vice-versa, so arcs and linear MDS codes are equivalent objects.

In [3], the author recently detailed a possible method to construct large arcs from small arcs. An alternative method to construct large arcs from small arcs as the intersection of the quadrics containing the small arc, was proposed by Glynn in [7]. In this note we firstly review Glynn's construction, then try and prove some kind of converse of Glynn's construction. We will construct a matrix from a small arc and show that for each vector of weight three in the column space of the matrix, there is a quadric containing the large arc to which the small arc extends.

The author acknowledges the support of the project MTM2014-54745-P of the Spanish *Ministerio de Economía y Competitividad*.
This article is for the proceedings of the *International Workshop on the Arithmetic of Finite Fields WAIFI 2016*, held in Ghent, Belgium, July 13–15, 2016.

S. Duquesne and S. Petkova-Nikova (Eds.): WAIFI 2016, LNCS 10064, pp. 95–102, 2016.
DOI: 10.1007/978-3-319-55227-9_8

The largest known arcs are of size $q+1$, unless $k \geqslant q+1$, in which case the largest arcs have size $k+1$, or $k=3$ or $q-1$ and q is even, in which case they have size $q+2$. Moreover, if $k \leqslant 3$ or $k \geqslant q-1$ then it is a simple matter to prove that there are no arcs larger than these largest known arcs.

The MDS conjecture, proposed as a question by Segre [9], is the following.

Conjecture 1.1. *If $4 \leqslant k \leqslant q-2$ then an arc of $\mathrm{PG}_{k-1}(\mathbb{F}_q)$ has size at most $q+1$.*

The MDS conjecture was proven for q prime in [1]. For q non-prime and $k \leqslant 2p-2$, where p is the prime for which q is a p-th power, the MDS conjecture was proven in [2]. As a consequence of these results and Tables 3.1, 3.2, 3.3, 3.4 and 3.7 from [8], we have that the MDS conjecture is true for all $q \leqslant 27$ and for all $k \leqslant 7$, with the possible exceptions of $(k,q)=(6,81)$ and $(k,q)=(7,81)$.

For q odd and $k \leqslant q$, the largest known arcs have size $q+1$.

The normal rational curve,

$$S=\{(1,t,\ldots,t^{k-1}) \mid t \in \mathbb{F}_q\} \cup \{(0,\ldots,0,1)\}$$

is an arc of size $q+1$ in $\mathrm{PG}_{k-1}(\mathbb{F}_q)$, for all $k \leqslant q$.

Observe that S is contained the intersection of quadrics $V(\phi_{ij})$, where

$$\phi_{ij}(x)=x_i x_j - x_{i+1}x_{j-1},$$

and $1 \leqslant i \leqslant j-2$ and $3 \leqslant j \leqslant k$. These quadratic forms span a $\binom{k-1}{2}$-dimensional subspace of the vector space P_2, whose non-zero elements are the homogeneous polynomials of degree two.

The Glynn arc [6],

$$S=\{(1,t,t^2+\eta t^6,t^3,t^4) \mid t \in \mathbb{F}_q\} \cup \{(0,\ldots,0,1)\}$$

is an arc of size 10 in $\mathrm{PG}_4(\mathbb{F}_9)$, where η is fixed and satisfies $\eta^4=-1$.

As observed in [7], S is contained the intersection of quadrics,

$$\begin{aligned}
&\eta x_2^2 + x_4^2 - x_3x_5,\\
&x_3x_4 - \eta x_1x_2 - x_2x_5,\\
&x_2x_4 - x_1x_5,\\
&x_2x_3 - x_1x_4 - \eta x_4x_5,\\
&x_2^2 + \eta x_4^2 - x_1x_3.
\end{aligned}$$

These quadrics span a 5-dimensional subspace of the vector space P_2.

2 Glynn's Construction of Arcs as Intersection of Quadrics

Recall that P_2 denotes the vector space whose non-zero elements are the homogeneous polynomials in k variables of degree 2. The space P_2 is a $\binom{k+1}{2}$-dimensional vector space over $\mathbb{F}_q$ and a typical element of P_2 is

$$\psi(x)=a_1x_1^2+\cdots+a_kx_k^2+\sum_{i,j} b_{ij}x_ix_j,$$

where the sum runs over i and j such that $1 \leqslant i < j \leqslant k$.

For any subspace U of P_2, we denote by $V(U)$ the set of points of $PG_{k-1}(\mathbb{F}_q)$ which are zeros of ψ, for all $\psi \in U$.

As we have seen in the introduction, the normal rational curve is contained (in fact equal to) $V(U)$ for some $\binom{k-1}{2}$-dimensional subspace U in which every polynomial is irreducible. The following theorem is Theorem 3.1 from [6] and is a partial converse of this. Glynn goes on to conjecture that not only is $V(U)$ an arc, under the hypothesis, but is in fact a normal rational curve.

Theorem 2.1. *Let U be a $\binom{k-1}{2}$-dimensional subspace of P_2. If every polynomial in U is irreducible and $V(U)$ spans $PG_{k-1}(\mathbb{F}_q)$ then $V(U)$ is an arc of $PG_{k-1}(\mathbb{F}_q)$.*

Proof. Suppose that $V(U)$ is not an arc. Then there is a point u of $V(U)$ which is dependent on $r \leqslant k-1$ points of $V(U)$, say $e_1, \ldots, e_r$. Since $V(U)$ spans the space, there are points $e_{r+1}, \ldots, e_k$ of $V(U)$ such that $e_1, \ldots, e_k$ span $PG_{k-1}(\mathbb{F}_q)$. With respect to a suitable basis we can assume that e_i are the unit vectors, for $i = 1, \ldots, k$ and $u = (u_1, \ldots, u_{k-1}, 0)$, where $u_i u_j \neq 0$ for some $i \neq j$.

Let π be the hyperplane of $PG_{k-1}(\mathbb{F}_q)$ spanned by $e_1, \ldots, e_{k-1}$. The vector space of quadrics restricted to the hyperplane π has dimension $\binom{k}{2}$ and the subspace of this vector space of the quadrics which contains $e_1, \ldots, e_{k-1}$ and u has dimension $\binom{k}{2} - k = \frac{1}{2}k(k-3)$. Since U has dimension $\frac{1}{2}k(k-3)+1$ there must be two quadrics ψ_1 and ψ_2 in U whose restriction to π is the same. Therefore, $\psi_1 - \psi_2$ factorises into two linear forms, one of whose kernels is π, a contradiction. □

Given an arc G of $PG_{k-1}(\mathbb{F}_q)$, we define

$$QS(G) = \{\psi \in P_2 \mid \psi(x) = 0, \text{ for all } x \in G\},$$

a subspace of P_2.

The following theorem is Theorem 3.2(i) from [7].

Theorem 2.2. *If G is an arc of $PG_{k-1}(\mathbb{F}_q)$ of size $2k-1$ then $V(QS(G))$ is arc of $PG_{k-1}(\mathbb{F}_q)$ containing G.*

Proof. With respect to a suitable basis we can assume that G contains the unit vectors $e_1, \ldots, e_k$ and the points $(1, y_{\ell 2}, \ldots, y_{\ell k})$ for $\ell = 1, \ldots, k-1$. Quadrics in the subspace $QS(G)$ are of the form

$$\psi(x) = \sum_{i,j} b_{ij} x_i x_j,$$

where

$$(y_{\ell 2}, \ldots, y_{\ell k}, y_{\ell 2} y_{\ell 3}, \ldots, y_{\ell, k-2} y_{\ell, k-1}) \cdot (b_{12}, \ldots, b_{1k}, b_{23}, \ldots, b_{k-1,k}) = 0.$$

for all $\ell = 1, \ldots, k-2$. Here, $\cdot$ denotes the standard inner product of vectors. Since G is an arc, this system of equations has rank $k-2$ and so the dimension of $QS(G)$ is $\binom{k+1}{2} - (2k-1) = \binom{k-1}{2}$. Since $G \subseteq V(QS(G))$, the points $V(QS(G))$ span $PG_{k-1}(\mathbb{F}_q)$, so Theorem 2.1 applies. □

The following theorem is Theorem 3.2(ii) from [7].

Theorem 2.3. *If G is an arc of $\mathrm{PG}_{k-1}(\mathbb{F}_q)$ of size at most $2k-2$ then G does not extend to a larger arc within $V(\mathrm{QS}(G))$.*

Proof. As in the proof of Theorem 2.2, each point of G imposes a condition on the quadrics in $\mathrm{QS}(G)$. Hence, the dimension of $\mathrm{QS}(G)$ is $\binom{k+1}{2} - |G|$. Any point of $V(\mathrm{QS}(G)) \setminus G$ which extends G to a larger arc would impose a further condition on $\mathrm{QS}(G)$, thus implying that its dimension is smaller than it is. □

Observe that if we take a larger arc G, of size $2k$ or $2k+1$ say, then $\mathrm{QS}(G)$ will, in general, get smaller and then $V(\mathrm{QS}(G))$ larger. We will not then have that $V(\mathrm{QS}(G))$ is necessarily an arc. However, we should be able to find a larger arc extending G within $V(\mathrm{QS}(G))$. Alternatively, we can take a subspace U of $\mathrm{QS}(G)$. Although Theorem 2.1 will not apply, so we cannot conclude that $V(U)$ is an arc, it may be that there is a large arc containing G contained in $V(U)$. This is precisely what happens if we take 9 points G of Glynn arc, as described in the previous section. The subspace of quadrics $\mathrm{QS}(G)$ is a 6-dimensional subspace and it contains a five dimensional subspace U such that the Glynn arc of size 10 is contained in $V(U)$. Observe that for the subspace U spanned the five quadrics detailed in the previous section, $V(U)$ also contains the point $(0,0,1,0,0)$, so is not an arc itself.

3 Small Arcs Which Extend to Large Arcs

Although Theorem 3.4, Theorem 3.5 and Theorem 3.6 have not explicitly appeared before, the results in this section are based on the articles [3–5].

Let $\mathrm{V}_k(\mathbb{F}_q)$ denote the k-dimensional vector space over $\mathbb{F}_q$.

Let $\det(v_1,\ldots,v_k)$ denote the determinant of the matrix whose i-th row is v_i, a vector of $\mathrm{V}_k(\mathbb{F}_q)$. If $C=\{p_1,\ldots,p_{k-1}\}$ is an ordered set of $k-1$ vectors then we write

$$\det(u,C)=\det(u,p_1,\ldots,p_{k-1}),$$

where we evaluate the determinant with respect to a fixed canonical basis.

Throughout this section S will be an arbitrarily ordered arc of size $q+k-1-t$ of $\mathrm{V}_k(\mathbb{F}_q)$. In other words, we will consider S as a set of vectors of $\mathrm{V}_k(\mathbb{F}_q)$, as opposed to a set of points of $\mathrm{PG}_{k-1}(\mathbb{F}_q)$.

For each $(k-1)$-subset C of S, there is a non-zero element $\alpha_C \in \mathbb{F}_q$, such that the following lemma holds, see [4, Lemma 2.2].

Lemma 3.1. *Suppose that q is odd. Let E be a subset of S of size $k+t-1$. For any subset D of E of size $k-3$,*

$$\sum \alpha_C \prod_{z\in(E\cup\{x\})\setminus C} \det(z,C)^{-1}=0,$$

where the sum runs over the subsets C of E of size $k-1$ containing D, for all $x\in S\setminus E$.

Let G be an arc of $V_k(\mathbb{F}_q)$ and let $n \leqslant |G| + 1 - k$ be a non-negative integer. We define a matrix F_n whose rows are indexed by the $(k-1)$-subsets of G. The columns of F_n are indexed by pairs (D, E), where D is a subset of E of size $k-3$ and E is a subset of G of size $|G| - n$. The $(C, (D, E))$ entry of F_n is

$$\prod_{u \in G \setminus E} \det(u, C),$$

if $D \subset C \subset E$ and zero otherwise.

Let $v(x)$ be the row vector whose coordinates are indexed by the $(k-1)$-subsets C of G and whose C entry is

$$\alpha_C \det(x, C)^{-1} \prod_{z \in G \setminus C} \det(z, C)^{-1}.$$

Note that all the coordinates in $v(x)$ are non-zero.

Lemma 3.2. *If G extends to an arc S of size $q+2k+n-|G|-2$ then $v(x)F_n = 0$ for all $x \in S \setminus G$.*

Proof. The (D, E) coordinate in the vector $v(x)F_n$ is

$$\sum \alpha_C \det(x, C)^{-1} \prod_{z \in G \setminus C} \det(z, C)^{-1} \prod_{u \in G \setminus E} \det(u, C),$$

where the sum runs over the $(k-1)$-subsets C with the property that $D \subset C \subset E$. By Lemma 3.1, this sum is zero. □

Lemma 3.2 allows us to prove various theorems depending on vectors appearing in the column space of F_n.

Theorem 3.3. *If there is a vector w of weight one in the column space of F_n then G cannot be extended to an arc of size $q + 2k + n - |G| - 2$.*

Proof. Suppose G extends to an arc of size $q + 2k + n - |G| - 2$. By Lemma 3.2, $v(x) \cdot w = 0$, where $\cdot$ denotes the standard inner product of vectors. Since w has weight one, this equation implies that one of the coordinates of $v(x)$ is zero, which it is not. □

Theorem 3.4. *If there is a vector w of weight two in the column space of F_n and $|G| < \frac{1}{2}(q+k+n-1)$ then G cannot be extended to an arc of size $q+2k+n-|G|-2$.*

Proof. Suppose G extends to an arc S of size $q+2k+n-|G|-2$. By Lemma 3.2, $v(x) \cdot w = 0$, where $\cdot$ denotes the standard inner product of vectors. Since w has weight two, there are two $(k-1)$-subsets C_1 and C_2 of G such that

$$\beta_1 \det(x, C_1)^{-1} + \beta_2 \det(x, C_2)^{-1} = 0,$$

for some $\beta_1, \beta_2 \in \mathbb{F}_q$, for all $x \in S \setminus G$. This implies that all the vectors of $S \setminus G$ are in the kernel of the linear form

$$\alpha(x) = \beta_1 \det(x, C_2) + \beta_2 \det(x, C_1).$$

Note that α is not zero since C_1 and C_2 do not span the same hyperplane of $\mathrm{V}_k(\mathbb{F}_q)$. Since the kernel of a linear form is a hyperplane and a hyperplane of $\mathrm{V}_k(\mathbb{F}_q)$ contains at most $k-1$ vectors of an arc of $\mathrm{V}_k(\mathbb{F}_q)$, we have that $|S \setminus G| \leqslant k-1$. Therefore,

$$q + 2k + n - |G| - 2 - |G| \leqslant k - 1,$$

contradicting the hypothesis on the size of G. □

Theorem 3.5. *Suppose G can be extended to an arc S of size $q+2k+n-|G|-2$. For each vector w of weight three in the column space of F_n there is a quadratic form ψ_w such that $S \setminus G \subseteq V(\psi_w)$.*

Proof. By Lemma 3.2, $v(x) \cdot w = 0$, where $\cdot$ denotes the standard inner product of vectors. Since w has weight three, there are three $(k-1)$-subsets C_1, C_2 and C_3 of G such that

$$\beta_1 \det(x, C_1)^{-1} + \beta_2 \det(x, C_2)^{-1} + \beta_3 \det(x, C_3)^{-1} = 0,$$

for some $\beta_1, \beta_2, \beta_3 \in \mathbb{F}_q$, for all $x \in S \setminus G$.

This implies that

$$S \setminus G \subseteq V(\psi_w),$$

where

$$\psi_w(x) = \beta_1 \det(x, C_2)\det(x, C_3) + \beta_2 \det(x, C_1)\det(x, C_3) + \beta_3 \det(x, C_1)\det(x, C_2).$$

□

Theorem 3.6. *Let S be an arc of $\mathrm{V}_k(\mathbb{F}_q)$ of size $q+k-t-1$ and suppose that C_1, C_2, C_3 are $(k-1)$-subsets of S with the property that no element of $C_1 \cup C_2 \cup C_3$ is in exactly one of the C_i, for some $i = 1, 2, 3$. Suppose that there is a 2-subset T of S such that for each $x \in S \setminus (T \cup C_1 \cup C_2 \cup C_3)$, there is a subset G of $S \setminus (T \cup \{x\})$ with the property that there is a vector with support $\{C_1, C_2, C_3\}$ in the column space of $\mathrm{F}_n(G)$, where $n = |G| - k - t + 1$. Then*

$$S \subseteq V(\psi),$$

for some quadratic form

$$\psi(x) = \beta_1 \det(x, C_2)\det(x, C_3) + \beta_2 \det(x, C_1)\det(x, C_3) + \beta_3 \det(x, C_1)\det(x, C_2),$$

where $\beta_1, \beta_2, \beta_3$ are non-zero elements of $\mathbb{F}_q$.

Proof. For each $x \in S \setminus (T \cup C_1 \cup C_2 \cup C_3)$, there is, by hypothesis, a subset G of $S \setminus (T \cup \{x\})$ such that Theorem 3.5 implies the existence of non-zero elements $\beta_1, \beta_2, \beta_3$ of $\mathbb{F}_q$ (possibly dependent on G which is dependent on x) such that

$$T \cup \{x\} \subseteq V(\psi),$$

where

$$\psi(X) = \beta_1 \det(X, C_2) \det(X, C_3) + \beta_2 \det(X, C_1) \det(X, C_3) + \beta_3 \det(X, C_1) \det(X, C_2).$$

Since $\psi(a) = 0$ and $\psi(b) = 0$ where $T = \{a, b\}$, the elements $\beta_1, \beta_2, \beta_3$ are determined (up to scalar factor) by T. Therefore, they do not depend on x and we have that $S \setminus (C_1 \cup C_2 \cup C_3) \subseteq V(\psi)$.

By hypothesis, C_1, C_2, C_3 are $(k-1)$-subsets of S with the property that no element of $C_1 \cup C_2 \cup C_3$ is in exactly one of the C_i, for some $i = 1, 2, 3$. Hence, $C_1 \cup C_2 \cup C_3 \subseteq V(\psi)$. □

The following example is essentially what is used to classify arcs of $\mathrm{V}_k(\mathbb{F}_q)$ of size $q+1$, for $k \leqslant p$ and $k \leqslant \frac{1}{2}(q-1)$ in [1]. Recall, p is the prime for which q is a p-th power.

Example 3.7. Let S be an arc of $\mathrm{V}_k(\mathbb{F}_q)$ of size $q+1$, where $k \leqslant p$ and $k \leqslant \frac{1}{2}(q-1)$. Let K be a subset of S of size k and suppose $C_i = K \setminus \{e_i\}$, for $i = 1, 2, 3$, for any $e_1, e_2, e_3 \in K$. Then $\mathrm{F}_1(G)$ contains a vector with support $\{C_1, C_2, C_3\}$, for all $(2k-2)$-subsets G of S containing K. Since $2k-2 \leqslant |S|-3$, we can fix a 2-subset T of S and choose G to be a subset of $S \setminus (T \cup \{x\})$. Then Theorem 3.6 implies that

$$S \subseteq V(\psi),$$

where, with respect to the basis K of $\mathrm{V}_k(\mathbb{F}_q)$,

$$\psi(x) = \beta_1 x_2 x_3 + \beta_2 x_1 x_3 + \beta_3 x_1 x_2,$$

for some $\beta_1, \beta_2, \beta_3$, non-zero elements of $\mathbb{F}_q$. As proven in [1], this is sufficient to prove that S (when viewed as points in the corresponding projective space) is a normal rational curve.

References

1. Ball, S.: On sets of vectors of a finite vector space in which every subset of basis size is a basis. J. Eur. Math. Soc. **14**, 733–748 (2012)
2. Ball, S., De Beule, J.: On sets of vectors of a finite vector space in which every subset of basis size is a basis II. Des. Codes Cryptogr. **65**, 5–14 (2012)
3. Ball, S.: Extending small arcs to large arcs. arXiv:1603.05795 (2016)
4. Ball, S., De Beule, J.: On subsets of the normal rational curve. arXiv:1603.06714 (2016)
5. Chowdhury, A.: Inclusion matrices and the MDS conjecture. arXiv:1511.03623v2 (2015)

6. Glynn, D.G.: The non-classical 10-arc of $PG(4,9)$. Discret. Math. **59**, 43–51 (1986)
7. Glynn, D.G.: On the construction of arcs using quadrics. Austral. J. Combin. **9**, 3–19 (1994)
8. Hirschfeld, J.W.P., Storme, L.: The packing problem in statistics, coding theory and finite projective spaces: update 2001. In: Blokhuis, A., Hirschfeld, J.W.P., Jungnickel, D., Thas, J.A. (eds.) Finite Geometries. Developments in Mathematics, vol. 3, pp. 201–246. Springer, Heidelberg (2001). doi:10.1007/978-1-4613-0283-4_13
9. Segre, B.: Introduction to Galois geometries. Atti Accad. Naz. Lincei Mem. **8**, 133–236 (1967)

Applications to Cryptography

A Generalised Successive Resultants Algorithm

James H. Davenport[1], Christophe Petit[2], and Benjamin Pring[1(✉)]

[1] University of Bath, Bath BA2 7AY, UK
{J.H.Davenport,b.i.pring}@bath.ac.uk
[2] University of Oxford, Oxford OX2 6GG, UK
Christophe.Petit@maths.ox.ac.uk

Abstract. The Successive Resultants Algorithm (SRA) is a root-finding algorithm for polynomials over $\mathbb{F}_{p^n}$ and was introduced at ANTS in 2014 [19]. The algorithm is efficient when the characteristic p is small and $n > 1$. In this paper, we abstract the core SRA algorithm to arbitrary finite fields and present three instantiations of our general algorithm, one of which is novel and makes use of a series of isogenies derived from elliptic curves with sufficiently smooth order.

Keywords: Root finding · Finite fields · Algorithms · Elliptic curves

1 Introduction

The factorization of polynomials over finite fields is an important problem in computer algebra, both from theoretical and practical points of view [11]. An important subcase of this problem is the root-finding problem, which given a polynomial over a finite field, asks for one, several or all roots of this polynomial over the field. It is well-known that factoring polynomials is deterministically reducible to root-finding [3], so in this paper we will mostly focus on the root-finding problem.

1.1 Finding Roots of Polynomials over Finite Fields

From now on in this paper, let $\mathbb{F}_{p^n}$ be a finite field of size p^n and f a polynomial of degree d with coefficients in $\mathbb{F}_{p^n}$. As it is often the case in the literature, we will assume that f is entirely split over $\mathbb{F}_{p^n}$ and that it has no repeated roots. One can reduce the general case to this one by computing $\gcd(f(x), x^{q^n} - x)$, for example using a variant of the square and multiply algorithm [10]. We allow the notation x and $f(x)$ to denote the variable and polynomial in $\mathbb{F}_{p^n}[x]$ with $\hat{x} \in \mathbb{F}_{p^n}$ and $f(\hat{x})$ to represent the evaluation of the polynomial f at $\hat{x}$.

Since the seminal work of Berlekamp in the seventies [3], the root-finding problem can be solved in probabilistic polynomial time (in the degree of f and in $n \log p$). Significant practical and theoretical improvements have been made since then, with the current best probabilistic algorithm for the general factorization of a polynomial being due to Kedlaya and Umans [16]. In practice, one will often

S. Duquesne and S. Petkova-Nikova (Eds.): WAIFI 2016, LNCS 10064, pp. 105–124, 2016.
DOI: 10.1007/978-3-319-55227-9_9

use either Berlekamp's trace algorithm [3] or Cantor-Zassenhaus algorithm [5], depending on the parameters.

Berlekamp's trace algorithm in fact provides a polynomial time reduction from the root-finding problem over $\mathbb{F}_{p^n}$ to the root-finding problem over $\mathbb{F}_p$. The reduction can be made deterministic, leading to a polynomial time deterministic algorithm for fields of small characteristic.

In contrast, Shoup's algorithm [23] is still the best unconditional deterministic algorithm over $\mathbb{F}_p$ today, with a complexity in $\tilde{O}(d^2\sqrt{p})$. Designing a deterministic polynomial time algorithm in that setting is an important open problem, even in the case of degree 2 polynomials. Evdokimov has provided a quasi-polynomial time algorithm when a quadratic non-residue is provided together with the field as an input to the algorithm [8]. Under the Generalised Riemann Hypothesis (GRH), this element can be computed in polynomial time, removing the need for an extra input. Polynomial time algorithms have also been suggested under additional assumptions on the polynomial [21], other conjectures [1,9] or for specific families of primes [2,21,24], still under GRH.

In 2014, Petit introduced a new algorithm in the small characteristic case, called the Successive Resultants Algorithm [19].

1.2 Our Contributions

In this paper, we introduce a generalisation of the Successive Resultants Algorithm to arbitrary finite fields. Our generalisation covers both the original SRA algorithm for finite fields $\mathbb{F}_{p^n}$ with small characteristic and the generalised Graeffe transform approach of Grenet et al. [12] when $p^n - 1$ is smooth. We also present a third instance using an elliptic curve with smooth order over $\mathbb{F}_{p^n}$, leading to a new algorithm of independent interest.

Our initial observation is that the linearized polynomials used in SRA can be replaced by any set of polynomials, and in fact even rational maps K_i, such that the image of the composed map $K_t \circ K_{t-1} \circ \ldots \circ K_2 \circ K_1$ under a restricted domain is sufficiently small. Similar generalisations were made in different contexts in [20].

Like the original SRA, our generalisation reduces the root-finding problem for large degree polynomials to the same problem for "small" degree polynomials. The original SRA has two stages, a resultant stage and a gcd stage. We show how to adapt both stages to the case of arbitrary rational maps, and how to overcome the technical difficulties introduced by the denominators of the maps.

Recently, De Feo, Petit and Quisquater showed that the Successive Resultants Algorithm and Berlekamp's celebrated Trace Algorithm (BTA) [3] are in a certain sense dual of each other [18]. We remark that the generalised Graeffe transform algorithm mentioned above can similarly be seen as a dual of Shoup's algorithm when $p-1$ is smooth [24], and our new algorithm as a dual of a slight variant of an algorithm due to Ronyai [21, Sect. 7] (See Table 1).

In the algorithm of Sect. 3.3, we have used Icart's embeddings [15] to map $\mathbb{F}_p$ elements to the x-coordinates of $\mathbb{F}_p$-rational points of a smooth order elliptic curve over $\mathbb{F}_p$, where Ronyai's algorithm would map them to a smooth curve

Table 1. Special instances of SRA and corresponding "dual" algorithms

	p small	p^n-1 smooth	Elliptic curves
Resultant-based	[19]	[12]	Section 3
GCD-based	[3]	[24]	[21, Sect. 4]

over $\mathbb{F}_{p^2}$. Our approach has some efficiency advantages and more importantly it leads to a larger set of suitable parameters in the algorithm.

We remark that all our algorithms can be made deterministic for certain parameters after some precomputation dependent upon the field and assuming the Generalised Riemann Hypothesis. These deterministic versions can be seen in the continuity of [2,21,24], providing polynomial time deterministic algorithms under GRH for special fields.

Proof of concept code in SageMath [7] for all three instantiations may be found at https://www.github.com/bip20/SRA.

2 A Generalised Form of the Successive Resultants Algorithm

The Successive Resultants Algorithm (SRA) is a root finding algorithm which exploits the properties of an ordered set of rational mappings in order to extract roots by computing the roots of polynomials of small degree. As explained in the introduction, we will be considering the problem of finding the roots of a polynomial $f \in \mathbb{F}_{p^n}[x]$ whose splitting field is $\mathbb{F}_{p^n}$.

The generation of the rational maps is a key factor in the efficiency and utility of the SRA algorithm. These maps may be considered as input to the algorithm and the existence of a useful set of maps currently depends upon the structure of $\mathbb{F}_{p^n}$. We note that the rational maps are independent of f and may be performed as precomputation.

Given a polynomial f of degree d and a sequence of rational maps $K_1, \ldots, K_t$ the SRA algorithm involves computing finite sequences of length $j \leq t+1$ obtained by successively transforming the roots of f by application of the rational maps. In other words, the sequences $(x_1, \ldots, x_j)$ of length 1 to $t+1$ fulfilling

$$\begin{cases} f(x_1) & = 0 \\ K_i(x_i) = \frac{a_i(x_i)}{b_i(x_i)} & = x_{i+1} \quad \text{for } i = 1, \ldots, j \end{cases} \tag{1}$$

where $K_i : \mathbb{F}_{p^n} \setminus N_i \rightarrow \mathbb{F}_{p^n}$, $N_i := \{x_i \in \mathbb{F}_{p^n} \ : \ b_i(x_i) = 0\}$ and $a_i, b_i \in \mathbb{F}_{p^n}[X]$ such that $\gcd(a_i(x_i), b_i(x_i))$ is trivial.
We define the composed map

$$\begin{aligned} K^{[j]}(x_1) &: \mathbb{F}_{p^n} \setminus N^{[j]} \rightarrow \mathbb{F}_{p^n} \\ K^{[j]}(x_1) &= K_j \circ \ldots \circ K_0(x_1) \end{aligned} \tag{2}$$

where $N^{[j]} := \{x_1 \in \mathbb{F}_{p^n} : \forall\, i \in \{1, \ldots, j\}\ K_i \circ \ldots \circ K_1(x_1) \notin N_i\}$. For notation purposes, we take $K^{[0]}$ to be the identity map $\mathrm{Id}_{\mathbb{F}_{p^n}} : \mathbb{F}_{p^n} \to \mathbb{F}_{p^n}$ and $N^{[0]} = \emptyset$.

Ideally, we will want a minimal number of rational maps of small degree with the property that Image($K^{[t]}$) is small. These points will improve the efficiency of the algorithm, as will become clear later in the paper.

The SRA algorithm consists of two separate stages, the Resultant Stage and the GCD stage. In the Resultant stage, a series of polynomials $f^{(1)}(x_1), \ldots, f^{(t)}(x_t)$ are computed with $f^{(i+1)}(x_{i+1})$ relying on $f^{(i)}(x_i)$ and $K_i(x_i)$. The roots of $f^{(i)}(x_i)$ lying in $\mathbb{F}_{p^n}$ correspond to the existence of a sequence $(x_1, \ldots, x_j)$ with the root as the ith value in the sequence. In the GCD stage, roots of $f^{(i+1)}(x_{i+1})$ are used to find the roots of $f^{(i)}(x_i)$ by computing roots of polynomials whose degree is constrained by the K_i maps.

Theorem 1. *Given the maps $K_i : \mathbb{F}_{p^n} \setminus N_i \to \mathbb{F}_{p^n}$ for $i = 1, \ldots, t$ we have that each distinct root of $f^{(1)}(x_1)$ produces a unique sequence $(x_1, \ldots, x_j)$ where $j \leq t+1$, obtained by computing $K_i(x_i) := x_{i+1}$ while $x_i \notin N_i$.*

Proof. This may be seen by successively applying the maps K_i to each root. If for some $j \in \{1, \ldots, t\}$ we have that $K_{j-1} \circ \ldots \circ K_0(x_1) \in N_j$, then the sequence is of length j. Otherwise the sequence is of length $t+1$. The sequence is unique for any distinct root of $f^{(1)}(x_1)$ by the fact that the root is the first value in any sequence. □

2.1 The Resultant Stage

We will use the basic result that the resultant possesses the property that for $f, g \in \mathbb{F}_{p^n}[x]$ we have that $\mathrm{Res}_x(f(x), g(x)) = 0$ if and only if the polynomials $f(x)$ and $g(x)$ share a common factor in $\mathbb{F}_{p^n}[x]$. The resultant of two polynomials $f, g \in \mathbb{F}_{p^n}[x]$ may be calculated via naively taking the determinant of the Sylvester matrix of f and g or a more specialised method depending upon the structure of the K_j maps. We will use the standard result [17, Definition 1.93] that

$$\mathrm{Res}_x(f, g) = \mathrm{lc}(f)^{\deg(g)} \prod_{x : f(x) = 0} g(x) \tag{3}$$

where the roots are taken over the splitting field of f and $\mathrm{lc}(f)$ is the leading coefficient of f.

We will use the resultant on polynomials in two variables x_i, x_{i+1}, with respect to the x_i variable to create a series of univariate polynomials, $f^{(i+1)}(x_{i+1})$, whose roots correspond to a non-empty subset of roots of $f^{(i)}(x_i)$.

In the Resultant stage we clear the denominator of the K_i rational function representation of the map and successively compute resultants of the resulting equation and the previously computed polynomial with respect to the variable x_i, defining $f^{(1)}(x_1) := f(x_1)$ as the first such polynomial. This results in the generation of an ordered list of polynomials $f^{(1)}(x_1), \ldots, f^{(t)}(x_t)$ via the procedure

$$\begin{cases} f^{(1)}(x_1) & := f(x_1) \\ f^{(i+1)}(x_{i+1}) & := \text{Res}_{x_i}(f^{(j)}(x_i),\ a_i(x_i) - b_i(x_i) \cdot x_{i+1}) \text{ for } i = 1, \ldots, t-1 \end{cases} \tag{4}$$

The roots of the resulting sequence of polynomials $f^{(1)}, \ldots, f^{(t)}$ encode the potential values any x_i may take for the sequences described in Theorem 1.

Theorem 2. *The polynomials $f^{(1)}, \ldots, f^{(t)}$ have the following properties:*

(i) If $f^{(1)}$ splits over $\mathbb{F}_{p^n}$ then all $f^{(i)}$ split over $\mathbb{F}_{p^n}$.
(ii) The degree of any $f^{(i+1)}$ is less than or equal to the degree of $f^{(i)}$.
(iii) The degree of $f^{(i+1)}$ is strictly less than the degree of $f^{(i)}$ if and only if the gcd of $f^{(i)}$ and b_i is non-trivial.

Proof. We have that by formula (3), that

$$\begin{aligned} f^{(i+1)}(x_{i+1}) &= \text{Res}_{x_i}(f^{(i)}(x_i), a_i(x_i) - b_i(x_i) \cdot x_{i+1}) \\ &= \text{lc}(f^{(i)})^{\deg(a_i(x_i) - b_i(x_i) \cdot x_{i+1})} \prod_{x_i : f^{(i)}(x_i) = 0} (a_i(x_i) - b_i(x_i) \cdot x_{i+1}) \end{aligned} \tag{5}$$

For (i) we note that for any polynomials $a_i, b_i \in \mathbb{F}_{p^n}[x]$ and $\hat{x}_i \in \mathbb{F}_{p^n}$ we have that $a_i(\hat{x}_i), b_i(\hat{x}_i) \in \mathbb{F}_{p^n}$. For (ii) & (iii) as $f^{(i)}$ splits over $\mathbb{F}_{p^n}$ it is clear that the degree of $f^{(i+1)}(x_{i+1}) \leq d$ with equality holding if and only if $b_i(x_i)$ and $f^{(i)}(x_i)$ share no roots in $\mathbb{F}_{p^n}$. □

We note that in the case of polynomial maps, we have that $\deg(f^{(i)}) = d$ as $b_i(x_i) = 1$ and therefore possesses no roots.

Theorem 3. *If $f^{(1)}(x_1)$ splits over $\mathbb{F}_{p^n}$ then the union of all i^{th} values for valid sequences $(x_1, \ldots, x_j)$ as produced in Theorem 1 is equal to the set of roots of each $f^{(i)}(x_i)$.*

Proof. We use induction to prove the result. As $f^{(1)}$ splits over $\mathbb{F}_{p^n}$, we have that the roots of $f^{(1)}(x_1)$ comprise exactly the first values of the sequences as we have defined $f^{(1)}(x_1) := f(x_1)$. Assuming that the set of roots of $f^{(i)}$ is equal to the set of possible x_i sequence values, we have that, by the computation of the resultant and Eq. (5), the roots of $f^{(i+1)}(x_{i+1})$ are those values such that $f^{(i)}(x_i) = 0$ and $a_i(x_i) - b_i(x_i) \cdot x_{i+1} = 0$. If $x_i \in N_i$ then by the product equation of the resultant as in (5) we have that if $b_i(x_i) = 0$, there is no root of $f^{(i+1)}(x_{i+1})$ corresponding to a solution of $a_i(x_i) - b_i(x_i) \cdot x_{i+1} = 0$. If $x_i \notin N_i$, we have that $b_i(x_i) \neq 0$ and so the root x_{i+1} satisfies both $f^{(i)}(x_i) = 0$ and $\frac{a_i(x_i)}{b_i(x_i)} = x_{i+1}$ and is therefore the corresponding point in a sequence. □

We note that if some $f^{(i)}$ does not split over $\mathbb{F}_{p^n}$ then if x_i is a root of an irreducible factor of $f^{(i)}$ in the splitting field of $f^{(i)}$ we have the potential for $f^{(i)}(x_i) = 0$ and $a_i(x_i) - b_i(x_i) \cdot x_{i+1} = 0$ with $x_{i+1} \in \mathbb{F}_{p^n}$. This would lead to sequences of the form $(x_{i+1}, \ldots, x_j)$, but as we assume that $f^{(i)}$ splits over $\mathbb{F}_{p^n}$ we have that all $f^{(i)}$ split over $\mathbb{F}_{p^n}$ by Theorem 1.

Theorem 4. *The roots of $f^{(i)}$ lie in the image of $K^{[i-1]}$.*

Proof. By the property of the resultant, the roots of $f^{(i)}(x_i)$ possess the property that $\frac{a_{i-1}(x_{i-1})}{b_{i-1}(x_{i-1})} = x_i$. This successively constrains the potential values that the roots of each $f^{(i)}(x_i)$ may take. □

By Theorem 4, we have that by sensible choice of the K_i maps, we may obtain a small set which contains our potential x_{t+1} values. This will be useful in the GCD stage.

It is clear that as the roots are successively constrained by Theorem 4, the sequences may be considered as a series of trees of depth $j \leq t+1$ with the $\hat{x}_j$ forming the root nodes and the distinct subsets of the roots of f corresponding to sequences of length j forming the leaves. The SRA algorithm may be considered as a means of generating this tree with relation to the K_i rational maps by first encoding the information concerning the nodes at each level with the Resultant stage and then computing the root and children nodes with the GCD stage.

2.2 The GCD Stage

In the GCD stage, the $f^{(1)}, \ldots, f^{(t)}$ polynomials computed by the Resultant stage to are used to locate the final values of all sequences $(x_1, \ldots, x_j)$ as created by the procedure in Theorem 1. Once we have the final value, we recursively determine the sequence by application of the gcd algorithm and by iteratively computing roots of polynomials, whose degree is bounded by the degree of the K_i maps. Once we have obtained the first value in all sequences, we naturally have found the roots of $f^{(1)}(x_1) = f(x_1)$.

Theorem 5. *Given any $\hat{x}_{i+1} \in \mathbb{F}_{p^n}$ which forms part of a sequence as computed by successive application of the K_i maps on the roots of f, we may compute all i^{th} values of sequences which possess $\hat{x}_{i+1}$ as the $i+1^{th}$ value.*

Proof. If $\hat{x}_{i+1} \in \mathbb{F}_{p^n}$ is such that there exists a sequence $(x_1, \ldots, x_i, \hat{x}_{i+1}, \ldots, \hat{x}_j)$ we have that all values of x_i such that $K_i(x_i) = \frac{a_i(x_i)}{b_i(x_i)} = \hat{x}_{i+1}$ are contained in the roots of $f^{(i)}(x_i)$ by Theorem 3. As $\mathsf{Res}_{x_i}(f^{(i)}(x_i), a_i(x_i) - b_i(x_i) \cdot \hat{x}_{i+1}) = 0$, we have that

$$g^{(i)}_{\hat{x}_{i+1}}(x_i) := \gcd(f^{(i)}(x_i), a_i(x_i) - b_i(x_i) \cdot \hat{x}_{i+1}) \tag{6}$$

is non-trivial and as $f^{(i)}(x_i)$ is split over $\mathbb{F}_{p^n}$, we have that (6) is a product of linear factors whose roots are exactly those such that $f^{(i)}(x_i) = 0$ and $K_i(x_i) = \frac{a_i(x_i)}{b_i(x_i)} = \hat{x}_{i+1}$. We may therefore extract all x_i values whose next value in the sequence is $\hat{x}_{i+1}$ by finding the roots of a split polynomial of degree bounded by $\deg g^{(i)}_{\hat{x}_{i+1}} \leq \max\{\deg a_i, \deg b_i\}$. □

Theorem 6. *(i) We may detect that there exists a sequence* $(\hat{x}_1, \ldots, \hat{x}_j)$ *of length* $j < t+1$ *and may compute* $\hat{x}_j \in \mathbb{F}_{p^n}$ *by computing the roots of a polynomial of degree no larger than* $\deg b_j$.
(ii) Given $\text{Image}(K^{[t]})$ *we may detect that there exists a sequence* $(\hat{x}_1, \ldots, \hat{x}_{t+1})$ *of length* $t+1$ *and its final value* $\hat{x}_{t+1} \in \mathbb{F}_{p^n}$ *by computing the roots of* $|\text{Image}(K^{[t]})|$ *polynomials of degree no larger than* $\max\{\deg a_t, \deg b_t\}$.

Proof. (i) If a sequence is of length $j < t+1$, then $x_j \in N_j$. If this is the case, then it is detected by observing that $\deg f^{(j+1)} < \deg f^{(j)}$ as in Theorem 2. As $x_j \in N_j$, we have that $b_j(x_j) = 0$ and $f^{(j)}(x_j) = 0$, so we may compute

$$g^{(j)}(x_j) := \gcd(f^{(j)}(x_j), b_j(x_j)) \tag{7}$$

whose degree is $\leq \deg b_j$ and whose roots are the final values of all sequences of length j. For sequences of length t there will be no indication of degree drop and so we must compute $g^{(t)}(x_t)$. For *(ii)*, in the case where a sequence of length $t+1$ exists, we have by Theorem 4 that the potential values of x_t lie in the roots of $f^{(t)}(x_t)$ and that $b_t(x_t) \neq 0$. The x_t for which $K_t(x_t) = \frac{a_t(x_t)}{b_t(x_t)} = \hat{x}_{t+1}$ may therefore be extracted by computing the roots of

$$g^{(t)}_{\hat{x}_{t+1}}(x_t) := \gcd(f^{(t)}(x_t), a_t(x_t) - b_t(x_t) \cdot \hat{x}_{t+1}) \tag{8}$$

for each $\hat{x}_{t+1} \in \text{Image}(K^{[t]})$. In the case where no such sequence exists for a chosen $\hat{x}_{i+1}$ then by Theorem 4 we have that (8) is trivial. □

Taken together with a sensible choice of mappings to constrain each $|\text{Image}(K^{[i]})|$ for $i = 1, \ldots, t$, Theorems 4, 5 and 6 allow us to find the length and final value of all sequences. For sequences of length $j < t+1$, we may use Theorem 4, whilst for sequences of length $t+1$ Theorem 5 constrains the possible values which x_{t+1} may take. Ideally, we will wish $\text{Image}(K^{[t]})$ to contain only one element.

We note that SRA may also be used to explicity extract roots possessing certain properties. The algorithm may specifically pick out only roots corresponding to sequences of specific length or roots which intersect with ${K^{[j]}}^{-1}(B)$ for $B \subseteq \text{Image}(K^{[j]})$.

Together the Resultant stage and the GCD stage give us the generalised SRA.

2.3 Generic Complexity Analysis

We allow the notation that $h(n)$ is $O(g(n))$ if there exists some $C \in \mathbb{R}$ and $N \in \mathbb{N}$ such that for all $n \geq N$ we have that $|h(n)| \leq C|g(n)|$. We also make use of the notation that $h(n)$ is $\tilde{O}(g(n))$ if $h(n)$ is $O(g(n) \log^c g(n))$. Complexity is given in terms of basic operations in $\mathbb{F}_p$. The cost of these operations is $O(\log p)$ for addition, $O((\log p)^2)$ for multiplication and inversion with classical arithmetic or $\tilde{O}(\log p)$ for addition, multiplication and $\tilde{O}((\log p)^2)$ for exponentiation with fast FFT-style arithmetic, such as via the Schönhage-Strassen algorithm [10].

Algorithm 1. The generalised Successive Resultants Algorithm

Data: $f \in \mathbb{F}_{p^n}[x]$ – the polynomial whose roots we wish to find
$\frac{a_1}{b_1}, \ldots, \frac{a_t}{b_t}$ – a set of rational maps
$B \subseteq \mathbb{F}_{p^n}$ – a set of points contained in Image$(K^{[t]})$
ParSeq $\in \{\mathrm{T}, \mathrm{F}\}$ – whether to extract roots not in ${K^{[t]}}^{-1}(\mathbb{F}_{p^n})$

Result:
The roots of f in ${K^{[t]}}^{-1}(B)$ and optionally all roots not in ${K^{[t]}}^{-1}(\mathbb{F}_{p^n})$.

begin
- $f^{(1)}(x_1) \longleftarrow f(x_1)$
- **for** $i = 1, \ldots, t-1$ **do**
  - $f^{(i+1)}(x_{i+1}) \longleftarrow \mathsf{Resultant}_{x_i}(f^{(i)}(x_i), a_i(x_i) - b_i(x_i) \cdot x_{i+1})$
- CandidateRoots $\longleftarrow B$
- **for** $i = t, \ldots, 1$ **do**
  - TempRoots $\longleftarrow \{\}$
  - **for** $y \in$ *CandidateRoots* **do**
    - $g_y(x_i) \longleftarrow \mathsf{gcd}(f^{(t)}(x_i), a_i(x_i) - b_i(x_i) \cdot y)$
    - TempRoots $\longleftarrow$ TempRoots $\cup$ $\mathsf{Roots}(g_y(x_i))$
  - **if** *PartialSeq* **and** $(i < t$ **and** $\deg(f^{(i+1)}) < \deg(f^{(i)})$ **or** $i == t)$ **then**
    - $g_b(x_i) \longleftarrow \mathsf{gcd}(f^{(i)}(x_i), b_i(x_i))$
    - TempRoots $\longleftarrow$ TempRoots $\cup$ $\mathsf{Roots}(g_b(x_i))$
  - CandidateRoots $\longleftarrow$ TempRoots
- **return** *CandidateRoots*

We write $\mathbf{a}(n)$ and $\mathbf{m}(n)$ to represent the cost of addition and multiplication over $\mathbb{F}_{p^n}$. We let $\mathbf{A}(d)$, $\mathbf{M}(d)$ and $\mathbf{G}(d)$ respectively represent the cost of performing the addition, multiplication and taking the gcd of two polynomials of degree d in $\mathbb{F}_{p^n}[x]$. We allow $\mathbf{R}(d)$ to be the cost of computing the resultant of two polynomials in $\mathbb{F}_{p^n}[y][x]$ with respect to x where the maximum degree of either polynomials in x is d and the maximum degree of y is 1. We represent the cost of finding the roots in $\mathbb{F}_{p^n}$ of a degree d polynomial to be $\mathbf{P}(d)$.

The following table summarises the cost of performing these with regard to $\mathbb{F}_{p^n}$ in terms of basic operations over $\mathbb{F}_p$ [10]. We provide a cost for $\mathbf{P}(d)$ in terms of the Berlekamp Trace Algorithm, though other methods including standard formulae for quadratics, cubics and quartics may be used (Table 2).

Table 2. Cost of operations over $\mathbb{F}_{p^n}$ in terms of basic operations over $\mathbb{F}_p$.

	$\mathbf{a}(n)$	$\mathbf{m}(n)$	$\mathbf{A}(d)$	$\mathbf{M}(d)$	$\mathbf{G}(d)$	$\mathbf{R}(d)$	$\mathbf{P}(d)$
Classical	$O(n)$	$O(n^2)$	$O(dn)$	$O(d^2n^2)$	$O(d^2n^2 \log d)$	$O(d^3n^2 \log d)$	$O(d^2n^3)$
Fast	$O(n)$	$\tilde{O}(n)$	$O(dn)$	$\tilde{O}(dn)$	$\tilde{O}(dn)$	$\tilde{O}(d^2n))$	$\tilde{O}(dn^2)$

We assume that we are attempting to find the roots of the polynomial $f \in \mathbb{F}_{p^n}[x]$ of degree d such that f possesses d distinct roots, all defined over

$\mathbb{F}_{p^n}$. Whilst the complexity of the SRA algorithm depends upon the choice of K_i rational maps, we may assume there are t maps and that these have been provided by a precomputation. We assume that the maximum degree of any $a_i, b_i \in \mathbb{F}_{p^n}[x]$ is B, that $|\text{Image}(K^{[t]})| = L$ and that $d \geq \max\{B, L\}$. Computing purely the linear factors of a polynomial of degree d may be done at a cost of $O(\mathbf{m}(n)\mathbf{M}(d) \log d \log(p^n d))$ operations in $\mathbb{F}_p$ [10]. We also assume that the maps K_i have been precomputed and exclude this cost.

We have that the resultant stage will consist of taking t resultants of bivariate polynomials where the maximum degree of x is d and the degree of y is always 1. After noting if $\deg f^{(i)} < \deg f^{(i+1)}$, we may compute the square-free part by a cost bounded by $O(\mathbf{M}(n) \log d)$ - this cost is overwhelmed by the computation of the first resultant. We therefore have that the resultant stage generically costs $O(td^3 n^2 \log d)$ with classical arithmetic or $\tilde{O}(td^2 n)$ with fast arithmetic.

Each of the t steps in the GCD consists of taking a maximum of d gcds of maximum degree d resulting in polynomials of degree bounded by B which require solving. At any stage, at most $\frac{d}{2}$ of these will be of degree ≥ 2 and will require solving. This results in a total cost of $O(td^3 n^2 \log d + td\mathbf{P}(B))$ for classical arithmetic and $\tilde{O}(td^2 n + td\mathbf{P}(B))$ for fast arithmetic.

We therefore have the generic cost of SRA is $O(td^3 n^2 \log d + td\mathbf{P}(B))$ with classical arithmetic or $\tilde{O}(td^2 n + td\mathbf{P}(B))$ with fast arithmetic. As the $f^{(i)}$ are square-free, we may obtain the linear factors by using equal-degree splitting for a cost of $O(\mathbf{m}(n)\mathbf{M}(B) \log d \log(p^n B))$. This gives us a total complexity cost of $\tilde{O}(td^3 n^2 + tBd^3 n^5)$ for classical arithmetic or $\tilde{O}(td^2 n + tBd^2 n^3)$ with fast arithmetic. We note that optimized versions of the separate stages may be constructed by exploiting the structure of the K_i maps for specific instantiations of SRA [13,19].

3 Instantiations of SRA

In this section we present three instantiations of our generalised SRA algorithm. We first show how the original SRA algorithm fits into our generalised version. We then present an algorithm for efficiently determining the roots of a polynomial $f \in \mathbb{F}_p[x]$ when $p - 1$ is smooth; this algorithm is equivalent to the one in [12]. We conclude with a description of a method to transform polynomials with roots in $\mathbb{F}_p^*$ into a polynomial whose roots correspond to x-coordinates of an elliptic curve defined over $\mathbb{F}_p$ when $p = 2 \bmod 3$. This method demonstrates a procedure with which we can exploit the structure of $\mathbb{F}_p$ to generate the required rational maps. Proof of concept code written in SageMath [7] for all three cases may be found at https://github.com/bip20/SRA.

3.1 SRA over $\mathbb{F}_{p^n}$ with p Small

In this section we show how the original SRA algorithm fits into the framework of our generalised algorithm. The original SRA algorithm was designed for extension fields $\mathbb{F}_{p^n}$. It uses n polynomial maps K_i of degree p, hence the algorithm is only efficient for small characteristic fields. The polynomials K_i are chosen as follows. For any $\{v_1, \ldots, v_n\}$ a basis of $\mathbb{F}_{p^n}$ over $\mathbb{F}_p$, we define the system of *linearized polynomials*

$$\begin{cases} L_0(z) &= z \\ L_i(z) &= \prod_{i \in \mathbb{F}_p} L_{i-1}(z - iv_i) \qquad \text{for } i = 1, \ldots, n \end{cases} \tag{9}$$

The system

$$K_i(x_i) = x_i^p - c_i x_i = x_{i+1} \qquad \text{for } i = 1, \ldots, n \tag{10}$$

may be derived from system (9) by means of the setting $x_i = L_i(z)$ and the c_i may be precomputed for a cost of $O(n^4)$ with classical arithmetic or $\tilde{O}(n^3)$ with fast arithmetic.

From system (9) it may be deduced that $\text{Image}(K^{[n]}) = \{0\} \subset \mathbb{F}_{p^n}$ and we may call the SRA algorithm as described in Algorithm 2 with the precomputed K_i polynomials and $B = \{0\}$. As the maps are polynomials, all sequences will be of full length, so there is no need to check for partial sequences.

A straight forward application of the SRA algorithm as described in Sect. 2 would lead to a cost of $\tilde{O}(d^3n^3 + pd^3n^6)$ with classical arithmetic and $\tilde{O}(d^2n^2 + pd^2n^4)$ for fast arithmetic. The algorithm possesses several optimizations for both the Resultant stage and the GCD stage which utilise the structure of the polynomial maps (10), fast arithmetic, multipoint evaluation and reuse calculations. This leads to a cost of $O(d^2n^3)$ with classical arithmetic or $\tilde{O}(dn^2)$ with fast arithmetic excluding the precomputation of the c_i values [19]. These optimizations have allowed it outperform traditional algorithms such as Berlekamp's Trace Algorithm for certain parameters [19].

3.2 SRA Maps for $\mathbb{F}_p^*$ of Smooth Order

In this section we explore the use of SRA in the field $\mathbb{F}_p$ where $|\mathbb{F}_p^*|$ is smooth. This algorithm is equivalent to one based on *Generalised Graeffe transforms* [12,13].

Definition 1. *For any integer n we will denote the* smoothness function $S : \mathbb{N} \longrightarrow \mathbb{N}$ *by* $S(n) = \max\{p \;:\; p \textit{ is a prime factor of } n\}$.
We say that an integer $n \in \mathbb{N}$ is B-smooth *if* $S(n) \leq B$.

We will assume that $f \in \mathbb{F}_p[x]$ is of degree d and that f splits over $\mathbb{F}_p$ and is square-free. From Fermat's little theorem we have that $x^{p-1} - 1 = 0$ for $x \in \mathbb{F}_p^*$. As we have that $p - 1 = n_1 \cdots n_t$ we exploit this structure to create the following system of maps

$$K_i(x_i) = x_i^{n_i} = x_{i+1} \qquad \text{for } i = 1, \ldots, t \tag{11}$$

with $K^{[t]}(x_1) = K_t \circ \cdots \circ K_1(x_1) = x_1^{n_1 \cdots n_t} = x_1^{p-1}$ and $\text{Image}(K^{[t]}) = \{0, 1\}$. We may therefore use the K_i maps and $\{0, 1\}$ as input to the SRA algorithm. As with the original SRA, the maps are polynomials and so all sequences will be of length $t + 1$.

An Optimized Resultant Stage. We may assume that $t \leq \log p$ and that $B = S(p-1)$. A straightforward adaptation of the algorithm would cost $O(Bd^3)$ with classical arithmetic or $\tilde{O}(Bd^2)$ with fast arithmetic.

After checking whether 0 is a root of f, we may use an improved method of computing the resultant which requires the precomputation of a root of unity. Full details of procedure is described in [13, Sect. 2.5]. This method replaces taking each resultant and is based upon the result that

$$f^{(i+1)}(x_{i+1}) := \mathsf{Res}_{x_i}(f^{(i)}(x_i), x_i^{n_i} - x_{i+1}) \tag{12}$$
$$f^{(i+1)}(x_{i+1}^{n_i}) = \prod_{k \in \{1,\ldots,n_i\}} f^{(i)}(\zeta_{n_i}^k x_{i+1})$$

where ζ_{n_i} is an $n_i{}^{\text{th}}$ root of unity. After computing $f^{(i+1)}(x_{i+1}^{n_i})$ this way and shifting the coefficients to obtain $f^{(i+1)}(x_{i+1})$ the total cost of taking each resultant costs $O(d \log p \log B)$ with fast arithmetic. The optimisation proposed in [13, Sect. 2.5] results in a total cost for the Resultant stage of $\tilde{O}(d \log^2 p \log B)$ with fast arithmetic.

Complexity Analysis. We assume that we wish to find the roots of a polynomial $f \in \mathbb{F}_p[x]$ of degree d lying in $\mathbb{F}_p$ and that $S(p-1) = B$. Using the standard GCD stage and the optimized resultant gives us a complexity of $\tilde{O}(d^2B^2 + Bd^3)$ with classical arithmetic and $\tilde{O}(d \log^2 p + d \log^2 pB)$ with fast arithmetic and the improvements from [13]. We note that the algorithm of [13] also utilises an improved equivalent to the GCD stage and their algorithm is more efficient, with a total complexity $\tilde{O}(B^{\frac{1}{2}} d \log^2 p)$ for certain parameters.

3.3 SRA Map in Conjunction with Hashing to an Elliptic Curve

We now generalise the previous instance by working with the group of rational points of an elliptic curve over $\mathbb{F}_p$ instead of the multiplicative group $\mathbb{F}_p^*$. This generalisation is analogous to Lenstra's generalisation of Pollard's $p-1$ factorization method as the elliptic curve factorization method.

We assume that the field $\mathbb{F}_p$ is provided with an elliptic curve $E_{a,b}$ (in reduced Weierstass coordinates $Y^2 = X^2 + aX + b$) of smooth order $N = \prod_{i=1}^t n_i$ over $\mathbb{F}_p$ and a sequence of isogenies $\varphi_i : E_i \to E_{i+1}$, where $E_1 = E_{a,b}$ and φ_i has degree n_i. It is well-known that φ_i can be defined as $\varphi_i(x,y) = \left(\frac{\xi_i(x)}{\psi_i^2(x)}, y\frac{\omega_i(x)}{\psi_i^3(x)}\right)$ where ξ_i, ω_i, ψ_i are polynomials [28]. From this data we define the rational maps $K_i : \mathbb{F}_p \to \mathbb{F}_p : x \to \frac{\xi_i(x)}{\psi_i^2(x)}$. The composition map $K^{[t]}(x) = K_t \circ \ldots \circ K_1$ clearly maps to infinity all $\mathbb{F}_p$ elements that are the x-coordinate of some $P \in E_{a,b}(\mathbb{F}_p)$. By Hasse's theorem, this set covers roughly half of the elements in $\mathbb{F}_p$ [25].

Applying the SRA algorithm with those maps on f, we would not be able to separate roots that are not the x-coordinates of a point in E. At this point, one could try to split the remaining factor g by applying SRA again on $g(x-\alpha)$, for α randomly chosen in $\mathbb{F}_p$. This algorithm, somehow reminiscent of Berlekamp's trace algorithm, would probably work well in practice but it is not clear how it could be rigorously analyzed. An alternative approach would be to assume that E has a smooth order over $\mathbb{F}_{p^2}$ instead of a smooth order over $\mathbb{F}_p$. This approach would be more satisfactory from a theoretical point of view but it would also put more severe restrictions on the set of parameters, requiring that the curve order is smooth over $\mathbb{F}_{p^2}$ instead of $\mathbb{F}_p$. This is essentially the approach taken by Ronyai [21] for a different algorithm.

In this paper, we use recent progress on hashing into elliptic curves [15] to solve this problem in a different way, which moreover fits nicely within our generalised SRA framework. We first recall the following results from Icart [15].

Lemma 1 [15]. *Let $p = 2 \bmod 3$ be an odd prime. For any $z \in \mathbb{F}_p$, there is a unique cube root of z defined over $\mathbb{F}_p$, which we write $z^{1/3}$. For any $a, b \in \mathbb{F}_p$ let $E_{a,b}$ be the elliptic curve defined by the equation $y^2 = x^3 + ax + b$. The map $f_{a,b} : \mathbb{F}_p \to E_{a,b}$ sending 0 to the point at infinity and $u \in \mathbb{F}_p^*$ to $(x, y) \in E_{a,b}(\mathbb{F}_p)$ where*

$$x = \left(v^2 - b - \frac{u^6}{27}\right)^{\frac{1}{3}} + \frac{u^6}{3}, \qquad y = ux + v, \qquad v = \frac{3a - u^4}{6u}, \tag{13}$$

is a well-defined surjective map. Reciprocally, if $P = (x, y)$ is a point on the curve $E_{a,b}$, then the solutions u_s of $f_{a,b}(u_s) = P$ are the solutions of the polynomial equation $u^4 - 6u^2x + 6uy - 3a = 0$.

The map $f_{a,b}$ defined in Lemma 1 is in fact an algebraic map as $z^{1/3} = z^{(2p-1)/3}$.

Let $K_{a,b} : \mathbb{F}_p \to \mathbb{F}_p$, where

$$u \to \left(v^2 - b - \frac{u^6}{27}\right)^{\frac{1}{3}} + \frac{u^6}{3} \tag{14}$$

be the composition of $f_{a,b}$ with a projection on the x-coordinate of the curve. Lemma 1 implies that the modified composition map $K'(x) = K_t \circ \ldots \circ K_1 \circ K_{a,b}$ maps all $\mathbb{F}_p$ elements to infinity. However, the degree of $K_{a,b}$ is prohibitively large to run SRA efficiently. We therefore modify our algorithm as follows.

In the first resultant step instead of computing $\mathrm{Res}_u(f(u), x - K_{a,b}(u))$, we compute

$$f^{(1)}(x) = \mathrm{Res}_u(f(u), \tilde{K}_{a,b}(u, x))$$

where

$$\tilde{K}_{a,b}(u, x) := (u^4 - 6u^2x - 3a)^2 - 36u^2(x^3 + ax + b).$$

Note that $\deg f^{(1)} = 3 \deg f$ as $\tilde{K}_{a,b}$ has degree 3 with respect to variable x. In fact, for every root u of f the three values

$$\xi^i \left(v^2 - b - \frac{u^6}{27}\right)^{\frac{1}{3}} + \frac{u^6}{3}, \qquad i = 1, 2, 3,$$

where ξ is a primitive cube root of unity, are roots of the polynomial $f^{(1)}$. Since only one value in each triple is defined over $\mathbb{F}_p$, one can eliminate the other "parasitic" roots by replacing $f^{(1)}(x)$ by $\gcd(f^{(1)}(x), x^p - x)$ at this stage.

Alternatively, one can just ignore this issue and work with bigger polynomials, and eventually the SRA algorithm will only produce $\mathbb{F}_p$ roots anyway. The choice of computing a gcd or not may depend on the parameters; we will not explore this further here.

After this conversion is completed we may call the original SRA algorithm to find the roots of $f^{(1)}(x)$ via the rational maps derived from the isogenies. As we know that the composed map of isogenies maps all elements in $E_{a,b}(\mathbb{F}_p)$ to the point at infinity, we know that all roots of $f^{(1)}(x)$ will result in partial sequences. Therefore we do not have to calculate or supply SRA with the set of points in the image of $K^{[t]}$. Once the roots are returned from the SRA algorithm, we compute the potential corresponding y coordinates on $E_{a,b}(\mathbb{F}_p)$ for each root x and use the final equation from Lemma 1 to recover the roots by taking gcds with our original polynomial.

We conclude the section with a comment on the existence and computation of suitable parameters for this variant of SRA. The existence of a curve of order N over $\mathbb{F}_p$ is equivalent to the existence of an integer solution to the equation

$$(N + 1 - p)^2 - Df^2 = 4N$$

with $D < 0$ (see [4, Eq. 4.3]). Once this solution is known, the curve can be constructed using the complex multiplication algorithm [4, p. 30] provided that the *reduced discriminant* D is small enough. Finally, computing small degree isogenies can be done efficiently with Vélu's formulae [27]. In order to find suitable parameters for the algorithm of this section, one can therefore for example first fix D small, then choose N randomly among a set of numbers of the desired smoothness, and finally solve the above equation for p and f using Cornacchia's algorithm [6].

Algorithm 2. SRA in conjunction with Icart's map

Data: $f \in \mathbb{F}_p[u]$ – the polynomial whose roots we wish to find
$M := \{\frac{a_1}{b_1}, \ldots, \frac{a_t}{b_t}\}$ – rational map representation of the isogenies
$a, b \in \mathbb{F}_p$ – description the smooth order curve $E_{a,b}$
Result: The roots of f
begin
  $f^{(1)}(x_1) \longleftarrow \mathsf{Res}_u(f(u), (u^4 - 6u^2x_1 - 3a)^2 - 36u^2(x_1^3 + ax_1 + b))$
  $f^{(1)}(x_1) \longleftarrow \mathsf{gcd}(f^{(1)}(x_1), x_1^p - x_1)$
  UnconvertedRoots $\longleftarrow$ $\mathsf{SRA}(f^{(1)}(x_1), M, \{\}, True)$
  Roots $\longleftarrow \{\}$
  for $x_1 \in$ *UnconvertedRoots* **do**
    $y_1 \longleftarrow \mathsf{Sqrt}(x_1^3 + ax + b)$
    $y_2 \longleftarrow -y_1$
    Roots $\longleftarrow$ Roots $\cup\, \mathsf{gcd}(f(u), u^4 - 6u^2x_1 + 6uy_1 - 3a)$
    Roots $\longleftarrow$ Roots $\cup\, \mathsf{gcd}(f(u), u^4 - 6u^2x_1 + 6uy_2 - 3a)$
  return *Roots*

Complexity Analysis. The first step requires that we take one resultant of two bivariate polynomials, where the degree of u is d and the degree of x_1 is 3. We must then normalise the resulting polynomial $f^{(1)}(x_1)$ by removing the irreducible factors. This step costs $O(d\mathbf{M}(d)\log d + \mathbf{M}(d)\log p)$ to compute the resultant and take the gcd using a square-and-multiply algorithm. SRA is called with $f^{(1)}$ and the rational maps derived from the isogenies. There will be a maximum of $\lceil \log p \rceil$ rational maps with their degree bounded by the smoothness of $|E_{a,b}|$ and denoted B. Finally the roots must be converted back, costing $O(\mathbf{G}(d))$. We therefore have the total complexity of the algorithm is $\tilde{O}(d^2 \log p + Bd^3)$ with classical arithmetic and $\tilde{O}(d \log p + Bd^2)$ with fast arithmetic.

4 Conclusions and Open Problems

In this paper, we provided a framework to extend the Successive Resultants Algorithm of [19] to arbitrary finite fields. As it stands we have three sets of maps for which the SRA algorithm works in an efficient manner. The maps for SRA in the $p = 2 \bmod 3$ case exploit the structure of the rational map framework introduced in Sect. 2 and additionally require Icart's map to transform the polynomial before converting our solutions back into roots of our original polynomial. We believe that the creation of suitable maps for specific finite fields exploiting the structure of the rational map framework remains an interesting open problem.

We remark that the SRA algorithm may be used to solve the problem of deterministic root finding for polynomials in $\mathbb{F}_p[x]$ under the *Generalised Riemann Hypothesis* (GRH), which gives us the result that finding a generator of may be done in time $O(\ln^6 p)$. Provided with such a generator, we may use deterministic algorithms to find square roots [22,26] and cubic roots [14] allowing us to exploit the standard formula for finding roots of quadratic, cubic and quartic equations. As long as $\max\{\deg a_i, \deg b_i\} \leq 4$, we may therefore use SRA in a deterministic manner.

Given the various ways that the algorithm may operate, in terms of choosing either to include elements which produce partial sequences or not and restriction of the input $B \subseteq \mathrm{Image}(K^{[t]})$ to contain specific elements, there also exists the possibility of using SRA to extract only roots with specific properties as defined by the K_i maps.

Acknowledgements. Christophe Petit is supported by a GCHQ research grant and Benjamin Pring is supported by an EPSRC doctoral research grant. The authors would like to thank the anonymous reviewers both for their time and for their helpful advice, much of which was incorporated into the final paper.

A Example of SRA with $|\mathbb{F}_p^*|$ Smooth

We demonstrate the $p - 1$ instantiation of SRA with a toy example. We use the finite field $\mathbb{F}_{37}$, where $36 = 2 \cdot 2 \cdot 3 \cdot 3$ is 3-smooth. Precomputation for $\mathbb{F}_{37}$ gives us the series of rational maps (in fact polynomials)

$$\begin{cases} K_1(x_1) &= x_1^2 = x_2 \\ K_2(x_2) &= x_2^2 = x_3 \\ K_3(x_3) &= x_3^3 = x_4 \\ K_4(x_4) &= x_4^3 = x_5 \end{cases} \tag{15}$$

The composed map is $K^{[4]}(x_1) = x_1^{36}$, which gives us $B = \text{Image}(K^{[t]}) = \{0, 1\}$. We wish to find the roots of

$$f(x) = x^{10} + 21x^9 + 22x^8 + 7x^7 + 12x^6 + 25x^5 + 35x^4 + 4x^3 + 25x \tag{16}$$

We first compute $f^{(i+1)}(x_{i+1}) = \mathsf{Res}_{x_i}(f^{(x_i)}, x_i^{n_i} - x_{i+1})$ with $f^{(1)}(x_1) = f(x_1)$.

$$f^{(1)}(x_1) = x_1^{10} + 19x_1^9 + 25x_1^8 + 6x_1^7 + 22x_1^6 + 32x_1^5 + 13x_1^4 + 32x_1^3 + 6x_1^2 + 24x_1$$
$$f^{(2)}(x_2) = x_2^{10} + 22x_2^9 + 34x_2^8 + 22x_2^7 + 27x_2^6 + 32x_2^5 + 21x_2^4 + x_2^3 + 17x_2^2 + 16x_2$$
$$f^{(3)}(x_3) = x_3^{10} + 28x_3^9 + 20x_3^8 + 23x_3^7 + 36x_3^6 + 36x_3^4 + 22x_3^3 + 35x_3^2 + 3x_3$$
$$f^{(4)}(x_4) = x_4^{10} + 28x_4^9 + 28x_4^8 + 25x_4^7 + 18x_4^6 + 18x_4^5 + 21x_4^4 + 28x_4^3 + 28x_4^2 + 27x_4$$

We then compute $g^{(i)}(x_i) = \mathsf{gcd}(f^{(i)}(x_i), x_i^{n_i} - \hat{x}^{i+1})$ for $i = 4, 3, 2, 1$, where $\hat{x}_5 \in B$ for $i = 4$ and $\hat{x}^{i+1}$ is a root of $g^{(i+1)}(x_i)$ for $i = 3, 2, 1$.
We will note the solutions of these polynomials to the right of each equation.

$$g^{(4)}(x_4) = \mathsf{gcd}(f^{(4)}(x_4), x_4^3 - 0) \quad = x_4 \qquad \{0\}$$
$$g^{(4)}(x_4) = \mathsf{gcd}(f^{(4)}(x_4), x_4^3 - 1) \quad = x_4^3 + 36 \qquad \{1, 10, 26\}$$

giving us the values for $\hat{x_4} : \{0, 1, 10, 26\}$. We use these roots to calculate

$$g^{(3)}(x_3) = \mathsf{gcd}(f^{(3)}(x_3), x_3^3 - 0) \quad = x_3, \qquad \{0\}$$
$$g^{(3)}(x_3) = \mathsf{gcd}(f^{(3)}(x_3), x_3^3 - 1) \quad = x_3^3 - 1, \qquad \{10\}$$
$$g^{(3)}(x_3) = \mathsf{gcd}(f^{(3)}(x_3), x_3^3 - 10) \quad = x_3^2 + 9x_3 + 7, \qquad \{7, 33, 34\}$$
$$g^{(3)}(x_3) = \mathsf{gcd}(f^{(3)}(x_3), x_3^3 - 26) \quad = x_3^2 + 16x_3 + 34, \qquad \{9, 12\}$$

giving us the values for $\hat{x}_3 : \{0, 7, 9, 10, 12, 33, 34\}$. We use these roots to calculate

$$g^{(2)}(x_2) = \mathsf{gcd}(f^{(2)}(x_2), x_2^2 - 0) \quad = x_2, \qquad \{0\}$$
$$g^{(2)}(x_2) = \mathsf{gcd}(f^{(2)}(x_2), x_2^2 - 7) \quad = x_2 + 28, \qquad \{9\}$$
$$g^{(2)}(x_2) = \mathsf{gcd}(f^{(2)}(x_2), x_2^2 - 9) \quad = x_2 + 34, \qquad \{3\}$$
$$g^{(2)}(x_2) = \mathsf{gcd}(f^{(2)}(x_2), x_2^2 - 10) \quad = x_2 + 26, \qquad \{11\}$$

$$g^{(2)}(x_2) = \mathsf{gcd}(f^{(2)}(x_2), x_2^2 - 12) = x_2 + 30, \quad \{7\}$$
$$g^{(2)}(x_2) = \mathsf{gcd}(f^{(2)}(x_2), x_2^2 - 33) = x_2 + 25, \quad \{12\}$$
$$g^{(2)}(x_2) = \mathsf{gcd}(f^{(2)}(x_2), x_2^2 - 34) = x_2 + 16, \quad \{21\}$$

giving us the values for $\hat{x}_2 : \{0, 3, 7, 9, 11, 12, 21\}$. We use these roots to calculate

$$g^{(1)}(x_1) = \mathsf{gcd}(f^{(1)}, x_1^2 - 0) = x_1, \quad \{0\}$$
$$g^{(1)}(x_1) = \mathsf{gcd}(f^{(1)}, x_1^2 - 3) = x_1^2 + 34, \quad \{15, 22\}$$
$$g^{(1)}(x_1) = \mathsf{gcd}(f^{(1)}, x_1^2 - 7) = x_1 + 9, \quad \{28\}$$
$$g^{(1)}(x_1) = \mathsf{gcd}(f^{(1)}, x_1^2 - 9) = x_1 + 34, \quad \{3\}$$
$$g^{(1)}(x_1) = \mathsf{gcd}(f^{(1)}, x_1^2 - 11) = x_1^2 + 26, \quad \{14, 23\}$$
$$g^{(1)}(x_1) = \mathsf{gcd}(f^{(1)}, x_1^2 - 12) = x_1^2 + 25, \quad \{7, 30\}$$
$$g^{(1)}(x_1) = \mathsf{gcd}(f^{(1)}, x_1^2 - 21) = x_1 + 13, \quad \{24\}$$

whose union is the set of roots $\hat{x}_1 : \{0, 3, 7, 14, 15, 22, 23, 24, 28, 30\}$ which are the roots of of our original polynomial $f(x)$.

B Example of SRA with $p = 2 \bmod 3$

We provide the toy example for the case of finding solutions for $h(u) \in \mathbb{F}_{41}[x]$, which fulfils our initial condition that $p = 41 = 2 \bmod 3$.

We first perform the precomputation stage for the given p. A value of N is computed so that a large enough proportion of N is smooth and allows a suitable curve to be constructed. We find that $N = 32$ is a such a value and compute the auxiliary curve

$$E_{1,0}(\mathbb{F}_{41}) := \{(x, y) \in \mathbb{F}_{41} \times \mathbb{F}_{41} \ : \ y^2 = x^3 + x\} \tag{17}$$

whose rational points we will convert our points in $\mathbb{F}_{41}$ to via Icart's map [15] as in Eq. 18,

$$\begin{aligned} K_0 &: \mathbb{F}_{41}[u, x] \to E_{1,0}(\mathbb{F}_{41}) \\ (u, x) &\mapsto -8u^8 + 14u^6x - u^4x^2 + u^2x^3 + 7u^4 + 10 \end{aligned} \tag{18}$$

The final step of the precomputation is to compute suitable elliptic curves and successive isogenies between them such that their degree is bounded by our smoothness bound. We will only use the rational map representations of the x-coordinate for these maps. The following series of isogenies with their rational-map representations of the mappings from x-coordinate to x-coordinates give rise to the following system of equations

$$
\begin{aligned}
&K_1(x) : E_{1,0}(\mathbb{F}_p) \rightarrow E_{37,0}(\mathbb{F}_p), && \frac{x_1^2+1}{x_1} &&= x_2 \\
&K_2(x) : E_{37,0}(\mathbb{F}_p) \rightarrow E_{38,11}(\mathbb{F}_p) && \frac{x_2^2-2x_2+8}{x_2-2} &&= x_3 \\
&K_3(x) : E_{38,11}(\mathbb{F}_p) \rightarrow E_{25,8}(\mathbb{F}_p) && \frac{x_3^2+12x_3+19}{x_3+12} &&= x_4 \\
&K_4(x) : E_{25,8}(\mathbb{F}_p) \rightarrow E_{34,7}(\mathbb{F}_p) && \frac{x_4^2+17x_4-10}{x_4+17} &&= x_5 \\
&K_5(x) : E_{34,7}(\mathbb{F}_p) \rightarrow E_{1,0}(\mathbb{F}_p) && \frac{x_5^2+16x_5-18}{x_5+16} &&= x_6
\end{aligned} \tag{19}
$$

After this precomputation is completed, we may begin the process of calculating the roots of $h(u)$. We seek to find the roots of the polynomial

$$h(u) = u^5 + 19u^4 + 6u^3 + 37u^2 + 38u + 30 \tag{20}$$

We first use the Icart map K_0 to create our polynomial $f^{(1)}(x_1)$, whose roots represent solutions of both $h(u)$ and $K_0(u, x)$ by means of taking the resultant with regards to u.

$$
\begin{aligned}
f(x) &= \mathsf{Res}_u(h(u), -8u^8 + 14u^6x - u^4x^2 + u^2x^3 + 7u^4 + 10) \\
&= 39x^{15} + x^{14} + 22x^{13} + 30x^{12} + 4x^{11} + 33x^{10} + 33x^9 \\
&\quad + 32x^8 + 9x^7 + 4x^6 + 33x^5 + 40x^4 + 12x^3 + x + 2
\end{aligned} \tag{21}
$$

we note that we now have a polynomial three times the degree of our original one, but we are only interested in linear factors hence we may obtain

$$
\begin{aligned}
f(x) &= \gcd(x^p - x, f^{(1)}(x)) \\
f(x) &= x^4 - 15x^3 - 5x^2 + 14x + 14
\end{aligned} \tag{22}
$$

which is of degree bounded by *deg*(h). We then compute the roots of f using the SRA algorithm with the maps $M = \{K_i\}_{i=1}^5$, the set $B = \emptyset$ and the flag ParSeq = True.

As described in the generic case of SRA, we now apply the resultant stage to obtain our $f^{(2)}(x_2), f^{(3)}(x_3), f^{(4)}(x_4), f^{(5)}(x_5)$ polynomials using the map structure we have derived from the rational maps of the isogenies as described in system 19. To do this we successively compute

$$f^{(i+1)}(x) = \mathsf{Res}_u(f^{(i)}(x_i), a_i(x_i) - x_4 + 17 \cdot x_{i+1}) \qquad \text{for } i = 1, \dots, t-1. \tag{23}$$

This results in the series of polynomials

$$
\begin{aligned}
f^{(1)}(x_1) &= x_1^4 - 15x_1^3 - 5x_1^2 + 14x_1 + 14 \\
f^{(2)}(x_2) &= 14x_2^4 + 9x_2^3 - 13x_2^2 - 5x_2 + 11 \\
f^{(3)}(x_3) &= -14x_3^4 - 4x_3^3 - 16x_3^2 - 13x_3 - 16 \\
f^{(4)}(x_4) &= -3x_4^4 + 7x_4^3 - x_4^2 - 11x_4 + 11 \\
f^{(5)}(x_5) &= -2x_5^4 - 5x_5^3 + 3x_5^2 - 9x_5 + 5
\end{aligned} \tag{24}
$$

We may then begin the gcd stage of the algorithm. We note that at each stage we must repeatedly extract those values in the kernel as these values are not picked up by the root merging process. Our first set of roots is therefore calculated via

$$g^{(5)}(x_5) = \gcd(-2x_5^4 - 5x_5^3 + 3x_5^2 - 9x_5 + 5, x_5 + 16)$$

giving us the candidate roots $\hat{x}_5$: $\{25\}$. We use this to compute the polynomial

$$\begin{aligned} g^{(4)}(x_4) &= \gcd(-3x_4^4 + 7x_4^3 - x_4^2 - 11x_4 + 11, x_4^2 + 17x_4 - 10 - (x_4 + 17) \cdot 25) \\ &= x_4^2 + 33x_4 + 16 \end{aligned}$$

These give us $\hat{x}_4 : \{4\}$. We perform the same procedure to compute

$$\begin{aligned} g^{(3)}(x_3) &= \gcd(-14x_3^4 - 4x_3^3 - 16x_3^2 - 13x_3 - 16, x_3^2 + 12x_3 + 19 - (x_3 + 12) \cdot 4) \\ &= x_3^2 + 8x_3 + 12 \end{aligned}$$

Giving us the candidate solutions $\hat{x}_3 : \{35, 39\}$. We perform the same procedure to compute

$$\begin{aligned} g^{(2)}(x_2) &= \gcd(14x_2^4 + 9x_2^3 - 13x_2^2 - 5x_2 + 11, x_2^2 - 2x_2 + 8 - (x_2 - 2) \cdot 35) \\ &= x_2 + 9 \\ g^{(2)}(x_2) &= \gcd(14x_2^4 + 9x_2^3 - 13x_2^2 - 5x_2 + 11, x_2^2 - 2x_2 + 8 - (x_2 - 2) \cdot 39) \\ &= x_2^2 + 4 \end{aligned}$$

Giving us the candidate solutions $\{32\}$ and $\{18, 23\}$ respectively. Finally we compute

$$\begin{aligned} g^{(1)}(x_1) &= \gcd(x_1^4 - 15x_1^3 - 5x_1^2 + 14x_1 + 14, x_1^2 + 1 - (x_1) \cdot 18)) \\ &= x_1 + 21 \\ g^{(1)}(x_1) &= \gcd(x_1^4 - 15x_1^3 - 5x_1^2 + 14x_1 + 14, x_1^2 + 1 - (x_1) \cdot 23)) \\ &= x_1^2 + 18x_1 + 1 \\ g^{(1)}(x_1) &= \gcd(x_1^4 - 15x_1^3 - 5x_1^2 + 14x_1 + 14, x_1^2 + 1 - (x_1) \cdot 32)) \\ &= x_1 + 28 \end{aligned}$$

Solving these provides us with solutions $\{20\}, \{2, 21\}, \{13\}$ for $f(x)$.

We then must convert these back into solutions for $h(u)$. We now possess x-coordinate solutions and may retrieve the corresponding y-coordinates via substitution of x into the auxiliary curve and taking square roots. These lead to the solutions

$$(13, 18), (13, 23), (21, 4), (21, 37), (2, 16), (2, 25), (20, 5), (20, 36)$$

each of which we substitute into the precomputed map $L(x, y) = u^4 - 6xu^2 + 6uy - 3$ and take the gcd with $h(u)$ to obtain the list of equations whose roots are precisely those of $h(u)$ (excluding 0, which may be specially checked for).

$(13, 18)$	1	$\{\}$
$(13, 23)$	$u + 17$	$\{24\}$
$(21, 4)$	1	$\{\}$
$(21, 37)$	$u^2 + 22u + 23$	$\{34, 26\}$
$(2, 16)$	1	$\{\}$
$(2, 25)$	$u + 1$	$\{40\}$
$(20, 5)$	$u + 20$	$\{21\}$
$(20, 36)$	1	$\{\}$

The roots of $h(u)$ are therefore $\{21, 24, 26, 34, 40\}$.

References

1. Arora, M., Ivanyos, G., Karpinski, M., Saxena, N.: Deterministic polynomial factoring and association schemes. Electron. Colloq. Computat. Complex. **19**, 68 (2012)
2. Bach, E., von zur Gathen, J., Lenstra, H.: Deterministic factorization of polynomials over special finite fields. University of Wisconsin-Madison, Computer Sciences Department (1988)
3. Berlekamp, E.: Factoring polynomials over large finite fields. Math. Comput. **111**, 713–735 (1970)
4. Bröker, R.: Constructing elliptic curves of prescribed order. Ph.D. thesis, University of Leiden (2006)
5. Cantor, D.G., Zassenhaus, H.: A new algorithm for factoring polynomials over finite fields. Math. Comput. **36**(154), 587–592 (1981)
6. Cornacchia, G.: Su di un metodo per la risoluzione in numeri interi dell' equazione $\sum_{h=0}^{n} c_h x^{n-h} y^h = p$. Giornale di Matematiche di Battaglini **46**, 33–90 (1903)
7. The Sage Developers: SageMath, the Sage Mathematics Software System (Version 6.8) (2015). http://www.sagemath.org
8. Evdokimov, S.: Factorization of polynomials over finite fields in subexponential time under GRH. In: Adleman, L.M., Huang, M.-D. (eds.) ANTS 1994. LNCS, vol. 877, pp. 209–219. Springer, Heidelberg (1994). doi:10.1007/3-540-58691-1_58
9. Gao, S.: On the deterministic complexity of factoring polynomials. J. Symb. Comput. **31**, 19–36 (2001)
10. von zur Gathen, J., Gerhard, J.: Modern Computer Algebra. Cambridge University Press, Cambridge (2013)
11. von zur Gathen, J., Panario, D.: Factoring polynomials over finite fields: a survey. J. Symb. Comput. **31**(1/2), 3–17 (2001)
12. Grenet, B., van der Hoeven, J., Lecerf, G.: Randomized root finding over finite FFT-fields using tangent Graeffe transforms. In: Proceedings of ISSAC, pp. 197–204. ACM (2015)
13. Grenet, B., van der Hoeven, J., Lecerf, G.: Deterministic root finding over finite fields using Graeffe transforms. Appl. Algebra Eng. Commun. Comput. **27**(3), 237–257 (2016)
14. Harasawa, R., Sueyoshi, Y., Aichi, K.: Root computation in finite fields. IEICE Trans. Fundam. Electron. Commun. Comput. Sci. **96**(6), 1081–1087 (2013)

15. Icart, T.: How to hash into elliptic curves. In: Halevi, S. (ed.) CRYPTO 2009. LNCS, vol. 5677, pp. 303–316. Springer, Heidelberg (2009). doi:10.1007/978-3-642-03356-8_18
16. Kedlaya, K.S., Umans, C.: Fast polynomial factorization and modular composition. SIAM J. Comput. **40**(6), 1767–1802 (2011)
17. Lidl, R., Niederreiter, H.: Finite Fields, vol. 20. Cambridge University Press, Cambridge (1997)
18. De Feo, L., Petit, C., Quisquater, M.: Application of the affine geometry of $GF(q^n)$ to root finding. Poster presented at International Symposium on Symbolic and Algebraic Computation (2015)
19. Petit, C.: Finding roots in $GF(p^n)$ with the successive resultant algorithm. LMS J. Comput. Math. (Spec. Issue ANTS XI) **17A**, 203–217 (2014)
20. Petit, C., Kosters, M., Messeng, A.: Algebraic approaches for the elliptic curve discrete logarithm problem over prime fields. In: Cheng, C.-M., Chung, K.-M., Persiano, G., Yang, B.-Y. (eds.) PKC 2016. LNCS, vol. 9615, pp. 3–18. Springer, Heidelberg (2016). doi:10.1007/978-3-662-49387-8_1
21. Rónyai, L.: Galois groups and factoring polynomials over finite fields. SIAM J. Disc. Math. **5**(3), 345–365 (1992)
22. Shanks, D.: Five number-theoretic algorithms. In: Proceedings of the Second Manitoba Conference on Numerical Mathematics, pp. 51–70 (1972)
23. Shoup, V.: On the deterministic complexity of factoring polynomials over finite fields. Inf. Process. Lett. **33**(5), 261–267 (1990)
24. Shoup, V.: Smoothness and factoring polynomials over finite fields. Inf. Process. Lett. **38**(1), 39–42 (1991)
25. Silverman, J.H.: Heights and elliptic curves. In: Cornell, G., Silverman, J.H. (eds.) Arithmetic Geometry, pp. 253–265. Springer, Heidelberg (1986)
26. Tonelli, A.: Bemerkung über die Auflösung quadratischer Congruenzen. Nachrichten von der Königl. Gesellschaft der Wissenschaften und der Georg-Augusts-Universität zu Göttingen 1891, pp. 344–346 (1891)
27. Vélu, J.: Isogénies entre courbes elliptiques. Communications de l'Académie royale des Sciences de Paris. CR Acad. Sci. Paris Sér. AB **273**, A238–A241 (1971)
28. Washington, L.C.: Elliptic Curves: Number Theory and Cryptography. CRC Press, Boca Raton (2008)

Distribution and Polynomial Interpolation of the Dodis-Yampolskiy Pseudo-Random Function

Thierry Mefenza[1,2](✉) and Damien Vergnaud[1]

[1] ENS, CNRS, INRIA, and PSL, Paris, France
mefenza@di.ens.fr
[2] Department of Mathematics, University of Yaounde 1, Yaounde, Cameroon

Abstract. We give some theoretical support to the security of the cryptographic pseudo-random function proposed by Dodis and Yampolskiy in 2005. We study the distribution of the function values over general finite fields and over elliptic curves defined over prime finite fields. We also prove lower bounds on the degree of polynomials interpolating the values of these functions in these two settings.

Keywords: Dodis-Yampolskiy pseudo-random function · Discrepancy · Polynomial interpolation · Finite fields · Elliptic curves

1 Introduction

A cryptographic pseudo-random function family is a collection of functions that can be evaluated in polynomial-time using a secret key but for which no polynomial-time algorithm can distinguish (with significant advantage) between a function chosen randomly from the family and a truly random function (*i.e.* whose outputs are sampled uniformly and independently at random). In 2005, Dodis and Yampolskiy [DY05] proposed an efficient pseudo-random function family which takes inputs in $\{1, \dots, d\}$ (for some parameter $d \in \mathbb{N}$) and outputs an element in a group $\mathbb{G}$ (multiplicatively written) of prime order t with generator g. The secret key is a scalar $x \in \mathbb{Z}_t^*$ and the pseudo-random function is defined by:

$$\begin{aligned} V_x : \{1, \dots, d\} &\longrightarrow \mathbb{G} \\ m &\longmapsto V_x(m) = g^{\frac{1}{x+m}} \quad \text{if } x + m \neq 0 \bmod t \text{ and } 1_{\mathbb{G}} \text{ otherwise.} \end{aligned}$$

The Dodis-Yampolskiy pseudo-random function family has found numerous applications in cryptography (e.g., for compact e-cash [CHL05] or anonymous authentication [CHK+06]). Dodis and Yampolskiy showed that their construction has some very attractive security properties, provided that some assumption about the hardness of breaking the so-called *Decision Diffie-Hellman Inversion* problem holds in $\mathbb{G}$ [DY05]. This assumption is non-standard and Cheon [Che10] proved that it is stronger than the classical discrete logarithm assumption in $\mathbb{G}$.

S. Duquesne and S. Petkova-Nikova (Eds.): WAIFI 2016, LNCS 10064, pp. 125–140, 2016.
DOI: 10.1007/978-3-319-55227-9_10

In practice, two interesting choices for the group $\mathbb{G}$ are a subgroup of the multiplicative group of any finite field (in particular, for the so-called *verifiable* Dodis-Yampolskiy pseudo-random function in groups equipped with a bilinear map [DY05]) or a subgroup of points of an elliptic curve defined over a prime finite field. Very few results supporting the Decision Diffie-Hellman Inversion assumption hardness were proven in these settings (contrary to the Naor-Reingold pseudo-random function family [NR04] for which numerous results are known, e.g. distribution [LSW14], linear complexity [GGI11] and non-linear complexity [BGLS00]). This paper deals with the distribution of the Dodis-Yampolskiy pseudo-random function over finite fields and over elliptic curves and proves lower bounds on the degree of polynomials which interpolate these functions.

Contributions of the Paper. As a first contribution, we prove that for almost all values of parameters, the Dodis-Yampolskiy pseudo-random function produces a uniformly distributed sequence. This simple result is based on some recent bounds on character sums with exponential functions. Shparlinski [Shp11] has obtained in 2011 an explicit bound for exponential sums with consecutive modular roots over a prime finite field. Ostafe and Shparlinski [OS11] obtained an analoguous result for exponential sums over multiples of a point on an elliptic curve defined over a prime finite field. Following the method from [Shp11], we obtain readily a bound for such sums over any extension of a prime finite field (Proposition 1). This new bound allows us to give results on the distribution of the Dodis-Yampolskiy pseudo-random functions over finite fields (Theorem 1). We use the bounds from [OS11] to give results on the distribution of the Dodis-Yampolskiy pseudo-random functions over elliptic curves (Theorem 2).

In order to break the security of the Dodis-Yampolskiy pseudo-random function, it would be sufficient to have a polynomial over a finite field of low degree which reveals information on the function values. From the known lower bounds on the polynomial interpolation on the discrete logarithm in finite fields and elliptic curves (*e.g.* [CS00,LW02,KW06]), one can prove that a low-degree univariate polynomial cannot reveal the secret key x when evaluated at $V_x(m)$ (for some integer $m \in \{1,\dots,d\}$) for all x. However, the security of the Dodis-Yampolskiy pseudo-random function would also be broken if such low-degree polynomial revealing a value $V_x(m')$ were proved to exist (for some integer $m' \in \{1,\dots,d\} \setminus \{m\}$ and many different keys x). Our main contribution is to prove lower bounds on the degree of polynomials interpolating the values of these functions over finite fields (Theorem 3) and elliptic curves (Theorem 4 and Theorem 5). These results can be regarded as first complexity lower bounds on the pseudo-randomness of the Dodis-Yampolskiy function families.

Both contributions are motivated by earlier results of the same flavour on the Naor-Reingold pseudo-random function family.

2 Auxiliary Results

In this section, we collect some statements about finite fields, exponential sums over finite fields and elliptic curves. We provide explicit upper-bounds for exponential sums with consecutive modular roots over a finite field and for analogous exponential sums over elliptic curves [Shp11,OS11]. The bound for exponential sums with consecutive modular roots over a general finite field is easily derived from [Shp11] and may be of independent interest.

2.1 Finite Fields and Exponential Sums

Let p be an odd prime number. We denote $\mathbb{F}_q = \mathbb{F}_{p^r}$ the finite field with $q = p^r$ elements ($r \geq 1$). For an integer t, denote by $\mathbb{Z}_t$ the residue ring modulo t and by $\mathbb{Z}_t^*$ the group of units of $\mathbb{Z}_t$. For an integer $m > 0$, we put $e_m(z) = \exp(2\pi i z/m)$. Let $g \in \mathbb{F}_{p^r}^*$ of order t (with $t \mid p^r - 1$), and ψ be a non-trivial character of $\mathbb{F}_{p^r}$. For $a \in \mathbb{F}_{p^r}^*$ and $b \in \mathbb{Z}_t$, we define the sum:

$$S_{a,b} = \sum_{n \in \mathbb{Z}_t^*} \psi(ag^{1/n}) e_t(bn).$$

Throughout the paper, the notation $U \ll V$ is equivalent to the inequality $|U| \leq cV$ with some constant $c>0$. In the following lemmas, the implied constants in the symbols "$\ll$" may occasionally depend on the integer parameters k, ℓ and are absolute otherwise.

In [BS08] Bourgain and Shparlinski proved, when $r = 1$, that for any $\varepsilon > 0$, there exists $\delta > 0$ such that for $t \geq p^\varepsilon$, we have the bound $S_{a,b} \ll t^{1-\delta}$. Shparlinski [Shp11] (Theorem 3.1) gave an explicit form of this result (again when $r = 1$) for relatively large values of t; in the case $t = p^{1+o(1)}$, it takes the form $S_{a,b} \ll t^{127/128+o(1)}$. Using Shparlinski's methods, we generalize this bound on $S_{a,b}$ for any $r \geq 1$ (see Appendix A for a proof which follows [Shp11]):

Proposition 1. *For any integers $k \geq 2$, $\ell \geq 1$ we have for $t \geq q^{1/2}(\log q)^2$:*

$$S_{a,b} \leq t^{1-\alpha_{k,\ell}} q^{\beta_{k,\ell}+o(1)},$$

where $\alpha_{k,\ell} = \frac{1}{2(2k+\ell)} - \frac{1}{4k\ell}$ *and* $\beta_{k,\ell} = \frac{1}{4(2k+\ell)}$.

2.2 Elliptic Curves and Exponential Sums

We will also consider the setting of an elliptic curve E defined over $\mathbb{F}_p$ (where p is a prime number), that is a rational curve given by the following Weierstrass equation $y^2 = x^3 + Ax + B$ with $A, B \in \mathbb{F}_p$ and $4A^3 + 27B^2 \neq 0$. The set $E(\mathbb{F}_p)$ of the points of the curve defined over $\mathbb{F}_p$ (including the special point O at infinity) has a group structure (denoted additively) with an appropriate composition rule where O is the neutral element. Given P a point of the curve E with prime order ℓ (with $\ell \mid |E(\mathbb{F}_p)|$), we denote $[n]P$ the scalar multiplication, i.e. in fact the adding of the point P to itself n times (for $n \geq 0$).

Let E be an elliptic curve and $G \in E(\mathbb{F}_p)$ be a point of order $t \geq 1$. For $a \in \mathbb{F}_p^*$ and $b \in \mathbb{Z}_t$, we define the sum:

$$\hat{S}_{a,b} = \sum_{n \in \mathbb{Z}_t^*} e_p\left(aX\left(\left[\frac{1}{n}\right]G\right)\right) e_t(bn),$$

where $X(P)$ denotes the abscissa of a point $P \in E(\mathbb{F}_p)$.

In [OS11, Theorem 6], Ostafe and Shparlinski obtained an upper-bound on $\hat{S}_{a,b}$ (with $H(X) = X^{-1}$ following the notation from [OS11]):

Proposition 2 [OS11]. *For any integers $k \geq 2$, $\ell \geq 1$ we have for $t \geq q^{1/2}(\log q)^2$:*

$$\hat{S}_{a,b} \leq t^{1-\alpha_{k,\ell}} p^{\beta_{k,\ell}+o(1)},$$

where $\alpha_{k,\ell} = \frac{1}{2(4k+\ell)} - \frac{1}{4k\ell}$ *and* $\beta_{k,\ell} = \frac{1}{4(4k+\ell)}$.

2.3 Division Polynomials Over Elliptic Curves

In this section, we recall some basic facts on division polynomials of elliptic curves (see [Was08, BSS99]). The *division polynomials* $\psi_m(X, Y) \in \mathbb{F}_p[X, Y]/(Y^2 - X^3 - AX - B)$, $m \geq 0$, are recursively defined by:

$$\begin{aligned}
\psi_0 &= 0 \\
\psi_1 &= 1 \\
\psi_2 &= 2Y \\
\psi_3 &= 3X^4 + 6AX^2 + 12BX - A^2 \\
\psi_4 &= 4Y(X^6 + 5AX^4 + 20BX^3 - 5A^2X^2 - 4ABX - 8B^2 - A^3) \\
\psi_{2m+1} &= \psi_m + 2\psi_m^3 - \psi_{m-1}\psi_{m+1}^3, \quad m \geq 2 \\
\psi_{2m} &= \psi_m(\psi_{m+2}\psi_{m-1}^2 - \psi_{m-2}\psi_{m+1}^2)/\psi_2, \quad m \geq 3,
\end{aligned}$$

where ψ_m is an abbreviation for $\psi_m(X, Y)$. If m is odd, then $\psi_m(X, Y) \in \mathbb{F}_p[X]$ is univariate and if m is even then $\psi_m(X, Y) \in 2Y\mathbb{F}_p[X]$. Therefore, we have $\psi_m^2(X, Y) \in \mathbb{F}_p[X]$ and $\psi_{m-1}(X, Y)\psi_{m+1}(X, Y) \in \mathbb{F}_p[X]$. In particular, we may write $\psi_{2m+1}(X)$ and $\psi_m^2(X)$.

The division polynomials can be used to calculate multiples of a point on the elliptic curve E. Let $P = (x, y) \in E$ with $P \neq O$, then the abscissa of $[m]P$ is given by $\theta_m(x)/\psi_m^2(x)$ where $\theta_m(X) = X\psi_m^2 - \psi_{m-1}\psi_{m+1}$. The zeros of the denominator $\psi_m^2(X)$ are exactly the first coordinates of the non-trivial m-torsion points, i.e., the points $Q = (x, y) \in \overline{\mathbb{F}_p}^2 \setminus \{O\}$ on E with $[m]Q = O$. Note, that these points occur in pairs $Q = (x, y)$ and $-Q = (x, -y)$, which coincide only if $2Q = O$, i.e., if x is a zero of $\psi_2^2(X)$.

We recall that the group of m-torsion points $E[m]$, for an elliptic curve E defined over a field of characteristic p, is isomorphic to $(\mathbb{Z}/m\mathbb{Z})^2$ if $p \nmid m$ and to a proper subgroup of $(\mathbb{Z}/m\mathbb{Z})^2$ if $p \mid m$. If m is a power of p then $E[m]$ is either isomorphic to $(\mathbb{Z}/m\mathbb{Z})$ or to $\{O\}$. Accordingly, the degree of $\psi_m^2(X)$ is $m^2 - 1$ if $p \nmid m$ and strictly less than $m^2 - 1$ otherwise. In particular, for $p = 2$ and m a power of 2 we have $\deg(\psi_m^2) = m - 1$ if E is not supersingular and $\deg(\psi_m^2) = 0$ otherwise. By induction one can show that $\theta_m(X) \in \mathbb{F}_p[X]$ is monic of degree m^2.

3 Distribution of the Dodis-Yampolskiy Pseudo-Random Functions

For a real z, we use the notation $e(z) = exp(2\pi iz)$. For a sequence of N points $\Gamma = (\gamma_{0,n}, \dots, \gamma_{s-1,n})_{n\in\{1,\dots,N\}}$ in the s-dimensional unit cube, we denote its *discrepancy* by D_Γ:

$$D_\Gamma = \sup_{B\subseteq[0,1)^s} \left| \frac{T_\Gamma(B)}{N} - |B| \right|,$$

where $T_\Gamma(B)$ denotes the number of points of the sequence Γ in a box B (i.e. a polyhedron $[\alpha_0, \beta_0)\times\cdots\times[\alpha_{s-1}, \beta_{s-1}) \subseteq [0,1)^s$) of volume $|B|$ and the supremum is taken over all such boxes. For an integer vector $a = (a_0, \dots, a_{s-1}) \in \mathbb{Z}^s$, we define $|a| = max_{\nu\in\{0,\dots,s-1\}}|a_\nu|$ and $r(a) = \prod_{\nu=0}^{s-1} \max\{|a_\nu|, 1\}$.

In order to show that a sequence Γ is uniformly distributed, we need to show that its discrepancy D_Γ is very small (i.e. tends to 0). The following lemma is our main tool for finding non-trivial upper bound for the discrepancy. It is a slightly weaker form of the Koksma-Szüsz inequality [DT97, Theorem 1.21]. The implied constant in the symbol "$\ll$" depends on the integer s.

Lemma 1. *For any integer $L > 1$ and any sequence Γ of N points, we have*

$$D_\Gamma \ll \frac{1}{L} + \frac{1}{N} \sum_{0<|a|<L} \frac{1}{r(a)} \left| \sum_{n=1}^{N} e\left(\sum_{\nu=0}^{s-1} a_\nu \gamma_{\nu,n}\right) \right|,$$

where the sum is taken over all integer vectors $a \in \mathbb{Z}^s$ with $0 < |a| < L$.

We also need the well-known orthogonality relation:

$$\sum_{\eta=0}^{m-1} e_m(\eta\lambda) = \begin{cases} 0 & \text{if} \lambda \neq 0 \bmod m \\ m & \text{otherwise} \end{cases} \tag{1}$$

and the inequality [[IK04], Bound (8.6)] (which holds for any integers m and M with $1 \leq M \leq m$):

$$\sum_{\eta=0}^{m-1} \left| \sum_{\lambda=1}^{M} e_m(\eta\lambda) \right| \ll m \log m. \tag{2}$$

3.1 Distribution of the Dodis-Yampolskiy Pseudo-Random Function Over Finite Fields

Let $q = p^r$ be a prime power for some integer $r > 1$, let $g \in \mathbb{F}_q^*$ be an element of prime order t. For $x \in \mathbb{Z}_t$ and $d \leq t$, we denote by $D_x(d)$ the discrepancy of the points $(V_{x,1}(n)/p, \dots, V_{x,r}(n)/p)$ for$1 \leq n \leq d$, where $V_x(n) = g^{\frac{1}{x+n}} \in \mathbb{F}_{p^r}$ and $V_x(n) = V_{x,1}(n)\beta_1 + \cdots + V_{x,r}(n)\beta_r$, where $\{\beta_1, \dots, \beta_r\}$ is an ordered basis of $\mathbb{F}_{p^r}$ over $\mathbb{F}_p$. We identify $\mathbb{F}_p$ with the set of integers $\{0, 1, \dots, p-1\}$.

Theorem 1. *For any $x \in \mathbb{Z}_t$, any integers $k \geq 2$, $\ell \geq 1$ and $1 \leq d \leq t$, we have:*

$$D_x(d) \leq \frac{t^{1-\alpha_{k,l}} q^{\beta_{k,l}+o(1)}}{d},$$

where $\alpha_{k,l} = \frac{1}{2(2k+l)} - \frac{1}{4kl}$ and $\beta_{k,l} = \frac{1}{4(2k+l)}$.

Proof. From Lemma 1, we derive

$$D_x(d) \ll \frac{1}{p} + \frac{1}{d} \sum_{0<|a|<p} \frac{1}{r(a)} \left| \sum_{n=1}^{d} e_p \left(\sum_{j=1}^{r} a_j V_{x,j}(n) \right) \right|,$$

where $a = (a_1, \ldots, a_r)$. Set

$$S_d(a) = \sum_{n=1}^{d} e_p(\sum_{j=1}^{r} a_j V_{x,j}(n)).$$

Let $\{\delta_1, \ldots, \delta_r\}$ be the dual basis of the given ordered basis $\{\beta_1, \ldots, \beta_r\}$. For $j \in \{1, \ldots, r\}$ and $n \in \{1, \ldots, d\}$, we have $V_{x,j}(n) = \mathrm{Tr}(\delta_j V_x(n))$, where Tr denotes the trace of $\mathbb{F}_{p^r}$ over $\mathbb{F}_p$ (namely $\mathrm{Tr}(x) = x + x^p + \cdots + x^{p^{r-1}}$). Therefore,

$$S_d(a) = \sum_{n=1}^{d} e_p \left(\mathrm{Tr} \left(\sum_{j=1}^{r} a_j \delta_j V_x(n) \right) \right) = \sum_{n=1}^{d} e_p(\mathrm{Tr}(\alpha_a V_x(n)))$$

where $\alpha_a = \sum_{j=1}^{r} a_j \delta_j \in \mathbb{F}_{p^r}$.

Let χ be defined by $\chi(z) = e_p(\mathrm{Tr}(z))$. Then χ is a non trivial additive character on $\mathbb{F}_{p^r}$. Since there exists $j \in \{1, \ldots, r\}$ such that $a_j \neq 0$, then $\alpha_a \neq 0$. We have:

$$S_d(a) = \sum_{n=1}^{d} \chi(\alpha_a V_x(n)) \text{ with } \alpha_a \neq 0.$$

We have

$$S_d(a) = \sum_{\substack{n=x+1 \\ n \in \mathbb{Z}_t^*}}^{x+d} \chi(\alpha_a g^{1/n}) = \frac{1}{t} \sum_{n \in \mathbb{Z}_t^*} \chi(\alpha_a g^{1/n}) \times \sum_{c=0}^{t-1} \sum_{\substack{v=x+1 \\ v \in \mathbb{Z}_t^*}}^{x+d} e_t(c(n-v))$$

$$= \frac{1}{t} \sum_{c=0}^{t-1} \left(\sum_{n \in \mathbb{Z}_t^*} \chi\left(\alpha_a g^{1/n}\right) e_t(cn) \right) \times \sum_{\substack{v=x+1 \\ v \in \mathbb{Z}_t^*}}^{x+d} e_t(-cv).$$

By applying Proposition 1 and (2), we obtain

$$S_d(a) \leq \frac{1}{t} \sum_{c=0}^{t-1} \left| \sum_{\substack{v=x+1 \\ v \in \mathbb{Z}_t^*}}^{x+d} e_t(-cv) \right| \times t^{1-\alpha_{k,\ell}} q^{\beta_{k,\ell}+o(1)} \leq t^{1-\alpha_{k,\ell}} q^{\beta_{k,\ell}+o(1)}.$$

By applying this bound to $D_x(d)$, we have

$$D_x(d) \ll \frac{1}{p} + \frac{t^{1-\alpha_{k,l}} q^{\beta_{k,\ell}+o(1)}}{d} \sum_{0<|a|<p} \frac{1}{r(a)} \ll \frac{1}{p} + \frac{t^{1-\alpha_{k,\ell}} q^{\beta_{k,\ell}+o(1)}}{d} \log^r p$$
$$\leq \frac{t^{1-\alpha_{k,\ell}} q^{\beta_{k,\ell}+o(1)}}{d}$$

□

With the choice $k = 4$, $l = 8$, $t = q^{1+o(1)}$ and $d = t^{\frac{127}{128}+\varepsilon}$, we obtain

$$D_x(d) \leq p^{r(-\varepsilon+o(1))} = q^{-\varepsilon+o(1)}.$$

3.2 Distribution of the Dodis-Yampolskiy Pseudo-Random Function Over Elliptic Curves

Let $E : y^2 = x^3 + Ax + B$, be an elliptic curve over $\mathbb{F}_p$. For $P \in E(\mathbb{F}_p)$ of prime order t, for $x \in \mathbb{Z}_t$, and for $1 \leq d \leq t$ we denote by $D_x(d)$ the discrepancy of the points $(X(V_x(n))/p)$ for $n \in \{1, \ldots, d\}$ where $V_x(n) = \left[\frac{1}{x+n}\right] P \in E(\mathbb{F}_p)$. We obtain the following theorem.

Theorem 2. *For any $x \in \mathbb{Z}_t$, any integers $k \geq 2$, $l \geq 1$ and $1 \leq d \leq t$, we have:*

$$D_x(d) \leq \frac{t^{1-\alpha_{k,\ell}} p^{\beta_{k,\ell}+o(1)}}{d},$$

where $\alpha_{k,\ell} = \frac{1}{2(4k+\ell)} - \frac{1}{4k\ell}$ *and* $\beta_{k,\ell} = \frac{1}{4(4k+\ell)}$.

Proof. From Lemma 1, we derive

$$D_x(d) \ll \frac{1}{p} + \frac{1}{d} \sum_{0<|a|<p} \frac{1}{|a|} \left| \sum_{n=1}^{d} e_p\left(aX(W_x(n))\right) \right|,$$

where a is an integer. Set $S_d(a) = \sum_{n=1}^{d} e_p(aX(W_x(n)))$, we have

$$S_d(a) = \sum_{\substack{n=x+1 \\ n \in \mathbb{Z}_t^*}}^{x+d} e_p\left(aX\left(\left[\frac{1}{n}\right] P\right)\right)$$
$$= \frac{1}{t} \sum_{n \in \mathbb{Z}_t^*} e_p\left(aX\left(\left[\frac{1}{n}\right] P\right)\right) \times \sum_{c=0}^{t-1} \sum_{\substack{v=x+1 \\ v \in \mathbb{Z}_t^*}}^{x+d} e_t\left(c(n-v)\right)$$
$$= \frac{1}{t} \sum_{c=0}^{t-1} \left(\sum_{n \in \mathbb{Z}_t^*} e_p\left(aX\left(\left[\frac{1}{n}\right] P\right)\right) e_t(cn) \right) \times \sum_{\substack{v=x+1 \\ v \in \mathbb{Z}_t^*}}^{x+d} e_t(-cv)$$

By applying Lemma 6 and (3), we obtain

$$\begin{aligned} S_d(a) &\leq \frac{1}{t}\sum_{c=0}^{t-1}\left|\sum_{\substack{v=x+1\\ v\in\mathbb{Z}_t^*}}^{x+d} e_t(-cv)\right| \times t^{1-\alpha_{k,\ell}}p^{\beta_{k,\ell}+o(1)} \\ &\leq t^{1-\alpha_{k,\ell}}p^{\beta_{k,\ell}+o(1)} \end{aligned}$$

By applying this bound to $D_x(d)$, we have

$$\begin{aligned} D_x(d) &\ll \frac{1}{p} + t^{1-\alpha_{k,\ell}}p^{\beta_{k,\ell}+o(1)} \times \frac{1}{d}\sum_{0<|a|<p}\frac{1}{|a|} \\ &\ll \frac{1}{p} + t^{1-\alpha_{k,\ell}}p^{\beta_{k,l}+o(1)} \times \frac{1}{d}\log p \\ &\leq t^{1-\alpha_{k,\ell}}p^{\beta_{k,\ell}+o(1)} \times \frac{1}{d} \end{aligned}$$

□

With the choice $k = 4$, $\ell = 16$, $t = p^{1+o(1)}$ and $d = t^{\frac{255}{256}+\varepsilon}$, we obtain $D_x(d) \ll p^{-\varepsilon+o(1)}$.

4 Polynomial Interpolation of the Dodis-Yampolskiy Pseudo-Random Function Over Finite Fields

Let $g \in \mathbb{F}_{p^r}^*$ for some integer $r > 1$, be an element of prime order $t \mid p^r - 1$. In this section, we prove a lower bound on the degree of univariate polynomial interpolation of the Dodis-Yampolskiy pseudo-random function over finite fields. We consider polynomials that interpolates values of the Dodis-Yampolskiy pseudo-random function for a fixed secret key $x \in \mathbb{F}_t^*$. The values considered are evaluation of the function at integers $n \in \{1, \ldots, d\}$ for some integer $1 \leq d \leq t$ and translates of these values by some fixed constants $\lambda \in \mathbb{N}$. This setting is interesting for applications in cryptography [CHL05, CHK+06]. Note that if one value n is larger than d then, the Dodis-Yampolskiy function is not necessarily defined at $n + \lambda$. In the following, we consider simple sets where all translates belong to the function domain but our method can be adapted to other settings.

Theorem 3. *Let λ be a fixed integer and let $A \subseteq \{1, \ldots, d\}$. For some $x \in \mathbb{F}_t^*$, let $F(X) \in \mathbb{F}_p[X]$ be such that $F(g^{\frac{1}{x+n}}) = g^{\frac{1}{x+n+\lambda}}$ for all $n \in A$. We have*

$$\deg(F) \geq \frac{t-2s}{4} \quad \textit{and} \quad w(F) \geq \left(\frac{t}{4s}\right)^{1/2} \quad \textit{where}\ \sharp A = t - s.$$

In the proof of Theorem 3, we use the following lemma [LW02] where the *weight* $w(F)$ (or sparsity) of a polynomial $F(X) \in \mathbb{F}_p[X]$ is the number of its non-zero coefficients.

Lemma 2 [LW02]. *Let $\gamma \in \mathbb{F}_p$ be an element of order ℓ and $F(X) \in \mathbb{F}_p[X]$ be a non-zero polynomial of degree at most $\ell - 1$ with at least b zeros of the form γ^x with $0 \le x \le \ell - 1$. The weight of $F(X)$ satisfies $w(F) \ge \ell/(\ell - b)$.*

Proof (Theorem 3). Let $R = \{(n + x) \bmod t : n \in A\}$. Then $R \subseteq \mathbb{F}_t$ and $\sharp R = t - s$. We have $F(g^{\frac{1}{n}}) = g^{\frac{1}{n+\lambda}}$ for all $n \in R$. Noticing that $\frac{1}{n+\lambda} = \frac{1}{\lambda}(1 - \frac{1}{\frac{\lambda}{n}+1})$, we obtain $F(g^{\frac{u}{\lambda}}) = g^{\frac{1}{\lambda}(1-\frac{1}{u+1})}$ for all $u = \frac{\lambda}{n}$, $n \in R$.

Let $R_0 = \{u = \frac{\lambda}{n} : n \in R \setminus \{0\}\}$ and $T = \{u \in R_0 : 2u + 1 \in R_0\}$. Since $\sharp R_0 = t - s$, we have $\sharp T \ge t - 2s$. Then

$$F\left(g^{\frac{2u+1}{\lambda}}\right) = g^{\frac{1}{\lambda}(1-\frac{1}{2u+2})} = g^{\frac{1}{\lambda}(\frac{1}{2}+\frac{1}{2}(1-\frac{1}{u+1}))} = g^{\frac{1}{2\lambda}} \times g^{\frac{1}{2\lambda}(1-\frac{1}{u+1})}$$

for all $u \in T$. We thus have

$$F^2\left(g^{\frac{2u+1}{\lambda}}\right) = g^{\frac{1}{\lambda}} \times g^{\frac{1}{\lambda}(1-\frac{1}{u+1})} = g^{\frac{1}{\lambda}} \times F(g^{\frac{u}{\lambda}}), \quad \text{for all } u \in T.$$

Let $H(X) = F^2(g^{\frac{1}{\lambda}}X^2) - g^{\frac{1}{\lambda}}F(X)$. The polynomial $H(X)$ is a non-zero polynomial and $\deg(H) \le 4\deg(F)$. Since $H(X)$ has at least $\sharp T = t - 2s$ zeros, we have $4\deg(F) \ge t - 2s$ and then $\deg(F) \ge \frac{t-2s}{4}$. Moreover, if $\deg(H) \le t - 1$, since the zeros of H are the powers of $g^{\frac{1}{\lambda}}$, then we have by Lemma 2, $w(H) \ge t/(t - (t - 2s))$, and since $w(H) \le 2(w(F))^2$, it follows that $w(F) \ge (t/4s)^{1/2}$. □

Remark 1. Theorem 3 is non-trivial only when $\sharp A > t/2$. It remains an open question to obtain non-trivial lowers bounds for smaller sets A.

5 Polynomial Interpolation of the Dodis-Yampolskiy Pseudo-Random Function Over Elliptic Curves

In this section, p is an odd prime number, E is an elliptic curve defined over $\mathbb{F}_p$ and P is a point of the curve $E(\mathbb{F}_p)$ with prime order t. We prove lower bounds on the degree of polynomial interpolation of the Dodis-Yampolskiy pseudo-random function over elliptic curves defined by $V_x(n) = X\left(\left[\frac{1}{x+n}\right]P\right)$ for a secret key $x \in \mathbb{F}_t^*$ and an integer $n \in \{1, \ldots, d\}$, with $1 \le d \le t$.

Theorem 4. *Let $S \subseteq \{1, \ldots, d\}$, $\sharp S = t - s$. We suppose $X(P) \ne 0$. For some $x \in \mathbb{F}_t^*$, let $F(X) \in \mathbb{F}_p[X]$ be such that $\psi_2^2(F(X(P))) \ne 0$ and $F(V_x(n)) = V_x(n+1)$ for all $n \in S$. We have*

$$\deg(F) \ge \frac{t - 2s}{176}.$$

Proof. Let $R = \{(n + x) \bmod t : n \in S\} \subseteq \mathbb{F}_t$. We have $\sharp R = t - s$. Let us denote $x_k = X([k]P)$ and $R_0 = \{\frac{1}{n} : n \in R\}$, then we have $F(x_u) = x_{1-\frac{1}{1+u}}$ for all $u \in R_0$. We consider the set $T = \{u \in R_0 : 2u + 1 \in R_0\}$, then $\sharp T \ge t - 2s$. For all $u \in T$, we have:

$$F(x_{2u+1}) = x_{1-\frac{1}{2(u+1)}} = x_{1/2+1/2(1-1/(u+1))} \quad \text{and} \quad F(x_u) = x_{1-1/(u+1)} \tag{3}$$

Using division polynomials (see Sect. 2.3), we can write:

$$x_{1+1-\frac{1}{(u+1)}} = \frac{\theta_2(F(x_{2u+1}))}{\psi_2^2(F(x_{2u+1}))} \tag{4}$$

Using the elliptic curve addition law, we have

$$x_{1+\alpha} = \frac{a(x_\alpha) - 2y_1y_\alpha}{(x_\alpha - x_1)^2} \text{ where } a(X) = x_1X^2 + (x_1^2 + A)X + Ax_1 + 2B,$$

and for any polynomial G of degree $m \geq 1$, we have

$$G(x_{1+\alpha}) = \frac{u(x_\alpha) - y_\alpha v(x_\alpha)}{(x_\alpha - x_1)^{2m}} \text{ and } lc(u) = G(x_1)$$

with uniquely determined polynomials $u(X)$ and $v(X)$ with $\deg(u) \leq 2m$ ($\deg(u) = 2m$ if $G(x_1) \neq 0$) and $\deg(v) \leq 2m - 2$ and where $lc(u)$ is the leading coefficient of the polynomial $u(X)$. Since $F(x_u) = x_{1-\frac{1}{u+1}}$, we can rewrite (4) as:

$$\frac{a(F(x_u)) - y_1y_{1-\frac{1}{u+1}}}{(F(x_u) - x_1)^2} = \frac{\theta_2(F(x_{2u+1}))}{\psi_2^2(F(x_{2u+1}))}.$$

Since the point $(x_{1-\frac{1}{u+1}}, y_{1-\frac{1}{u+1}}) \in E(\mathbb{F}_p)$ and $F(x_u) = x_{1-\frac{1}{u+1}}$, the polynomial $y_1^2(F(x_u)^3 + A \cdot F(x_u) + B)\psi_2^4(F(x_{2u+1}))$ is equal to the polynomial $[(F(x_u) - x_1)^2\theta_2(F(x_{2u+1})) - a(F(x_u))\psi_2^2(F(x_{2u+1}))]^2$. We thus obtain

$$y_1^2(F(x_u)^3 + A \cdot F(x_u) + B) \times \frac{p_1(x_{2u}) - y_{2u}p_2(x_{2u})}{(x_{2u} - x_1)^{12d_0}} = Q(x_u, x_{2u}, y_{2u}),$$

where $d_0 = \deg(F)$ and $Q(x_u, x_{2u}, y_{2u})$ denotes a polynomial of the form

$$\left[(F(x_u) - x_1)^2 \frac{p_3(x_{2u}) - y_{2u}p_4(x_{2u})}{(x_{2u} - x_1)^{8d_0}} - a(F(x_u)) \frac{p_5(x_{2u}) - y_{2u}p_6(x_{2u})}{(x_{2u} - x_1)^{6d_0}}\right]^2$$

such that $\deg(p_1) \leq 6d_0$, $\deg(p_2) \leq 6d_0 - 2$, $\deg(p_3) \leq 4d_0$, $\deg(p_4) \leq 4d_0 - 2$, $\deg(p_5) \leq 3d_0$ and $\deg(p_6) \leq 3d_0 - 2$. We obtain:

$$y_1^2(F(x_u)^3 + AF(x_u) + B)(x_{2u} - x_1)^{4d_0}(p_1(x_{2u}) - y_{2u}p_2(x_{2u})) = P(x_u, x_{2u}, y_{2u}),$$

where $P(x_u, x_{2u}, y_{2u}) = [(F(x_u) - x_1)^2 p_3(x_{2u}) - a(F(x_u))(x_{2u} - x_1)^{2d_0} p_5(x_{2u}) - y_{2u}((F(x_u) - x_1)^2 p_4(x_{2u}) - a(F(x_u))(x_{2u} - x_1)^{2d_0} p_6(x_{2u}))]^2$.

We then proceed as previously by trying to eliminate y_{2u}. We obtain an expression in function of x_u and x_{2u} and we replace x_{2u} by $\frac{\theta_2(x_u)}{\psi_2^2(x_u)}$. We finally obtain a rational function in x_u of the form:

$$\frac{Q(x_u)}{\psi_2^{40d_0}(x_u)} = 0, \text{ where } Q(X) \in \mathbb{F}_p[X] \text{ and } \deg(Q) \leq 88d_0.$$

Claim. $Q(X) \neq 0$ if $\psi_2^2(F(x_1)) \neq 0$ and $x_1 \neq 0$.

Proof (Claim). We have $\deg(P_5) = 3d_0$ iff $\psi_2^2(F(x_1)) \neq 0$. If $\deg(P_5) = 3d_0$, One can then verify that the leading coefficient of Q is the leading coefficient of the numerator of the rational function obtained from $[(F(x_u) - x_1)^2 p_3(x_{2u}) - a(F(x_u))(x_{2u} - x_1)^{2d_0} p_5(x_{2u})]^4$ after replacing x_{2u} by $\frac{\theta_2(x_u)}{\psi_2^2(x_u)}$.

Therefore, if $\deg(P_5) = 3d_0$, then the leading coefficient of Q is $(f^2 \times x_1 \times \psi_2^2(F(x_1)))^4$ which is non zero if $x_1 \neq 0$ since $\deg(P_5) = 3d_0$ iff $\psi_2^2(F(x_1)) \neq 0$, where f is the leading coefficient of F. Then if $\psi_2^2(F(x_1)) \neq 0$ and $x_1 \neq 0$, $Q(X)$ is a non-zero polynomial. □

If $\psi_2^2(F(x_1)) \neq 0$ and $x_1 \neq 0$, $Q(X)$ is a non-zero polynomial with at least $\sharp T/2$ different zeros. We thus have $88d_0 \geq (t-2s)/2$ and the claimed result. □

The condition $X(P) \neq 0$ in the statement of Theorem 4 holds obviously for almost all point P. The lower bound then holds if the group order $\sharp E(\mathbb{F}_p)$ is odd since in this case, the technical condition $\psi_2^2(F(X(P))) \neq 0$ is always satisfied. However, we obtain a weaker lower bound for the polynomial degree which holds for every curve E.

Theorem 5. *Let $1 \leq d \leq t$ be a fixed integer and let $A \subseteq \{1, \ldots, d\}$, $\sharp A = t - s$. For some $x \in \mathbb{F}_t^*$, let $F(X) \in \mathbb{F}_p[X]$ such that $F(V_x(n)) = V_x(n+1)$ for all $x \in A$. We have* $\deg(F) \geq (t-3s)^{1/2}/6$.

In the proof of Theorem 5, we use the following simple lemma:

Lemma 3. *Let $E: y^2 = x^3 + Ax + B$ be an elliptic curve over $\mathbb{F}_p$ with $A \neq 0$ and $B \neq 0$. Let $F(X) \in \mathbb{F}_p[X]$ be a non-constant polynomial with $F(X) \neq X$. Then there exists $\alpha \in \overline{\mathbb{F}_p}$ such that $\psi_2^2(F(\alpha)) = 0$ and $\psi_2^2(\alpha) \neq 0$.*

Proof. There are exactly three distinct zeros $\alpha_1, \alpha_2, \alpha_3 \in \overline{\mathbb{F}_p}$ of $\psi_2^2(X)$. For all index $i \in \{1,2,3\}$, there exists at least one $\beta_i \in \overline{\mathbb{F}_p}$ such that $F(\beta_i) = \alpha_i$, because F is not a constant polynomial. Since for all $i, j \in \{1,2,3\}$, $i \neq j$, we have $\alpha_i \neq \alpha_j$, then the system $F(X) = \alpha_i$ and $F(X) = \alpha_j$ has no solution. It follows that the polynomial $\psi_2^2(F(X))$ has at least three different zeros.

Let d denote the degree of F and let us suppose that there does not exist $\alpha \in \overline{\mathbb{F}_p}$ such that $\psi_2^2(F(\alpha)) = 0$ and $\psi_2^2(\alpha) \neq 0$. Then we have that $\psi_2^2(F(X))$ has exactly three zeros which are the zeros of $\psi_2^2(X)$. If $d = 1$, then it will imply that $F(X) = X$ which is impossible. If $d \geq 2$, for all $i \in \{1,2,3\}$, the equation $F(X) = \alpha_i$ has exactly one solution γ_i of multiplicity d which is one of $\{\alpha_1, \alpha_2, \alpha_3\}$. Then γ_1 and γ_2 are the zeros of the $(d-1)$-derivative of $F(X)$ which is of degree 1 and this is impossible because $\gamma_1 \neq \gamma_2$. Hence in all cases, we obtain a contradiction. So there exists $\alpha \in \overline{\mathbb{F}_p}$ such that: $\psi_2^2(F(\alpha)) = 0$ and $\psi_2^2(\alpha) \neq 0$.

Proof (Theorem). Let $R = \{(n+x) \bmod t : n \in A\}$. Then $R \subseteq \mathbb{F}_t$ and $\sharp R = t - s$. The equation $F(V_x(n)) = V_x(n+1)$ then becomes:

$$F\left(X\left(\left[\frac{1}{n}\right]P\right)\right) = X\left(\left[\frac{1}{n+1}\right]P\right),$$

for all $n \in R$. Denoting $x_k = X([k]P) = X([k \bmod t]P)$ and considering the set $T = \{n \in R/n/2, n+1 \in R\}$, we have

$$F\left(x_{\frac{2}{n}}\right) = F\left(x_{\frac{1}{n/2}}\right) = x_{\frac{1}{n/2+1}} = x_{\frac{2}{n+2}} = \frac{\theta_2(x_{\frac{1}{n+2}})}{\psi_2^2(x_{\frac{1}{n+2}})}$$
$$= \frac{\theta_2(F(x_{\frac{1}{n+1}}))}{\psi_2^2(F(x_{\frac{1}{n+1}}))}$$
$$= \frac{\theta_2(F(F(x_{\frac{1}{n}})))}{\psi_2^2(F(F(x_{\frac{1}{n}})))},$$

hence we have

$$F\left(\frac{\theta_2(x_{\frac{1}{n}})}{\psi_2^2(x_{\frac{1}{n}})}\right) = \frac{\theta_2(F(F(x_{\frac{1}{n}})))}{\psi_2^2(F(F(x_{\frac{1}{n}})))}, \text{ for all } n \in T.$$

Finally, we consider the polynomial

$$H(X) = \psi_2^{2d_0}(X)\psi_2^2(F(F(X)))\left(F\left(\frac{\theta_2(X)}{\psi_2^2(X)}\right) - \frac{\theta_2(F(F(X)))}{\psi_2^2(F(F(X)))}\right).$$

The polynomial $H(X)$ has at least $\sharp T/2$ zeros. We have $F(F(X)) \neq X$ and by Lemma 3, it will imply that there exists $\alpha \in \overline{\mathbb{F}_p}$ such that $\psi_2^2(F(F(\alpha))) = 0$ and $\psi_2^2(\alpha) \neq 0$. Hence, we have $H(\alpha) = -\theta_2(F(F(\alpha)))\psi_2^{2d_0}(\alpha) \neq 0$, since $\theta_2(X)$ and $\psi_2^2(X)$ have no common zeros. Therefore, $H(X)$ is a non-zero polynomial and $\deg(H) \leq 9d_0^2$. Then we get that $9d_0^2 \geq \sharp R/2$ and the result follows. □

6 Conclusion

We studied the distribution of the Dodis-Yampolskiy pseudo-random function values over finite fields and over elliptic curves. We also proved lower bounds on the degree of polynomials interpolating the values of these functions in this two settings of practical interest. As future works, it would be interesting to study the distribution of k-tuples $(V_x(m), \ldots, V_x(m+k))_m$ and to study the linear complexity and minimal polynomials of the sequence generated by the Dodis-Yampolskiy functions over finite fields and over elliptic curves.

Acknowledgments. The authors would like to thank the reviewers for their detailed comments and suggestions for the manuscript. The authors were supported in part by the French ANR JCJC ROMAnTIC project (ANR-12-JS02-0004) and by the Simons foundation Pole PRMAIS.

A Proof of Proposition 1

The classical Weil bound for exponential sums can be found in [Wei48, NW00].

Lemma 4. *Let $F(x)$ be a non constant polynomial in $\mathbb{F}_q[x]$ such that $F(x) \neq h(x)^p - h(x)$ for any $h(x) \in \overline{\mathbb{F}_q}(x)$. We have*

$$\left|\sum_{x\in\mathbb{F}_q} \psi(F(x))\right| \leq (\deg(F)-1)q^{1/2}$$

We deduce the following simple lemma:

Lemma 5. *For any pairwise distinct positive integers* $1 \le r_1, \dots, r_v \le R$, *we have*

$$\max_{\substack{(a_1,\dots,a_v)\in\mathbb{F}_{p^r}^v\\(a_1,\dots,a_v)\neq(0,\dots,0)}} \left|\sum_{n=1}^{t}\psi\left(\sum_{i=1}^{v}a_i g^{r_i n}\right)\right| \le Rq^{1/2}.$$

Proof. Let $s=(q-1)/t$. We have $g=\theta^s$, where θ is a primitive root in $\mathbb{F}_q$ and

$$\sum_{n=1}^{t}\psi\left(\sum_{i=1}^{v}a_i g^{r_i n}\right)=\sum_{n=1}^{t}\psi\left(\sum_{i=1}^{v}a_i\theta^{sr_i n}\right)=\frac{1}{s}\sum_{n=1}^{q-1}\psi\left(\sum_{i=1}^{v}a_i\theta^{sr_i n}\right)$$
$$=\frac{1}{s}\left(\sum_{x\in\mathbb{F}_q}\psi\left(\sum_{i=1}^{v}a_i x^{sr_i}\right)-1\right)$$

Applying Lemma 4, we obtain:

$$\max_{\substack{(a_1,\dots,a_v)\in\mathbb{F}_{p^r}^v\\(a_1,\dots,a_v)\neq(0,\dots,0)}} \left|\sum_{n=1}^{t}\psi\left(\sum_{i=1}^{v}a_i g^{r_i n}\right)\right| \le \frac{1}{s}((Rs-1)q^{1/2}+1)\le Rq^{1/2}.$$

□

Proof (Proposition 1). For any integer $k\ge 2$, we have

$${S_{a,b}}^k=\sum_{n_1,\dots,n_k\in\mathbb{Z}_t^*}\psi\left(a\sum_{j=1}^{k}g^{1/n_j}\right)e_t\left(b\sum_{j=1}^{k}n_j\right).$$

For $m\in\mathbb{Z}_t$, we collect together the terms with $n_1+\cdots+n_k\equiv m \bmod t$, getting:

$$|S_{a,b}|^k\le\sum_{m\in\mathbb{Z}_t}\left|\sum_{\substack{n_1,\dots,n_k\in\mathbb{Z}_t^*\\n_1+\cdots+n_k\equiv m \bmod t}}\psi\left(a\sum_{j=1}^{k}g^{1/n_j}\right)\right|.$$

By the Cauchy inequality, we can upper-bound $|S_{a,b}|^{2k}$ by

$$t\sum_{m\in\mathbb{Z}_t}\left|\sum_{\substack{n_1,\dots,n_k\in\mathbb{Z}_t^*\\n_1+\cdots+n_k\equiv m \bmod t}}\psi\left(a\sum_{j=1}^{k}g^{1/n_j}\right)\right|^2=t\sum_{(n_1,\dots,n_{2k})\in N_k}\psi\left(a\sum_{i=1}^{2k}(-1)^j g^{1/n_j}\right),$$

where the outside summation is taken over the set of vectors

$$N_k=\{(n_1,\dots,n_{2k})\in(\mathbb{Z}_t^*)^{2k}\ :n_1+\cdots+n_{2k-1}\equiv n_2+n_4+\cdots+n_{2k} \bmod t)\}.$$

One can see that for any $m \in \mathbb{N}$ with $\gcd(m,t)=1$, we have

$$\sum_{(n_1,\dots,n_{2k})\in N_k} \psi\left(a\sum_{j=1}^{2k}(-1)^j g^{1/n_j}\right) = \sum_{(n_1,\dots,n_{2k})\in N_k} \psi\left(a\sum_{j=1}^{2k}(-1)^j g^{m/n_j}\right).$$

Let us fix some parameter Q with $Q \geq 2\log t$. Let $\mathcal{Q}$ be the set of primes $m \leq Q$ with $\gcd(m,t)=1$. Averaging over all $m \in \mathcal{Q}$, we obtain

$$|S_{a,b}|^{2k} \leq \frac{t}{\sharp\mathcal{Q}} \sum_{m\in\mathcal{Q}} \sum_{(n_1,\dots,n_{2k})\in N_k} \psi\left(a\sum_{j=1}^{2k}(-1)^j g^{m/n_j}\right).$$

The number $w(t)$ of prime divisors of t satisfies $w(t) \leq (1+o(1))(\log t)/(\log\log t)$ (which can be seen from the trivial inequality $w(t)! \leq t$ and the Stirling formula). By the prime number theorem, we have (since $Q \geq 2\log t$):

$$\sharp\mathcal{Q} \geq (1+o(1))\frac{Q}{\log Q} - (1+o(1))\frac{\log t}{\log(\log t)} \geq 0.5\frac{Q}{\log Q},$$

provided that t is large enough. We have $\sharp N_k \leq t^{2k-1}$. Using the Hölder inequality and then extending the region of summation, we obtain that for any integer $\ell \geq 1$, we have:

$$\begin{aligned}
|S_{a,b}|^{4k\ell} &\leq \frac{t^{2\ell}}{\sharp\mathcal{Q}^{2\ell}}(\sharp N_k)^{2\ell-1} \sum_{n_1,\dots,n_{2k}\in\mathbb{Z}_t^*} \left|\sum_{m\in\mathcal{Q}} \psi\left(a\sum_{j=1}^{2k}(-1)^j g^{m/n_j}\right)\right|^{2\ell} \\
&\ll \frac{t^{4\ell-2k+1}\log^{2\ell} Q}{Q^{2\ell}} \sum_{n_1,\dots,n_{2k}=1}^{t} \left|\sum_{m\in\mathcal{Q}} \psi\left(a\sum_{j=1}^{2k}(-1)^j g^{mn_j}\right)\right|^{2\ell} \\
&= \frac{t^{4k\ell-2k+1}\log^{2\ell} Q}{Q^{2l}} \sum_{n_1,\dots,n_{2k}=1}^{t} \sum_{m_1,\dots,m_{2\ell}\in\mathcal{Q}} \psi\left(a\sum_{j=1}^{2k}\sum_{h=1}^{2\ell}(-1)^{j+h} g^{m_h n_j}\right) \\
&= \frac{t^{4k\ell-2k+1}\log^{2\ell} Q}{Q^{2\ell}} \sum_{m_1,\dots,m_{2\ell}\in\mathcal{Q}} \left|\sum_{n=1}^{t} \psi\left(a\sum_{h=1}^{2\ell}(-1)^h g^{m_h n}\right)\right|^{2k}.
\end{aligned}$$

For $O(\sharp\mathcal{Q}^\ell)=O(Q^\ell\log^{-\ell}Q)$ tuples $(m_1,\dots,m_{2\ell}) \in \mathcal{Q}^{2\ell}$ such that the tuple of the elements on the odd positions $(m_1,\dots,m_{2\ell-1})$ is a permutation of the elements on the even positions $(m_2,\dots,m_{2\ell})$, we estimate the inner sum trivially as t.

For the remaining $O((\sharp Q)^{2\ell}) = O(Q^{2\ell}(\log Q)^{-2\ell})$ tuples, we use the bound of Lemma 5. Therefore,

$$\begin{aligned}
|S_{a,b}|^{4k\ell} &\ll \frac{t^{4k\ell-2k+1}\log^{2\ell} Q}{Q^{2l}}(Q^\ell\log^{-\ell}Qt^{2k} + Q^{2\ell}\log^{-2\ell}Q(Qq^{1/2})^{2k}) \\
&= t^{4k\ell-2k+1}(Q^{-\ell}\log^\ell Qt^{2k} + Q^{2k}q^k).
\end{aligned}$$

Taking $Q = 2t^{2k/(2k+\ell)}q^{-k/(2k+\ell)}(\log q)^{\ell/(2k+\ell)}$ and if $t \geq q^{1/2}(\log q)^2$, one can see that $Q \geq 2\log t$ and we obtain

$$|S_{a,b}|^{4k\ell} \ll t^{4k\ell-(2k\ell-2k-\ell)/(2k+\ell)}q^{k\ell/(2k+\ell)}(\log q)^{\ell/(2k+\ell)}$$

and the result follows. □

References

[BGLS00] Banks, W.D., Griffin, F., Lieman, D., Shparlinski, I.E.: Non-linear complexity of the naor–reingold pseudo-random function. In: Song, J.S. (ed.) ICISC 1999. LNCS, vol. 1787, pp. 53–59. Springer, Heidelberg (2000). doi:10.1007/10719994_5

[BS08] Bourgain, J., Shparlinski, I.E.: Distribution of consecutive modular roots of an integer. Acta Arith. **134**(1), 83–91 (2008)

[BSS99] Blake, I.F., Seroussi, G., Smart, N.P.: Elliptic Curves in Cryptography. Cambridge University Press, Cambridge (1999)

[Che10] Cheon, J.H.: Discrete logarithm problems with auxiliary inputs. J. Cryptol. **23**(3), 457–476 (2010)

[CHK+06] Camenisch, J., Hohenberger, S., Kohlweiss, M., Lysyanskaya, A., Meyerovich, M.: How to win the clonewars: efficient periodic n-times anonymous authentication. In: ACM CCS 2006: 13th Conference on Computer and Communications Security, Alexandria, Virginia, USA, 30 October–3 November 2006, pp. 201–210. ACM Press (2006)

[CHL05] Camenisch, J., Hohenberger, S., Lysyanskaya, A.: Compact e-cash. In: Cramer, R. (ed.) EUROCRYPT 2005. LNCS, vol. 3494, pp. 302–321. Springer, Heidelberg (2005). doi:10.1007/11426639_18

[CS00] Coppersmith, D., Shparlinski, I.: On polynomial approximation of the discrete logarithm and the Diffie-Hellman mapping. J. Cryptol. **13**(3), 339–360 (2000)

[DT97] Drmota, M., Tichy, R.: Discrepancies and Applications. Springer, Berlin (1997)

[DY05] Dodis, Y., Yampolskiy, A.: A verifiable random function with short proofs and keys. In: Vaudenay, S. (ed.) PKC 2005. LNCS, vol. 3386, pp. 416–431. Springer, Heidelberg (2005). doi:10.1007/978-3-540-30580-4_28

[GGI11] Gómez, D., Gutierrez, J., Ibeas, A.: On the linear complexity of the Naor-Reingold sequence. Inf. Process. Lett. **111**(17), 854–856 (2011)

[IK04] Iwaniec, H., Kowalski, E.: Analytic Number Theory. American Mathematical Society, Providence (2004)

[KW06] Kiltz, E., Winterhof, A.: Polynomial interpolation of cryptographic functions related to Diffie-Hellman and discrete logarithm problem. Discrete Appl. Math. **154**(2), 326–336 (2006)

[LSW14] Ling, S., Shparlinski, I.E., Wang, H.: On the multidimensional distribution of the Naor-Reingold pseudo-random function. Math. Comput. **83**(289), 2429–2434 (2014)

[LW02] Lange, T., Winterhof, A.: Polynomial interpolation of the elliptic curve and XTR discrete logarithm. In: Ibarra, O.H., Zhang, L. (eds.) COCOON 2002. LNCS, vol. 2387, pp. 137–143. Springer, Heidelberg (2002). doi:10.1007/3-540-45655-4_16

[NR04] Naor, M., Reingold, O.: Number-theoretic constructions of efficient pseudo-random functions. J. ACM **51**(2), 231–262 (2004)

[NW00] Niederreiter, H., Winterhof, A.: Incomplete exponential sums over finite fields and their applications to new inversive pseudorandom number generators. Acta Arith. **93**(4), 387–399 (2000)

[OS11] Ostafe, A., Shparlinski, I.E.: Twisted exponential sums over points of elliptic curves. Acta Arith. **148**(1), 77–92 (2011)

[Shp11] Shparlinski, I.E.: Exponential sums with consecutive modular roots of an integer. Q. J. Math. **62**(1), 207–213 (2011)

[Was08] Washington, L.C.: Elliptic Curves: Number Theory and Cryptography, 2nd edn. Chapman and Hall/CRC, Boca Raton (2008)

[Wei48] Weil, A.: On some exponential sums. Proc. Natl. Acad. Sci. U.S.A. **34**, 204–207 (1948)

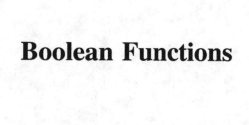

Boolean Functions

A Conjecture About Gauss Sums and Bentness of Binomial Boolean Functions

Jean-Pierre Flori(✉)

Agence nationale de la sécurité des systèmes d'information,
51 boulevard de La Tour-Maubourg, 75700 Paris 07 SP, France
jean-pierre.flori@ssi.gouv.fr

Abstract. In this note, the polar decomposition of binary fields of even extension degree is used to reduce the evaluation of the Walsh transform of binomial Boolean functions to that of Gauss sums. In the case of extensions of degree four times an odd number, an explicit formula involving a Kloosterman sum is conjectured, proved with further restrictions, and supported by extensive experimental data in the general case. In particular, the validity of this formula is shown to be equivalent to a simple and efficient characterization for bentness previously conjectured by Mesnager.

Keywords: Boolean functions · Bent functions · Walsh spectrum · Exponential sums · Gauss sums · Kloosterman sums

1 Introduction

Bent functions are Boolean functions defined over an extension of even degree and achieving optimal non-linearity. They are of both combinatorial and cryptographic interest. Unfortunately, characterizing bentness of an arbitrary Boolean function is a difficult problem, and even the less general question of providing simple and efficient criteria within infinite families of functions in a specific polynomial form is still challenging.

For a Boolean function f defined over $\mathbb{F}_{2^n}$ with $n = 2m$ and given in polynomial form, a classical characterization for bentness is that its Walsh transform $\widehat{\chi_f}$ values are only $2^{\pm m}$. Nevertheless, such a characterization is neither concise nor efficient: the best algorithm to compute the full Walsh spectrum has complexity $O(n2^n)$, which is asymptotically optimal. Whence the need to restrict to functions in a given form and to look for more efficient criteria. Unfortunately, only a few infinite families of Boolean functions with a simple and efficient criterion for bentness are known.

The most classical family is due to Dillon [7] and is made of monomial functions:

$$f_a(x) = \mathrm{Tr}_1^n \left(a x^{r(2^m-1)} \right),$$

where $n = 2m$, $a \in \mathbb{F}_{2^n}^*$ and r is co-prime with $2^m + 1$. Such functions are bent (and even hyper-bent: for any r coprime with $(2^n - 1)$ the function $f_a(x^r)$

S. Duquesne and S. Petkova-Nikova (Eds.): WAIFI 2016, LNCS 10064, pp. 143–159, 2016.
DOI: 10.1007/978-3-319-55227-9_11

is also bent) if and only if the Kloosterman sum $K_m(a)$ associated with a is equal to zero [3,7,15]. Not only does such a criterion gives a concise and elegant characterization for bentness, but using the connection between Kloosterman sums and elliptic curves [13,14] it also allows to check for bentness in polynomial time [1,16]. Further results on Kloosterman sums involving p-adic arithmetic [10,11,20] lead to even faster generation of zeros of Kloosterman sums and so of (hyper-)bent functions.

Mesnager [18,19] proved a similar criterion for a family Boolean functions in binomial form:

$$f_{a,b}(x) = \mathrm{Tr}_1^n\left(ax^{r(2^m-1)}\right) + \mathrm{Tr}_1^2\left(bx^{\frac{2^n-1}{3}}\right),$$

where $n = 2m$, $a \in \mathbb{F}_{2^n}^*$, $b \in \mathbb{F}_4^*$ and r is co-prime with $2^m + 1$ (but also $r = 3$ which divides 2^m+1 [17]). When the extension degree n is twice an odd number, that is when m is odd, $f_{a,b}$ is (hyper-)bent if and only if $K_m(a) = 4$. Moreover, (hyper-)bent functions in this family can be quickly generated as techniques used to generate zeros of Kloosterman sums can be transposed to the value 4 [9].

Unfortunately, the proof of the aforementioned characterization does not extend to the case where m is even. Nevertheless, it is easy to show that $K_m(a) = 4$ is still a necessary condition for $f_{a,b}$ to be bent in this latter case (but note that $f_{a,b}$ can no longer be hyper-bent). Further experimental evidence gathered by Flori et al. [9] supported the conjecture that it should also be a sufficient condition: for m up to 16, $f_{a,b}$ is bent if and only if $K_m(a) = 4$.

In this note, the polar decomposition of fields of even extension degree $n = 2^\nu m$ with m odd is used to reduce the evaluation of the Walsh transform of $f_{a,b}$ at $\omega \in \mathbb{F}_{2^n}^*$ to that of a Gauss sum of the form

$$\sum_{u\in U} \psi_n\left(b\mathrm{Tr}_m^n(\omega u)\right)\chi\left(\mathrm{Tr}_1^n\left(au^{2^{2^{\nu-1}m}-1}\right)\right), \tag{1}$$

where $\mathbb{F}_{2^n}^*$ is decomposed as $\mathbb{F}_{2^n}^* \simeq U \times \mathbb{F}_{2^m}^*$, ψ_n is a cubic multiplicative character and χ a quadratic additive character.

In the case of extensions of degree four times an odd number, that is when n is four times an odd number m, an explicit formula involving the Kloosterman sum $K_{n/2}(a)$ is proved for ω lying in the subfield $\mathbb{F}_{2^{n/2}}$, and conjectured and supported by extensive experimental evidence when $\omega \in \mathbb{F}_{2^n}$. In particular, the validity of this formula would prove the following conjecture for extensions of degree four times an odd number (and give hope to prove the conjecture for n of any 2-adic valuation):

Conjecture 1. Let $n = 4m$ with m odd, $a \in \mathbb{F}_{2^{n/2}}^*$ and $b \in \mathbb{F}_4^*$. The function $f_{a,b}$ is bent if and only if $K_{n/2}(a) = 4$.

2 Notation

2.1 Field Trace

Definition 1 (Field trace). *For extension degrees m and n such that m divides n, the field trace from $\mathbb{F}_{2^n}$ down to $\mathbb{F}_{2^m}$ is denoted by $\mathrm{Tr}_m^n(x)$.*

2.2 Polar Decomposition

Definition 2 (Extension degrees). *Let $n \geq 2$ be an even integer and $\nu \geq 1$ denote its 2-adic valuation. We denote by m_i for $0 \leq i \leq \nu$ the integer $n/2^i$, e.g. $m_0 = n$ and $m_\nu = m$ in the introduction.*

For $0 \leq i < \nu$, the multiplicative group $\mathbb{F}_{2^{m_i}}^*$ can be split using the so-called polar decomposition

$$\mathbb{F}_{2^{m_i}}^* \simeq U_{i+1} \times \mathbb{F}_{2^{m_{i+1}}}^*,$$

where $U_{i+1} \subset \mathbb{F}_{2^{m_i}}^*$ is the subgroup of $(2^{m_{i+1}} + 1)$-th roots of unity and $\mathbb{F}_{2^{m_{i+1}}}^*$ the subgroup of $(2^{m_{i+1}} - 1)$-th roots of unity. Repeating this construction yields the following decomposition.

Lemma 1 (Polar decomposition). *Let $\nu \geq 1$ and denote by U denote the image of $U_1 \times \cdots \times U_\nu$ within $\mathbb{F}_{2^{m_0}}^*$. Then $\mathbb{F}_{2^{m_0}}^*$ decomposes as*

$$\begin{aligned}\mathbb{F}_{2^{m_0}}^* &\simeq U_1 \times \cdots \times U_\nu \times \mathbb{F}_{2^{m_\nu}}^* \\ &\simeq U \times \mathbb{F}_{2^{m_\nu}}^*.\end{aligned}$$

2.3 Hilbert's Theorem 90

Definition 3. *For $1 \leq i \leq \nu$ and $j \in \mathbb{F}_2$, let $\mathcal{T}_{m_i}^j$ be the set*

$$\mathcal{T}_{m_i}^j = \{x \in \mathbb{F}_{2^{m_i}}, \mathrm{Tr}_1^{m_i}(x^{-1}) = j\}$$

of elements of $\mathbb{F}_{2^{m_i}}$ whose inverses have trace j (defining 0^{-1} to be 0).

Hilbert's Theorem 90 [8] implies that the function $x \mapsto x + x^{-1}$ is 2-to-1 from $U_i \backslash \{1\}$ to $\mathcal{T}_{m_i}^1$ and from $\mathbb{F}_{2^{m_i}}^* \backslash \{1\}$ to $\mathcal{T}_{m_i}^0 \backslash \{0\}$ (and both 0 and 1 are sent onto 0).

2.4 Dickson Polynomials

Definition 4. *We denote by D_3 the third Dickson polynomial of the first kind $D_3(x) = x^3 + x$.*

A notable property of D_3 is that $D_3(x + x^{-1}) = x^3 + x^{-3}$. It implies in particular that D_3 induces a permutation of $\mathcal{T}_{m_1}^0$ when m_1 is odd and of $\mathcal{T}_{m_1}^1$ when m_1 is even [8, Propositions 5, 6 and Theorem 7].

2.5 Characters

Definition 5 (Additive character). *Denote by χ the non-principal quadratic additive character of $\mathbb{F}_2$.*

Together with the field trace, χ can be used to construct all quadratic additive characters of $\mathbb{F}_{2^{m_i}}$ for any $0 \leq i \leq \nu$.

Definition 6 (Multiplicative character). *The non-principal cubic multiplicative character ψ_{m_i} of $\mathbb{F}_{2^{m_i}}$ for any $0 \leq i < \nu$ is defined for $x \in \mathbb{F}_{2^{m_i}}$ as*

$$\psi_{m_i}(x) = x^{\frac{2^{m_i}-1}{3}} .$$

Note that if x lies in a subextension, that is $x \in \mathbb{F}_{2^{m_{i+j}}}$ with $0 \leq i+j < \nu$, then

$$\psi_{m_i}(x) = \psi_{m_{i+j}}(x)^{2^j} .$$

Remark that 3 divides $2^{m_\nu}+1$ and is coprime with $2^{m_\nu}-1$ and $2^{m_i}+1$ for $0 \leq i < \nu$. Therefore the function $x \mapsto x^3$ is a permutation of $\mathbb{F}_{2^{m_\nu}}^*$ and U_i for $1 \leq i < \nu$, and 3-to-1 on U_ν. In particular, the multiplicative character ψ_{m_0} is trivial everywhere on $\mathbb{F}_{2^{m_0}}^*$ but on U_ν.

2.6 Walsh Transform

Definition 7. *The Walsh transform of a Boolean function f at $\omega \in \mathbb{F}_{2^{m_0}}$ is*

$$\widehat{\chi_f}(\omega) = \sum_{x \in \mathbb{F}_{2^{m_0}}} \chi\left(f(x) + \operatorname{Tr}_1^{m_0}(\omega x)\right) .$$

It is well-known that a Boolean function f is bent if and only if its Walsh transform only takes the values $2^{\pm m_1}$.

2.7 Kloosterman Sums

Definition 8. *For $a \in \mathbb{F}_{2^{m_1}}$, the Kloosterman sum $K_{m_1}(a)$ is*

$$K_{m_1}(a) = \sum_{x \in \mathbb{F}_{2^{m_1}}} \chi\left(\operatorname{Tr}_1^{m_1}(ax + x^{-1})\right) .$$

The following identities (proved using the map from Sect. 2.3) are well-known:

$$\begin{aligned}\sum_{u_1 \in U_1} \chi\left(\operatorname{Tr}_1^{m_0}(au_1)\right) &= 1 + 2 \sum_{t \in \mathcal{T}_{m_1}^1} \chi\left(\operatorname{Tr}_1^{m_1}(at)\right) \\ &= 1 - 2 \sum_{t \in \mathcal{T}_{m_1}^0} \chi\left(\operatorname{Tr}_1^{m_1}(at)\right) \\ &= 1 - K_{m_1}(a).\end{aligned}$$

2.8 Cubic Sums

Definition 9. *For $a, b \in \mathbb{F}_{2^{m_1}}$, the cubic sum $C_{m_1}(a,b)$ is*

$$C_{m_1}(a,b) = \sum_{x \in \mathbb{F}_{2^{m_1}}} \chi\left(\operatorname{Tr}_1^{m_1}(ax^3 + bx)\right) .$$

The possible values of $C_{m_1}(a,b)$ were determined by Carlitz [2] together with simple criteria involving a and b.

The most important consequence of Carlitz's results in our context is that $C_{m_1}(a,a) = \sum_{x\in\mathbb{F}_{2^{m_1}}} \chi(\mathrm{Tr}_1^{m_1}(aD_3(x))) = 0$ if and only if

- $\mathrm{Tr}_1^{m_1}(\alpha) = 0$ for $\alpha \in \mathbb{F}_{2^{m_1}}^*$ such that $a = \alpha^3$ when m_1 is odd (in that case a is always a cube),
- and when there exists $\alpha \in \mathbb{F}_{2^{m_1}}^*$ such that $a = \alpha^3$ (that is a is a cube or equivalently $\psi_{m_1}(a) = 1$) and $\mathrm{Tr}_2^{m_1}(\alpha) \neq 0$ (that is the cube root's half-trace is non zero) when m_1 is even.

Charpin et al. later deduced that both in the odd case [4] and in the even case [5,6] these conditions are equivalent to $K_{m_1}(a) \equiv 1 \pmod 3$.

For completeness, the other possible values for $C_{m_1}(a,a)$ when m_1 is even follow:

- When a is a cube and $\mathrm{Tr}_2^{m_1}(\alpha) = 0$, then $C_{m_1}(a,a) = 2^{m_2+1}\chi\left(\mathrm{Tr}_1^{m_1}(u_0^3)\right)$, where u_0 is any solution to $u^4 + u = \alpha^4$, that is $u_0 = \sum_{i=0}^{(m_2-3)/2} \alpha^{4^{2*i+2}} + \gamma$ for any $\gamma \in \mathbb{F}_4$.
- When a is not a cube, then $C_{m_1}(a,a) = -2^{m_2}\chi\left(\mathrm{Tr}_1^{m_1}(au_0^3)\right)$, where u_0 is the unique solution to $u^4 + u/a = 1$, that is $u_0 = \psi_{m_1}(a)\sum_{i=0}^{m_2-1} a^{4^i} a^{(4^i-1)/3}$.

Finally, Carlitz also proved the following result on $C_{m_1}(a,0)$ when $m_1 = 2m_2$ is even:

$$C_{m_1}(a,0) = \begin{cases} (-1)^{m_2+1}2^{m_2+1} & \text{if } \psi_{m_1}(a) = 1, \\ (-1)^{m_2}2^{m_2} & \text{if } \psi_{m_1}(a) \neq 1. \end{cases}$$

2.9 Binomial Functions

The binomial Boolean functions $f_{a,b}$ studied in this note are defined over $\mathbb{F}_{2^{m_0}}$.

Definition 10. *For $\nu \geq 1$, $a \in \mathbb{F}_{2^{m_0}}^*$ and $b \in \mathbb{F}_4^*$, we denote by $f_{a,b}$ the binomial function*

$$f_{a,b}(x) = \mathrm{Tr}_1^{m_0}\left(ax^{2^{m_1}-1}\right) + \mathrm{Tr}_1^2\left(b\psi_{m_0}(x)\right). \tag{2}$$

We also define $f_a = f_{a,0}$ (corresponding to Dillon's monomial) and $g_b(x) = \mathrm{Tr}_1^2(b\psi_{m_0}(x))$.

3 Preliminaries

3.1 Field of Definition of the Coefficients

First notice that it is enough to know how to evaluate the Walsh transform of functions $f_{a,b}$ for $a \in \mathbb{F}_{2^{m_1}}^*$.

Lemma 2. *Let $a \in \mathbb{F}_{2^{m_0}}^*$ be written as $a = \alpha\tilde{a}$ with $\alpha \in U_1$ and $\tilde{a} \in \mathbb{F}_{2^{m_1}}^*$ using the polar decomposition of $\mathbb{F}_{2^{m_0}}^*$. Let $\tilde{\alpha} \in U_1$ be a square root of α and $\beta \in \mathbb{F}_4^*$ be $\beta = \psi_{m_0}(\alpha)^{-1}$. Then*

$$\widehat{\chi_{f_{a,b}}}(\omega) = \widehat{\chi_{f_{\tilde{a},\beta b}}}(\tilde{\alpha}\omega).$$

Proof. Indeed, $x \mapsto \tilde{\alpha}x$ induces a permutation of $\mathbb{F}_{2^{m_0}}$, $\tilde{\alpha}^{2^{m_1}-1} = \tilde{\alpha}^{-2} = \alpha^{-1}$, and $\psi_{m_0}(\tilde{\alpha}) = \psi_{m_0}(\alpha)^{-1}$, so that

$$\begin{aligned}\widehat{\chi_{f_{a,b}}}(\omega) &= \sum_{x \in \mathbb{F}_{2^{m_0}}} \chi\left(f_{a,b}(x) + \operatorname{Tr}_1^{m_0}(\omega x)\right) \\ &= \sum_{x \in \mathbb{F}_{2^{m_0}}} \chi\left(f_{a,b}(\tilde{\alpha}x) + \operatorname{Tr}_1^{m_0}(\omega \tilde{\alpha} x)\right) \\ &= \sum_{x \in \mathbb{F}_{2^{m_0}}} \chi\left(f_{\tilde{a},\beta b}(x) + \operatorname{Tr}_1^{m_0}(\omega \tilde{\alpha} x)\right) \\ &= \widehat{\chi_{f_{\tilde{a},\beta b}}}(\tilde{\alpha}\omega).\end{aligned}$$

From now on we can suppose that $a \in \mathbb{F}_{2^{m_1}}^*$ without loss of generality.

3.2 Polar Decomposition

The polar decomposition yields the following expression for $f_{a,b}$.

Lemma 3. *For $\nu \geq 1$, $a \in \mathbb{F}_{2^{m_0}}^*$ and $b \in \mathbb{F}_4^*$, and $x \in \mathbb{F}_{2^{m_0}}^*$, $f_{a,b}(x)$ is*

$$f_{a,b}(x) = f_{a,b}(u) = f_a(u_1) + g_b(u_\nu). \tag{3}$$

Proof. Notice that $f_{a,b}(x) = f_a(x) + g_b(x)$. Moreover f_a is trivial on $\mathbb{F}_{2^{m_1}}^*$ and g_b is trivial everywhere but on U_ν as noted in Sect. 2.5.

We now split the sum expressing the Walsh transform of $f_{a,b}$ at $\omega \in \mathbb{F}_{2^{m_0}}$ using the polar decomposition of $\mathbb{F}_{2^{m_0}}^*$ as $\mathbb{F}_{2^{m_0}}^* \simeq U \times \mathbb{F}_{2^{m_\nu}}^*$. We write $x \in \mathbb{F}_{2^{m_0}}^*$ as $x = uy$ for $u \in U$, and $y \in \mathbb{F}_{2^{m_\nu}}^*$.

Lemma 4. *For $\nu \geq 1$, $a \in \mathbb{F}_{2^{m_1}}^*$ and $b \in \mathbb{F}_4^*$, the Walsh transform of $f_{a,b}$ at $\omega \in \mathbb{F}_{2^{m_0}}$ is, for $\omega = 0$:*

$$\widehat{\chi_{f_{a,b}}}(0) = 1 + (2^{m_\nu} - 1) \sum_{u \in U} \chi\left(f_{a,b}(u)\right), \tag{4}$$

and for $\omega \neq 0$:

$$\widehat{\chi_{f_{a,b}}}(\omega) = 1 - \sum_{u \in U} \chi\left(f_{a,b}(u)\right) + 2^{m_\nu} \sum_{u \in U, \operatorname{Tr}_{m_\nu}^{m_0}(\omega u) = 0} \chi\left(f_{a,b}(u)\right). \tag{5}$$

Proof. Using the polar decomposition, the Walsh transform of $f_{a,b}$ at $\omega \in \mathbb{F}_{2^{m_0}}$ can indeed be written

$$\begin{aligned}\widehat{\chi_{f_{a,b}}}(\omega) &= \sum_{x \in \mathbb{F}_{2^{m_0}}} \chi\left(f_{a,b}(x) + \operatorname{Tr}_1^{m_0}(\omega x)\right) \\ &= 1 + \sum_{x \in \mathbb{F}_{2^{m_0}}^*} \chi\left(f_{a,b}(x) + \operatorname{Tr}_1^{m_0}(\omega x)\right) \\ &= 1 + \sum_{(u,y) \in U \times \mathbb{F}_{2^{m_\nu}}^*} \chi\left(f_{a,b}(uy)\right) \chi\left(\operatorname{Tr}_1^{m_0}(\omega u y)\right).\end{aligned}$$

Note that 3 divides $2^{m_\nu}+1$ so that $\frac{2^{m_0}-1}{3} = (2^{m_\nu}-1)\frac{2^{m_\nu}+1}{3}\prod_{i=1}^{\nu-1}(2^{m_i}+1)$ and $f_{a,b}(uy) = f_{a,b}(u)$. Therefore

$$\widehat{\chi_{f_{a,b}}}(\omega) = 1 + \sum_{u\in U} \chi\left(f_{a,b}(u)\right) \sum_{y\in\mathbb{F}_{2^{m_\nu}}^*} \chi\left(\operatorname{Tr}_1^{m_\nu}\left(\operatorname{Tr}_{m_\nu}^{m_0}\left(\omega u\right) y\right)\right).$$

The sum ranging over $\mathbb{F}_{2^{m_\nu}}^*$ is equal to -1 when $\operatorname{Tr}_{m_\nu}^{m_0}(\omega u) \neq 0$ and $2^{m_1}-1$ when $\operatorname{Tr}_{m_\nu}^{m_0}(\omega u) = 0$. In particular, when $\omega = 0$, the trace is 0 for all $u \in U$.

To go further, the cases $\nu = 1$ and $\nu > 1$ have to be dealt with separately.

4 Odd Case

In this section, it is supposed that $\nu = 1$, i.e. m_1 is odd and $U = U_1$, which is the case that Mesnager settled [18,19] with the following proposition. We recall the main ingredients and results of her work as similar ideas will be used for the even case.

Proposition 1 [18,19]. *For $\nu = 1$, $a \in \mathbb{F}_{2^{m_1}}^*$ and $b \in \mathbb{F}_4^*$, the Walsh transform of $f_{a,b}$ at $\omega \in \mathbb{F}_{2^{m_0}}$ is, for $\omega = 0$:*

$$\widehat{\chi_{f_{a,b}}}(0) = \begin{cases} 1 + \frac{2^{m_1}-1}{3}\left(1 - K_{m_1}(a) - 4C_{m_1}(a,a)\right) & \text{if } b = 1, \\ 1 + \frac{2^{m_1}-1}{3}\left(1 - K_{m_1}(a) + 2C_{m_1}(a,a)\right) & \text{if } b \neq 1, \end{cases} \tag{6}$$

and for $\omega \neq 0$:

$$\widehat{\chi_{f_{a,b}}}(\omega) = \begin{cases} 1 + 2^{m_1}\chi\left(f_{a,b}(w_1^{-1})\right) + \frac{1}{3}\left(1 - K_{m_1}(a) - 4C_{m_1}(a,a)\right) & \text{if } b = 1, \\ 1 + 2^{m_1}\chi\left(f_{a,b}(w_1^{-1})\right) + \frac{1}{3}\left(1 - K_{m_1}(a) + 2C_{m_1}(a,a)\right) & \text{if } b \neq 1. \end{cases} \tag{7}$$

Proof. For $\omega \neq 0$, $\operatorname{Tr}_{m_1}^{m_0}(\omega u_1) = 0$ if and only if $u_1 = w_1^{-1}$, so that

$$\sum_{u_1\in U_1, \operatorname{Tr}_{m_1}^{m_0}(\omega u_1)=0} \chi\left(f_{a,b}(u_1)\right) = \chi\left(f_{a,b}(w_1^{-1})\right).$$

The only difficulty lies in the computation of $\sum_{u_1\in U_1}\chi\left(f_{a,b}(u_1)\right)$ which can be done by splitting the sum on U_1 according to the value of $\psi_{m_1}(u_1)$:

$$\begin{aligned}\sum_{u_1\in U_1}\chi\left(f_{a,b}(u_1)\right) &= \sum_{u_1\in U_1}\chi\left(f_a(u_1)\right)\chi\left(g_b(u_1)\right) \\ &= \sum_{u_1\in U_1, b\psi_{m_0}(u_1)=1}\chi\left(f_a(u_1)\right) - \sum_{u_1\in U_1, b\psi_{m_0}(u_1)\neq 1}\chi\left(f_a(u_1)\right) \\ &= 2\sum_{u_1\in U_1, b\psi_{m_0}(u_1)=1}\chi\left(f_a(u_1)\right) - \sum_{u_1\in U_1}\chi\left(f_a(u_1)\right).\end{aligned}$$

As noted in Sect. 2.7 the second sum is

$$\sum_{u_1\in U_1}\chi\left(f_a(u_1)\right) = 1 - K_{m_1}(a).$$

As far as the first one is concerned, let us denote it $S_1(a,b,\omega)$. As m_1 is odd, using properties of the Dickson polynomial D_3 given in Sect. 2.4, one can show that for $b=1$:

$$S_1(a,b,\omega) = \frac{1}{3}\left(1 - K_{m_1}(a) + 2C_{m_1}(a,a)\right).$$

As $S_1(a,b,\omega)$ takes the same value for both $b \neq 1$, one deduces that for $b \neq 1$:

$$S_1(a,b,\omega) = \frac{1}{3}\left(1 - K_{m_1}(a) - C_{m_1}(a,a)\right).$$

Results of Carlitz [2] on $C_{m_1}(a,a)$ when m_1 is odd yield a concise and easy to compute the Walsh transform of $f_{a,b}$ at any $\omega \in \mathbb{F}_{2^{m_0}}$.

Together with Charpin et al. results [5,6] and the Hasse–Weil bound on $K_{m_1}(a)$, these formulae prove that $f_{a,b}$ is (hyper-)bent if and only if $K_{m_1}(a) = 4$ as was noted by Mesnager [18,19].

Theorem 1 [18,19]. *For $\nu = 1$, $a \in \mathbb{F}_{2^m}^*$ and $b \in \mathbb{F}_4^*$, the function $f_{a,b}$ is bent if and only if $K_{m_1}(a) = 4$.*

5 Even Case

5.1 General Extension Degree

In this section, it is supposed that $\nu > 1$, i.e. both m_0 and m_1 are even. The main difference with the case $\nu = 1$ is that 3 does now divide $2^{m_1} - 1$ (in fact $2^{m_\nu} + 1$) rather than $2^{m_1} + 1$, and $\psi_{m_0}(u)$ does not depend on the value of u_1 (but only on that of u_ν).

In particular, the computation of $\sum_{u \in U} f_{a,b}(u)$ becomes straightforward.

Lemma 5. *For $\nu > 1$, $a \in \mathbb{F}_{2^{m_1}}^*$ and $b \in \mathbb{F}_4^*$,*

$$\sum_{u \in U} \chi\left(f_{a,b}(u)\right) = -\frac{2^{2^{\nu-1} m_\nu} - 1}{3\left(2^{m_\nu} - 1\right)}\left(1 - K_{m_1}(a)\right). \tag{8}$$

Proof. Splitting U as $U \simeq U_1 \times \cdots \times U_\nu$, the sum can be rewritten:

$$\begin{aligned}
\sum_{u \in U} \chi\left(f_{a,b}(u)\right) &= \prod_{j=2}^{\nu-1}\left(2^{m_j}+1\right) \sum_{u_1 \in U_1} \chi\left(f_a(u_1)\right) \sum_{u_\nu \in U_\nu} \chi\left(g_b(u_\nu)\right) \\
&= \prod_{j=2}^{\nu-1}\left(2^{m_j}+1\right) \frac{2^{m_\nu}+1}{3}\left(1 - K_{m_1}(a)\right) \sum_{c \in \mathbb{F}_4^*} \chi\left(\mathrm{Tr}_1^2(bc)\right) \\
&= -\prod_{j=2}^{\nu-1}\left(2^{m_j}+1\right) \frac{2^{m_\nu}+1}{3}\left(1 - K_{m_1}(a)\right).
\end{aligned}$$

Finally, using the identity $\left(2^{2^j m_\nu}+1\right)\left(2^{2^j m_\nu}-1\right)=\left(2^{2^{j+1} m_\nu}-1\right)$, the product of the $(2^{m_j}+1)$'s is

$$\prod_{j=2}^{\nu}(2^{m_j}+1)=\prod_{j=2}^{\nu}\left(2^{2^{\nu-j} m_\nu}+1\right)=\frac{2^{2^{\nu-1} m_\nu}-1}{2^{m_\nu}-1}.$$

The value of the Walsh transform at $\omega=0$ given by Eq. (4) can now be simplified.

Lemma 6. *For $\nu>1$, $a\in\mathbb{F}_{2^{m_1}}^*$ and $b\in\mathbb{F}_4^*$, the Walsh transform of $f_{a,b}$ at $\omega=0$ is*

$$\widehat{\chi_{f_{a,b}}}(0)=1-\frac{2^{m_1}-1}{3}\left(1-K_{m_1}(a)\right). \tag{9}$$

As noted by Mesnager [18,19], the Hasse–Weil bound on $K_{m_1}(a)$ implies that, if $f_{a,b}$ is bent, then $\widehat{\chi_{f_{a,b}}}(0)=2^{m_1}$ and $K_{m_1}(a)=4$.

Proposition 2 [18,19]. *For $\nu>1$, $a\in\mathbb{F}_{2^{m_1}}^*$ and $b\in\mathbb{F}_4^*$, if the function $f_{a,b}$ is bent, then $K_{m_1}(a)=4$.*

Finally, the value of the Walsh transform at $\omega\neq 0$ given by Eq. (5) is simplified as follows.

Lemma 7. *For $\nu>1$, $a\in\mathbb{F}_{2^{m_1}}^*$ and $b\in\mathbb{F}_4^*$, the Walsh transform of $f_{a,b}$ at $\omega\in\mathbb{F}_{2^{m_0}}^*$ is*

$$\widehat{\chi_{f_{a,b}}}(\omega)=1+\frac{2^{2^{\nu-1} m_\nu}-1}{3\left(2^{m_\nu}-1\right)}\left(1-K_{m_1}(a)\right)+2^{m_\nu}\sum_{u\in U,\mathrm{Tr}_{m_\nu}^{m_0}(\omega u)=0}\chi\left(f_{a,b}(u)\right). \tag{10}$$

5.2 Descending to an Odd Degree Extension

To simplify further Eq. (10), the sum over $u\in U$ can be split into smaller sums according to the extension $\mathbb{F}_{2^{m_i}}$ (with $1\leq i\leq\nu$) where $\mathrm{Tr}_{m_i}^{m_0}(u\omega)$ becomes 0, giving the following expression.

Proposition 3. *For $\nu>1$, $a\in\mathbb{F}_{2^{m_1}}^*$ and $b\in\mathbb{F}_4^*$, and $\omega\in\mathbb{F}_{2^{m_0}}^*$, denote by $S_\nu(a,b,\omega)$ the sum*

$$S_\nu(a,b,\omega)=\sum_{\mathrm{Tr}_{m_{\nu-1}}^{m_0}(u\omega)\neq 0,\mathrm{Tr}_{m_\nu}^{m_0}(u\omega)=0,\mathrm{hw}_{m_0}(u_\nu)=1}\chi\left(f_a(u_1)\right). \tag{11}$$

The Walsh transform of $f_{a,b}$ at $\omega\neq 0$ is

$$\begin{aligned}\widehat{\chi_{f_{a,b}}}(\omega)=1&-\frac{2\cdot 2^{\left(2^{\nu-1}-1\right)m_\nu}-1}{3}\left(1-K_{m_1}(a)\right)\\&-\frac{2\cdot 2^{\left(2^{\nu-1}-1\right)m_\nu}\left(2^{m_\nu-1}-1\right)}{3}\chi\left(f_a(w_1)\right)\\&+2^{m_\nu+1}S_\nu(a,b,\omega).\end{aligned} \tag{12}$$

Proof. The sum over U can be divided into subsums σ_i over U_i: $\sum_{u\in U,\mathrm{Tr}^{m_0}_{m_\nu}(\omega u)=0} \chi(f_{a,b}(u)) = \sum_{i=1}^{\nu} \sigma_i$ with

$$\sigma_i = \sum_{\substack{\mathrm{Tr}^{m_0}_{m_{i-1}}(u_1\cdots u_{i-1}w_1\cdots w_{i-1})\neq 0,\\ \mathrm{Tr}^{m_0}_{m_i}(u_1\cdots u_i w_1\cdots w_i)=0,\\ u_{i+1}\in U_{i+1},\ldots,u_\nu\in U_\nu}} \chi(f_a(u_1))\,\chi(g_b(u_\nu)) .$$

The first sum σ_1 can be simplified as Eq. (8):

$$\begin{aligned}\sigma_1 &= \prod_{j=2}^{\nu-1}(2^{m_j}+1)\,\chi\left(f_a(w_1^{-1})\right) \sum_{u_\nu\in U_\nu}\chi(g_b(u_\nu))\\ &= -\prod_{j=2}^{\nu-1}(2^{m_j}+1)\frac{2^{m_\nu}+1}{3}\chi\left(f_a(w_1^{-1})\right)\\ &= -\frac{2^{2^{\nu-1}m_\nu}-1}{3\,(2^{m_\nu}-1)}\chi\left(f_a(w_1^{-1})\right) .\end{aligned} \tag{13}$$

The last sum σ_ν can be split according to the value of $\psi_{m_0}(u_\nu)$ as in Sect. 4:

$$\sigma_\nu = 2\sum_{\substack{\mathrm{Tr}^{m_0}_{m_{\nu-1}}(u\omega)\neq 0,\\ \mathrm{Tr}^{m_0}_{m_\nu}(u\omega)=0,\\ b\psi_{m_0}(u_\nu)=1}} \chi(f_a(u_1)) - \sum_{\substack{\mathrm{Tr}^{m_0}_{m_{\nu-1}}(u\omega)\neq 0,\\ \mathrm{Tr}^{m_0}_{m_\nu}(u\omega)=0}} \chi(f_a(u_1)) , \tag{14}$$

where the first term is $2S_\nu(a,b,\omega)$ and the second term is

$$\begin{aligned}-\sum_{\substack{\mathrm{Tr}^{m_0}_{m_{\nu-1}}(u\omega)\neq 0,\\ \mathrm{Tr}^{m_0}_{m_\nu}(u\omega)=0}} \chi(f_a(u_1)) &= -\prod_{j=2}^{\nu-1}2^{m_j}\sum_{u_1\neq w_1^{-1}}\chi(f_a(u_1))\\ &= -\prod_{j=2}^{\nu-1}2^{m_j}\left(1-\chi\left(f_a(w_1^{-1})\right)-K_{m_1}(a)\right)\\ &= -2^{2(2^{\nu-2}-1)m_\nu}\left(1-\chi\left(f_a(w_1^{-1})\right)-K_{m_1}(a)\right) ,\end{aligned} \tag{15}$$

as the product of the 2^{m_j}'s is

$$\prod_{j=2}^{\nu-1}2^{m_j} = \prod_{j=2}^{\nu-1}2^{2^{\nu-j}m_\nu} = 2^{2^{\nu-2}\sum_{j=0}^{\nu-3}2^{-j}m_\nu} = 2^{2^{\nu-2}2(1-2^{-\nu+2})m_\nu} . \tag{16}$$

For $\nu > 2$, the intermediate sums σ_i for $2 < i < \nu$ are:

$$\begin{aligned}\sigma_i &= \prod_{j=2}^{i-1}2^{m_j}\prod_{j=i+1}^{\nu-1}(2^{m_j}+1)\sum_{u_1\neq w_1^{-1}}\chi(f_a(u_1))\sum_{u_\nu\in U_\nu}\chi(g_b(u_\nu))\\ &= -\prod_{j=2}^{i-1}2^{m_j}\prod_{j=i+1}^{\nu-1}(2^{m_j}+1)\frac{2^{m_\nu}+1}{3}\left(1-\chi\left(f_a(w_1^{-1})\right)-K_{m_1}(a)\right) .\end{aligned}$$

Fortunately, a simpler expression for the sum of the products of 2^{m_j}'s and $(2^{m_j}+1)$'s for $1 < j < \nu$ can be devised. Indeed, for $k \geq 3$ and any rational number m, the sum that we denote by $\Sigma(m,k)$ is

$$\Sigma(m,k) = \sum_{i=2}^{k-1}\left(\prod_{j=2}^{i-1} 2^{2^{k-j}m} \prod_{j=i+1}^{k}\left(2^{2^{k-j}m}+1\right)\right) = \frac{2^{2(2^{k-2}-1)m}-1}{2^m-1}. \tag{17}$$

The proof goes by induction on k. For $k=3$, the identity states $2^m+1 = \frac{2^{2m}-1}{2^m-1}$. Let us now suppose that Eq. (17) is verified up to some $k \geq 3$ for all rational numbers m's. The sum for $k+1$ is

$$\Sigma(m,k+1) = (2^m+1)\,\Sigma(2m,k) + (2^m+1)\prod_{j=2}^{k-1} 2^{2^{k-j}(2m)}$$

By induction and a variation of Eq. (16), the identity is proved for $k+1$:

$$\begin{aligned}\Sigma(m,k+1) &= (2^m+1)\,\frac{2^{2(2^{k-2}-1)(2m)}-1}{2^{2m}-1} + (2^m+1)\,2^{4(2^{k-2}-1)m}\\ &= \frac{2^{4(2^{k-2}-1)m}-1}{2^m-1} + \frac{(2^{2m}-1)\,2^{4(2^{k-2}-1)m}}{2^m-1}\\ &= \frac{2^{2(2^{k-1}-1)m}-1}{2^m-1}.\end{aligned}$$

Setting $k=\nu$ and $m=m_\nu$ in Eq. (17) yields

$$\sum_{i=2}^{\nu-1}\sigma_i = -\frac{2^{2(2^{\nu-2}-1)m_\nu}-1}{3\,(2^{m_\nu}-1)}\left(1-\chi\left(f_a(w_1^{-1})\right)-K_{m_1}(a)\right). \tag{18}$$

Note that for $\nu = 2$, both sides of the above equality are zero. Therefore, for any $\nu > 1$, Eqs. (13), (14), (15) and (18), lead to the following expression for the Walsh transform at $\omega \neq 0$:

$$\begin{aligned}\widehat{\chi_{f_{a,b}}}(\omega) = 1 &+ \frac{2^{2^{\nu-1}m_\nu}-1}{3\,(2^{m_\nu}-1)}\left(1-K_{m_1}(a)\right)\\ &- 2^{m_\nu}\frac{2^{2^{\nu-1}m_\nu}-1}{3\,(2^{m_\nu}-1)}\chi\left(f_a(w_1^{-1})\right)\\ &- 2^{m_\nu}\frac{2^{2(2^{\nu-2}-1)m_\nu}-1}{3\,(2^{m_\nu}-1)}\left(1-\chi\left(f_a(w_1^{-1})\right)-K_{m_1}(a)\right)\\ &- 2^{m_\nu}2^{2(2^{\nu-2}-1)m_\nu}\left(1-\chi\left(f_a(w_1^{-1})\right)-K_{m_1}(a)\right)\\ &+ 2^{m_\nu+1}S_\nu(a,b,\omega),\end{aligned}$$

which gives the announced expression by gathering independently the terms in $\chi\left(f_a(w_1^{-1})\right)$ and $\left(1-K_{m_1}(a)\right)$.

Unfortunately, making the remaining sum $S_\nu(a, b, \omega)$ explicit is a hard problem. Doing so is equivalent to evaluating a Gauss sum as in Eq. (1): an exponential sum involving a multiplicative character and an additive character. In the next section, we manage to tackle the case $\nu = 2$ when $\omega \in \mathbb{F}_{2^{m_1}}^*$ (that is $w_1 = 1$) and conjecture a partial formula when $\omega \notin \mathbb{F}_{2^{m_1}}^*$.

5.3 Four Times an Odd Number

From now on, it is supposed that $\nu = 2$, i.e. m_0 is four times the odd number m_2.

For $\operatorname{Tr}_{m_2}^{m_0}(u\omega)$ to be zero with $u_1 \neq w_1^{-1}$, u_2 must be the polar part of $\left(\omega_2 \operatorname{Tr}_{m_1}^{m_0}(u_1\omega_1)\right)^{-1}$ so that the sum of Eq. (11) becomes

$$S_2(a, b, \omega) = \sum_{u_1 \neq w_1^{-1}, \psi_{m_0}\left(w_2 \operatorname{Tr}_{m_1}^{m_0}(u_1 w_1)\right) = b} \chi\left(\operatorname{Tr}_1^{m_0}\left(a u_1^{-2}\right)\right). \tag{19}$$

The Subfield Case. We now restrict to the case $w_1 = 1$, that is $\omega \in \mathbb{F}_{2^{m_1}}^*$ rather than $\omega \in \mathbb{F}_{2^{m_0}}^*$.

Lemma 8. *For $a \in \mathbb{F}_{2^{m_1}}^*$ and $b \in \mathbb{F}_4^*$, and $\omega \in \mathbb{F}_{2^{m_1}}^*$, define $\gamma \in \mathbb{F}_4^*$ by $\gamma = b\psi_{m_1}(w_2)$. Then*

$$S_2(a, b, \omega) = 2 \sum_{t \in \mathcal{T}_{m_1}^1, \psi_{m_1}(t) = \gamma} \chi\left(\operatorname{Tr}_1^{m_1}(at)\right). \tag{20}$$

Proof. As $w_1 = 1$, both the multiplicative and additive characters act on the same inputs so that we can use the function $u_1 \mapsto u_1 + u_1^{-1}$ to transform the sum over U_1 of Eq. (19) into a sum over $\mathcal{T}_{m_1}^1$:

$$\begin{aligned} S_2(a, b, \omega) &= \sum_{u_1 \neq 1, \psi_{m_1}\left(u_1^2 + u_1^{-2}\right) = b\psi_{m_1}(w_2)} \chi\left(\operatorname{Tr}_1^{m_1}\left(a\left(u_1^{-2} + u_1^2\right)\right)\right) \\ &= \sum_{u_1 \neq 1, \psi_{m_1}\left(u_1 + u_1^{-1}\right) = b\psi_{m_1}(w_2)} \chi\left(\operatorname{Tr}_1^{m_1}\left(a\left(u_1 + u_1^{-1}\right)\right)\right) \\ &= 2 \sum_{t \in \mathcal{T}_{m_1}^1, \psi_{m_1}(t) = b\psi_{m_1}(w_2)} \chi\left(\operatorname{Tr}_1^{m_1}(at)\right). \end{aligned}$$

Remark that the sum in Eq. (20) can be seen as a first step toward generalizing the sum computed in Sect. 4 in the odd case: rather than involving u_1 directly, it involves its trace $t = \operatorname{Tr}_{m_1}^{m_0}(u_1)$.

As is customary, the sum over $\mathcal{T}_{m_1}^1$ can be evaluated using sums over all of $\mathbb{F}_{2^{m_1}}$:

$$S_2(a, b, \omega) = \sum_{x \in \mathbb{F}_{2^{m_1}}^*, \psi_{m_1}(x) = \gamma} \chi\left(\operatorname{Tr}_1^{m_1}(ax)\right) - \sum_{x \in \mathbb{F}_{2^{m_1}}^*, \psi_{m_1}(x) = \gamma} \chi\left(\operatorname{Tr}_1^{m_1}\left(ax + x^{-1}\right)\right). \tag{21}$$

The first sum is easily seen to be a cubic sum whereas the computation of the second sum is more involved.

Proposition 4. *For* $\nu = 2$, $a \in \mathbb{F}_{2^{m_1}}^*$ *and* $\gamma \in \mathbb{F}_4^*$. *Define* $\alpha \in \mathbb{F}_4^*$ *by* $\alpha = \psi_{m_1}(a)$. *The following equality holds:*

$$\sum_{x \in \mathbb{F}_{2^{m_1}}^*, \psi_{m_1}(x) = \gamma} \chi\left(\operatorname{Tr}_1^{m_1}(ax)\right) = \begin{cases} \frac{2^{m_2+1}-1}{3} & \text{if } \gamma = \alpha^{-1}, \\ \frac{-2^{m_2}-1}{3} & \text{if } \gamma \neq \alpha^{-1}. \end{cases} \tag{22}$$

Proof. Let $c \in \mathbb{F}_{2^{m_1}}^*$ be such that $\psi_{m_1}(c) = \gamma$. We make the change of variables $x = cx$ to transform the sum into a cubic sum:

$$\begin{aligned} \sum_{x \in \mathbb{F}_{2^{m_1}}^*, \psi_{m_1}(x) = \gamma} \chi\left(\operatorname{Tr}_1^{m_1}(ax)\right) &= \sum_{x \in \mathbb{F}_{2^{m_1}}^*, \psi_{m_1}(x) = \psi_{m_1}(c)} \chi\left(\operatorname{Tr}_1^{m_1}(ax)\right) \\ &= \sum_{x \in \mathbb{F}_{2^{m_1}}^*, \psi_{m_1}(x) = 1} \chi\left(\operatorname{Tr}_1^{m_1}(acx)\right) \\ &= \frac{1}{3} \sum_{x \in \mathbb{F}_{2^{m_1}}^*} \chi\left(\operatorname{Tr}_1^{m_1}(acx^3)\right) \\ &= \frac{1}{3}\left(C_{m_1}(ac, 0) - 1\right). \end{aligned}$$

Carlitz's results [2] give explicit values for this cubic sum when m_1 is even and m_2 is odd.

Proposition 5. *For* $\nu = 2$, $a \in \mathbb{F}_{2^{m_1}}^*$ *and* $\gamma \in \mathbb{F}_4^*$. *Define* $\alpha \in \mathbb{F}_4^*$ *by* $\alpha = \psi_{m_1}(a)$. *The following equality holds:*

$$\sum_{x \in \mathbb{F}_{2^{m_1}}^*, \psi_{m_1}(x) = \gamma} \chi\left(\operatorname{Tr}_1^{m_1}(ax + x^{-1})\right) = \begin{cases} \frac{1}{3}\left(2C_{m_1}(a, a) + K_{m_1}(a) - 1\right) & \text{if } \gamma = \alpha, \\ \frac{1}{3}\left(-C_{m_1}(a, a) + K_{m_1}(a) - 1\right) & \text{if } \gamma \neq \alpha. \end{cases} \tag{23}$$

Proof. First remark that summing over the three possible values of γ yields

$$\sum_{x \in \mathbb{F}_{2^{m_1}}^*} \chi\left(\operatorname{Tr}_1^{m_1}(ax + x^{-1})\right) = K_{m_1}(a) - 1.$$

Moreover, making the change of variable $x = (ax)^{-1}$ shows that the sum takes the same value for γ and $\alpha^{-1}\gamma^{-1}$. In particular, it takes the same value for $\alpha\beta$ and $\alpha\beta^2$, where $\beta \in \mathbb{F}_4^*$ is a primitive third root of unity, that is for the elements of $\mathbb{F}_4^*$ different from α, and this value can be deduced from the value for $\gamma = \alpha$ which we now compute.

Denote by r a square root of a. The change of variable $x = rx$ and properties of the Dickson polynomial D_3 when m_1 is even show that for $\gamma = \alpha = \psi_{m_1}(r^{-1})$:

$$\begin{aligned}
\sum_{x \in \mathbb{F}^*_{2^{m_1}}, \psi_{m_1}(x)=\alpha} \chi\left(\mathrm{Tr}_1^{m_1}(ax + x^{-1})\right) &= \sum_{x \in \mathbb{F}^*_{2^{m_1}}, \psi_{m_1}(x)=1} \chi\left(\mathrm{Tr}_1^{m_1}(r(x + x^{-1}))\right) \\
&= \frac{1}{3} \sum_{x \in \mathbb{F}^*_{2^{m_1}}} \chi\left(\mathrm{Tr}_1^{m_1}(r(x^3 + x^{-3}))\right) \\
&= \frac{1}{3} \sum_{x \in \mathbb{F}^*_{2^{m_1}}} \chi\left(\mathrm{Tr}_1^{m_1}(rD_3(x + x^{-1}))\right) \\
&= \frac{1}{3}\left(2 \sum_{t \in \mathcal{T}^0_{m_1}} \chi\left(\mathrm{Tr}_1^{m_1}(rD_3(t))\right) - 1\right) \\
&= \frac{1}{3}\left(2C_{m_1}(r,r) - 2 \sum_{t \in \mathcal{T}^1_{m_1}} \chi\left(\mathrm{Tr}_1^{m_1}(rt)\right) - 1\right) \\
&= \frac{1}{3}\left(2C_{m_1}(r,r) + K_{m_1}(r) - 1\right) \\
&= \frac{1}{3}\left(2C_{m_1}(a,a) + K_{m_1}(a) - 1\right).
\end{aligned}$$

Equations (22) and (23) give the following expression for $S_2(a,b,\omega)$.

Theorem 2. *For $\nu = 2$, $a \in \mathbb{F}^*_{2^{m_1}}$ and $b \in \mathbb{F}^*_4$, and $\omega \in \mathbb{F}^*_{2^{m_1}}$, let $\gamma = b\psi_{m_1}(w_2)$. Then the sum $S_2(a,b,\omega)$ is*

$$S_2(a,b,\omega) = \begin{cases} \frac{1}{3}\left(2^{m_2+1} - 2C_{m_1}(a,a) - K_{m_1}(a)\right) & \text{if } \gamma = \alpha \text{ and } \gamma = \alpha^{-1}, \\ \frac{1}{3}\left(-2^{m_2} - 2C_{m_1}(a,a) - K_{m_1}(a)\right) & \text{if } \gamma = \alpha \text{ and } \gamma \neq \alpha^{-1}, \\ \frac{1}{3}\left(2^{m_2+1} + C_{m_1}(a,a) - K_{m_1}(a)\right) & \text{if } \gamma \neq \alpha \text{ and } \gamma = \alpha^{-1}, \\ \frac{1}{3}\left(-2^{m_2} + C_{m_1}(a,a) - K_{m_1}(a)\right) & \text{if } \gamma \neq \alpha \text{ and } \gamma \neq \alpha^{-1}. \end{cases} \tag{24}$$

Carlitz's results [2] recalled in Sect. 2.8 can be used to make the cubic sum $C_{m_1}(a,a)$ explicit. In the particular case where $K_{m_1}(a) \equiv 1 \pmod 3$, which is equivalent to $C_{m_1}(a,a) = 0$ and implies that a is a cube, the expression for $S_2(a,b,\omega)$ gets very concise, as does Eq. (12) for the Walsh transform.

Corollary 1. *For $\nu = 2$, $a \in \mathbb{F}^*_{2^{m_1}}$ with $K_{m_1}(a) \equiv 1 \pmod 3$ and $b \in \mathbb{F}^*_4$, and $\omega \in \mathbb{F}^*_{2^{m_1}}$, let $\gamma = b\psi_{m_1}(w_2)$. Then the sum $S_2(a,b,\omega)$ is*

$$S_2(a,b,\omega) = \frac{2^{m_2+1} - K_{m_1}(a)}{3} - \mathrm{Tr}_1^2(\gamma)\, 2^{m_2}. \tag{25}$$

and the Walsh transform at $\omega \neq 0$ is

$$\widehat{\chi_{f_{a,b}}}(\omega) = \chi\left(\mathrm{Tr}_1^2(\gamma)\right) 2^{m_1} + \frac{4 - K_{m_1}(a)}{3}. \tag{26}$$

Note that Corollary 1 shows that for $\omega \in \mathbb{F}_{2^{m_1}}^*$ the Walsh transform of $f_{a,b}$ at ω is that of a bent function if and only if $K_{m_1}(a) = 4$.

5.4 A Conjectural General Formula

The techniques used in the previous section do not apply to the general case where $w_1 \neq 1$, i.e. $\omega \in \mathbb{F}_{2^{m_0}}^*$. The main reason being that the multiplicative and additive characters of $\mathbb{F}_{2^{m_1}}$ act on different values, e.g. $r = \mathrm{Tr}_{m_1}^{m_0}(w_1 u_1)$, or $s = \mathrm{Tr}_{m_1}^{m_0}(w_1^{-1} u_1)$, for one of them, and $t = \mathrm{Tr}_{m_1}^{m_0}(u_1)$ for the other one. Considering $v = \mathrm{Tr}_{m_1}^{m_0}(w_1)$, these values are related by $r + s = vt$. Moreover, the sum $S_2(a, b, \omega)$ takes the same value for w_1 and w_1^{-1}, so there is hope to introduce enough symmetry to reduce the case $w_1 \neq 1$ to the case $w_1 = 1$. Unfortunately, we could not devise a way to do so.

Yet, experimental evidence presented in more details in Sect. 5.5 suggests that the following conjecture, which relates the value of $S_2(a, b, \omega)$ for $w_1 = 1$ and $w_1 \neq 1$, is true.

Conjecture 2. For $\nu = 2$, $a \in \mathbb{F}_{2^{m_1}}^*$ with $K_{m_1}(a) \equiv 1 \pmod 3$ and $b \in \mathbb{F}_4^*$, and $\omega \in \mathbb{F}_{2^{m_0}}^*$, let $\gamma = b\psi_{m_1}(w_2)$. There exists a Boolean function $h_{a,b}(\omega)$ such that the sum $S_2(a, b, \omega)$ is

$$S_2(a,b,\omega) = \frac{2^{m_2+1} - K_{m_1}(a)}{3} - 2f_a(w_1^{-1})\frac{2^{m_2+1} - 1}{3} - h_{a,b}(\omega)\chi\left(f_a(w_1^{-1})\right)2^{m_2}. \tag{27}$$

The Walsh transform at $\omega \neq 0$ is then

$$\widehat{\chi_{f_{a,b}}}(\omega) = \chi\left(h_{a,b}(\omega) f_a(w_1^{-1})\right) 2^{m_1} + \frac{4 - K_{m_1}(a)}{3}. \tag{28}$$

In particular, this conjecture implies Conjecture 1: if $K_{m_1}(a) = 4$, then $f_{a,b}$ is bent. (And Corollary 1 already does so when $\omega \in \mathbb{F}_{2^{m_1}}^*$.)

5.5 Experimental Data

The computation of $S_2(a, b, \omega)$ was implemented in C and assembly[1], using AVX extensions for the arithmetic of $\mathbb{F}_{2^{m_0}}^*$, PARI/GP [21] to compute the Kloosterman sums $K_{m_1}(a)$, and Pthreads [12] for parallelization.

The computational cost of verifying Conjecture 2 can be somewhat leveraged using elementary properties of $S_2(a, b, \omega)$:

- it only depends on the cyclotomic class of $a \in \mathbb{F}_{2^{m_1}}^*$,
- it is the same for w_1 and w_1^{-1},
- the inner value can be computed at the same time for u_1 and u_1^{-1}.

[1] The source code is available at https://github.com/jpflori/expsums.

Whatsoever, there are:

- 3 values of $\gamma \in \mathbb{F}_4^*$,
- $\tilde{O}(2^{m_1})$ values of $a \in \mathbb{F}_{2^{m_1}}^*$,
- 2^{m_1-1} values of $w_1 \in U_1 \setminus \{1\}$,
- $\tilde{O}(2^{m_1})$ operations in $\mathbb{F}_{2^{m_0}}$ for each triple (γ, a, w_1).

Therefore, checking the conjectured formula for $S_2(a, b, \omega)$ over $\mathbb{F}_{2^{m_0}}$ has time complexity $\tilde{O}(2^{3m_1})$ which quickly becomes overcostly (and is comparable to that of computing the Walsh spectrum for every cyclotomic class of $a \in \mathbb{F}_{2^{m_1}}^*$ which has time complexity $\tilde{O}(2^{3m_1})$ as well but space complexity $\tilde{O}(2^{m_1})$).

Still, we checked Conjecture 2:

- completely for $m_2 = 3, 5, 7, 9$,
- for i up to 3405 where $a = z^i$ and z is a primitive element of $\mathbb{F}_{2^{m_1}}$ for $m_2 = 11$.

Finally, assuming $K_{m_1}(a) \equiv 1 \pmod 3$ and Conjecture 2 is correct, Parseval's equality yields the following relation:

$$\sum_{x \in \mathbb{F}_{2^{m_0}}^*} \chi\left(h_{a,b}(\omega) f_a(w_1^{-1})\right) = \frac{2^{m_1} - 1}{3} \left(K_{m_1}(a) - 1\right)$$
$$= \widehat{\chi_{f_{a,b}}}(0) - 1.$$

This is supported by experimental evidence that there are exactly $2^{m_1-1} + (5/6)\left(K_{m_1}(a) - 4\right) + 3$ (respectively $2^{m_1-1} - (1/6)\left(K_{m_1}(a) - 4\right)$) values of $w_1 \in U_1$ such that $h_{a,b}(\omega) f_a(w_1^{-1})$ is zero when $\gamma = 1$ (respectively $\gamma \neq 1$).

6 Further Research and Open Problems

Hopefully, Conjecture 2 can be proved using similar techniques as the ones used by Mesnager [18,19] and in this note. Otherwise, more involved techniques could be tried, e.g. considering a whole family of sums as a whole and their geometric structure. Another possibility would be to directly treat the general Gauss sums of Eqs. (1) and (11) without focussing on the case $\nu = 2$.

References

1. Ahmadi, O., Granger, R.: An efficient deterministic test for Kloosterman sum zeros. CoRR abs/1104.3882 (2011)
2. Carlitz, L.: Explicit evaluation of certain exponential sums. Math. Scand. **44**(1), 5–16 (1979)
3. Charpin, P., Gong, G.: Hyperbent functions, Kloosterman sums, and Dickson polynomials. IEEE Trans. Inf. Theory **54**(9), 4230–4238 (2008)
4. Charpin, P., Helleseth, T., Zinoviev, V.: The divisibility modulo 24 of Kloosterman sums on GF(2^m), m odd. J. Comb. Theory Ser. A **114**(2), 322–338 (2007). http://dx.doi.org/10.1016/j.jcta.2006.06.002

5. Charpin, P., Helleseth, T., Zinoviev, V.: Divisibility properties of Kloosterman sums over finite fields of characteristic two. In: IEEE International Symposium on Information Theory, ISIT 2008, pp. 2608–2612 (2008)
6. Charpin, P., Helleseth, T., Zinoviev, V.: Divisibility properties of classical binary Kloosterman sums. Discret. Math. **309**(12), 3975–3984 (2009)
7. Dillon, J.F.: Elementary Hadamard Difference Sets. ProQuest LLC, Ann Arbor, MI (1974). http://gateway.proquest.com/openurl?url_ver=Z39.88-2004&rft_val_fmt=info:ofi/fmt:kev:mtx:dissertation&res_dat=xri:pqdiss&rft_dat=xri:pqdiss:7501799, thesis (Ph.D.)–University of Maryland, College Park
8. Dillon, J.F., Dobbertin, H.: New cyclic difference sets with Singer parameters. Finite Fields Appl. **10**(3), 342–389 (2004)
9. Flori, J.-P., Mesnager, S., Cohen, G.: Binary Kloosterman sums with value 4. In: Chen, L. (ed.) IMACC 2011. LNCS, vol. 7089, pp. 61–78. Springer, Heidelberg (2011). doi:10.1007/978-3-642-25516-8_5
10. Göloğlu, F., Lisoněk, P., McGuire, G., Moloney, R.: Binary Kloosterman sums modulo 256 and coefficients of the characteristic polynomial. IEEE Trans. Inf. Theory **PP**(99), 1 (2012)
11. Göloğlu, F., McGuire, G., Moloney, R.: Binary Kloosterman sums using Stickelberger's theorem and the Gross-Koblitz formula. Acta Arith. **148**(3), 269–279 (2011). http://dx.doi.org/10.4064/aa148-3-4
12. Austin Group: Standard for Information Technology: Portable Operating System Interface (POSIX(R)) Base Specifications, Issue 7. IEEE Std 1003.1, 2013 Edition (incorporates IEEE Std 1003.1-2008, and IEEE Std 1003.1-2008/Cor 1-2013), pp. 1–3906, April 2013
13. Katz, N., Livné, R.: Sommes de Kloosterman et courbes elliptiques universelles en caractéristiques 2 et 3. C. R. Acad. Sci. Paris Sér. I Math. **309**(11), 723–726 (1989)
14. Lachaud, G., Wolfmann, J.: Sommes de Kloosterman, courbes elliptiques et codes cycliques en caractéristique 2. C. R. Acad. Sci. Paris Sér. I Math. **305**(20), 881–883 (1987)
15. Leander, G.: Monomial bent functions. IEEE Trans. Inf. Theory **52**(2), 738–743 (2006)
16. Lisoněk, P.: On the connection between Kloosterman sums and elliptic curves. In: Golomb, S.W., Parker, M.G., Pott, A., Winterhof, A. (eds.) SETA 2008. LNCS, vol. 5203, pp. 182–187. Springer, Heidelberg (2008). doi:10.1007/978-3-540-85912-3_17
17. Mesnager, S.: A new family of hyper-bent Boolean functions in polynomial form. In: Parker, M.G. (ed.) IMACC 2009. LNCS, vol. 5921, pp. 402–417. Springer, Heidelberg (2009). doi:10.1007/978-3-642-10868-6_24
18. Mesnager, S.: Bent and hyper-bent functions in polynomial form and their link with some exponential sums and Dickson polynomials. IEEE Trans. Inf. Theory **57**(9), 5996–6009 (2011)
19. Mesnager, S.: A new class of bent and hyper-bent Boolean functions in polynomial forms. Des. Codes Cryptogr. **59**(1–3), 265–279 (2011)
20. Moloney, R.: Divisibility properties of Kloosterman sums and division polynomials for Edward curves. Ph.D. thesis, University College Dublin (2011)
21. The PARI Group, Bordeaux: PARI/GP, version 2.7.0 (2014). http://pari.math.u-bordeaux.fr/

Generalized Bent Functions and Their Gray Images

Thor Martinsen[1], Wilfried Meidl[2], and Pantelimon Stănică[1](✉)

[1] Department of Applied Mathematics, Naval Postgraduate School, Monterey, CA 93943–5216, USA
{tmartins,pstanica}@nps.edu

[2] Johann Radon Institute for Computational and Applied Mathematics, Austrian Academy of Sciences, Altenbergerstrasse 69, 4040 Linz, Austria
meidlwilfried@gmail.com

Abstract. In this paper we prove that generalized bent (gbent) functions defined on $\mathbb{F}_2^n$ with values in $\mathbb{Z}_{2^k}$ are regular, and show connections between the (generalized) Walsh spectrum of these functions and their components. Moreover we analyze generalized bent and semibent functions with values in $\mathbb{Z}_{16}$ in detail, extending earlier results on gbent functions with values in $\mathbb{Z}_4$ and $\mathbb{Z}_8$.

1 Introduction

Let $\mathbb{V}_n$ be an n-dimensional vector space over $\mathbb{F}_2$ and for an integer q, let $\mathbb{Z}_q$ be the ring of integers modulo q. For a *generalized Boolean function* f from $\mathbb{V}_n$ to $\mathbb{Z}_q$ the *generalized Walsh-Hadamard transform* is the complex valued function

$$\mathcal{H}_f^{(q)}(\mathbf{u}) = \sum_{\mathbf{x}\in\mathbb{V}_n} \zeta_q^{f(\mathbf{x})}(-1)^{\langle \mathbf{u},\mathbf{x}\rangle}, \quad \zeta_q = e^{\frac{2\pi i}{q}},$$

where $\langle \mathbf{u},\mathbf{x}\rangle$ denotes a (nondegenerate) inner product on $\mathbb{V}_n$ (we shall use ζ, $\mathcal{H}_f$, instead of ζ_q, respectively, $\mathcal{H}_f^{(q)}$, when q is fixed). Throughout, we identify $\mathbb{V}_n$ with the vector space $\mathbb{F}_2^n$ of n-tuples over $\mathbb{F}_2$, and we use the regular scalar (inner) product $\langle \mathbf{u},\mathbf{x}\rangle = \mathbf{u}\cdot\mathbf{x}$. We denote the set of all generalized Boolean functions by $\mathcal{GB}_n^q$ and when $q=2$, by $\mathcal{B}_n$.

We recall that for $q=2$, where the generalized Walsh-Hadamard transform of f reduces to the conventional *Walsh-Hadamard transform*

$$\mathcal{W}_f(\mathbf{u}) = \sum_{\mathbf{x}\in\mathbb{V}_n} (-1)^{f(\mathbf{x})}(-1)^{\mathbf{u}\cdot\mathbf{x}},$$

a function f for which $|\mathcal{W}_f(\mathbf{u})| = 2^{n/2}$ for all $\mathbf{u}\in\mathbb{V}_n$ is called a *bent* function. Similarly, we say that function $f : \mathbb{V}_n \to \mathbb{Z}_q$ is a *generalized bent* (*gbent*) if

S. Duquesne and S. Petkova-Nikova (Eds.): WAIFI 2016, LNCS 10064, pp. 160–173, 2016.
DOI: 10.1007/978-3-319-55227-9_12

$|\mathcal{H}_f(\mathbf{u})| = 2^{n/2}$ for all $\mathbf{u} \in \mathbb{V}_n$. Further recall that $f \in \mathcal{B}_n$ is called *plateaued* if $|\mathcal{W}_f(\mathbf{u})| \in \{0, 2^{(n+s)/2}\}$ for all $\mathbf{u} \in \mathbb{V}_n$ for a fixed integer s depending on f (we also call f then *s-plateaued*). If $s = 1$ (n must then be odd), or $s = 2$ (n must then be even), we call f semibent. With this notation, a semibent function is an s-plateaued Boolean function with smallest possible $s > 0$. Accordingly we call a function $f \in \mathcal{GB}_n^q$, with $q = 2^k$, $k > 1$, *generalized semibent* (*gsemibent*, for short) if $|\mathcal{H}_f(\mathbf{u})| \in \{0, 2^{(n+1)/2}\}$ for all $\mathbf{u} \in \mathbb{V}_n$, and more general, *generalized s-plateaued* if $|\mathcal{H}_f(\mathbf{u})| \in \{0, 2^{(n+s)/2}\}$ for all $\mathbf{u} \in \mathbb{V}_n$.

Let $f : \mathbb{V}_m \to \mathbb{Z}_q$. If $2^{k-1} < q \leq 2^k$, we associate a unique sequence of Boolean functions $a_i : \mathbb{V}_m \to \mathbb{F}_2$, $1 \leq i \leq k$, such that (the addition below is in $\mathbb{Z}_q$)

$$f(\mathbf{x}) = a_1(\mathbf{x}) + \cdots + 2^{k-1}a_k(\mathbf{x}), \text{ for all } \mathbf{x} \in \mathbb{V}_m.$$

If $q = 2^k$, following Carlet [1], we further define the *generalized Gray map* $\psi(f) : \mathcal{GB}_n^q \to \mathcal{B}_{n+k-1}$ by $\psi(f)(\mathbf{x}, y_1, \ldots, y_{k-1}) = \bigoplus_{i=1}^{k-1} a_i(\mathbf{x})y_i \oplus a_k(\mathbf{x})$.

Generalized bent functions were introduced in [7] in connection with applications in CDMA systems, and lately have attracted increasing attention, see e.g. [2,3,8,9].

In [8,9] gbent functions $f(\mathbf{x}) = a_1(\mathbf{x}) + 2a_2(\mathbf{x})$ in $\mathcal{GB}_n^4$ and $f = a_1(\mathbf{x}) + 2a_2(\mathbf{x}) + 2^2a_3(\mathbf{x})$ in $\mathcal{GB}_n^8$ were completely characterized in terms of properties of the Boolean functions $a_i(\mathbf{x})$. In particular, relations between gbentness of f and bentness of associated Boolean functions have been investigated. In [9] it was moreover shown that $f \in \mathcal{GB}_n^4$, with $f(\mathbf{x}) = a_1 + 2a_2(\mathbf{x})$, $a_1, a_2 \in \mathcal{B}_n$, is gbent if and only if the Gray image $\psi(f)$ is bent if n is odd, or semibent and the associated a_2 and $a_1 \oplus a_2$ have complementary autocorrelation if n is even (see [9] for the details). Currently one observes a lot of research activities on gbent functions for general k, and we expect that many more general results will be discovered in the near future. Our generalizations in Sect. 2, where we analyze gbent functions in terms of their components, are some first results. In particular we show that a gbent function in even dimension is an affine space of bent functions and we decompose a gbent function in $\mathcal{GB}_n^{2^k}$ into two gbent functions in $\mathcal{GB}_n^{2^{k-1}}$. In Sect. 3 we analyze gbent functions in $\mathcal{GB}_n^{16}$ in detail.

2 Gbent Functions and Their Components

In accordance with the terminology for bent functions, we call a gbent function $f \in \mathcal{GB}_n^q$ *regular*, if $\mathcal{H}_f(\mathbf{u}) = 2^{n/2}\zeta_q^{f^*(\mathbf{u})}$ for some function $f^* \in \mathcal{GB}_n^q$. We start with a theorem about the regularity of gbent functions, which is also of independent interest. We prove the result by modifying a method of Kumar et al. [6].

Theorem 1. *All gbent functions $f \in \mathcal{GB}_n^{2^k}$ are regular, except for n odd and $k = 2$, in which case we have $\mathcal{H}_f^4(\mathbf{u}) = 2^{\frac{n-1}{2}}(\pm 1 \pm i)$.*

Proof. If $k = 1$, the result is known, as we are dealing with classical bent functions. Let $k \geq 2$. Let $\zeta = e^{\frac{2\pi i}{2^k}}$ be a 2^k-primitive root of unity. It is known that

$\mathbb{Z}[\zeta]$ is the ring of algebraic integers in the cyclotomic field $\mathbb{Q}(\zeta)$. We recall some facts from [6] (we change the notations slightly). The decomposition for the ideal generated by 2 in $\mathbb{Z}[\zeta]$ has the form $\langle 2\rangle = P^{2^{k-1}}$, where $P = \langle 1-\zeta\rangle$ is a prime ideal in $\mathbb{Z}[\zeta]$. The decomposition group

$$G_2 = \{\sigma \text{ in the Galois group of } \mathbb{Q}(\zeta)/\mathbb{Q} \mid \sigma(P) = P\}$$

contains also the conjugation isomorphism $\sigma^*(z) = \overline{z^{-1}}$ (Proposition 2 in [6]). Observe that $\mathcal{H}_f^{(2^k)}(\mathbf{u})\overline{\mathcal{H}_f^{(2^k)}(\mathbf{u})} = 2^n$. Now, as in Property 7 of [6], observing that our generalized Walsh transform is simply $S(f, 2^{k-1}\mathbf{u})$ (in the notations of Kumar et al. [6]; $\mathbf{u}$ is a binary vector in our case), then $\mathcal{H}_f^{(2^k)}(\mathbf{u})$ and $\overline{\mathcal{H}_f^{(2^k)}(\mathbf{u})}$ will generate the same ideal in $\mathbb{Z}[\zeta]$ and so, $2^{-n}(\mathcal{H}_f^{(2^k)}(\mathbf{u}))^2$ is a unit, and consequently, $2^{-n/2}\mathcal{H}_f^{(2^k)}(\mathbf{u})$ is an algebraic integer. Therefore, by Proposition 1 of [6] (which, in fact, is an old result of Kronecker from 1857), $2^{-n/2}\mathcal{H}_f^{(2^k)}(\mathbf{u})$ must be a root of unity. That alone would still not be enough to show regularity since this root of unity may be in a cyclotomic field outside $\mathbb{Q}(\zeta)$, however, that is not the case here, since the Gauss quadratic sum $G(2^k) = \sum_{i=0}^{2^k-1} \zeta^{i^2} = 2^{k/2}(1+i)$ and so, $\sqrt{2} \in \mathbb{Q}(\zeta)$, unless $k = 2$ (since then $1 + i \notin \mathbb{Q}(\zeta)$). The first assertion is shown for n even, as well as for n odd with $k \geq 3$.

When n is odd and $k = 2$, then $\mathcal{H}_f^{(4)}(\mathbf{u}) = \sum_{\mathbf{u}\in\mathbb{V}_n} i^{f(\mathbf{u})}(-1)^{\mathbf{u}\cdot\mathbf{x}} = a + bi$, for some integers a, b. Since f is gbent, with $|\mathcal{H}_f^{(4)}(\mathbf{u})|^2 = 2^n$ we get the diophantine equation $a^2 + b^2 = 2^n$. If n is even, the only solutions are $(a, b) = (\pm 2^{n/2}, 0)$, or $(0, \pm 2^{n/2})$. If n is odd, the solutions are $(a, b) = (\pm 2^{\lfloor n/2\rfloor}, \pm 2^{\lfloor n/2\rfloor})$ (independent choices of signs). The theorem is shown.

From the definition of a Boolean bent function via the Walsh-Hadamard transform we immediately obtain the following equivalent definition, where we denote the support of a Boolean function f by $\mathrm{supp}(f) := \{\mathbf{x} \in \mathbb{V}_n : f(\mathbf{x}) = 1\}$: A Boolean function $f : \mathbb{V}_n \to \mathbb{F}_2$ is bent if and only if for every $\mathbf{u} \in \mathbb{V}_n$ the function $f_{\mathbf{u}}(\mathbf{x}) := f(\mathbf{x}) \oplus \mathbf{u}\cdot\mathbf{x}$ satisfies $|\mathrm{supp}(f_{\mathbf{u}})| = 2^{n-1} \pm 2^{n/2}$. Our next target is to show an analog description for gbent functions. We use the following lemma.

Lemma 1. *Let $q = 2^k$, $k > 1$, $\zeta = e^{2\pi i/q}$. If $\rho_l \in \mathbb{Q}$, $0 \leq l \leq q-1$ and $\sum_{l=0}^{q-1} \rho_l\zeta^l = r$ is rational, then $\rho_j = \rho_{2^{k-1}+j}$, for $1 \leq j \leq 2^{k-1}-1$.*

Proof. Since $\zeta^{2^{k-1}+l} = -\zeta^l$ for $0 \leq l \leq 2^{k-1}-1$, we can write every element z of the cyclotomic field $\mathbb{Q}(\zeta)$ as

$$z = \sum_{l=0}^{2^{k-1}-1} \lambda_l\zeta^l,\ \lambda_l \in \mathbb{Q}, 0 \leq l \leq 2^{k-1}-1.$$

As $[\mathbb{Q}(\zeta):\mathbb{Q}] = \varphi(q) = 2^{k-1}$ (φ is Euler's totient function), the set $\{1, \zeta, \ldots, \zeta^{2^{k-1}-1}\}$ is a basis of $\mathbb{Q}(\zeta)$. Since

$$0 = \sum_{l=0}^{q-1} \rho_l \zeta^l - r = (\rho_0 - \rho_{2^{k-1}} - r) + \sum_{l=1}^{2^{k-1}-1} (\rho_j - \rho_{2^{k-1}+j})\zeta^l,$$

the assertion of the lemma follows.

Proposition 1. *Let $n = 2m$ be even, and for a function $f : \mathbb{V}_n \to \mathbb{Z}_{2^k}$ and $\mathbf{u} \in \mathbb{V}_n$, let $f_\mathbf{u}(\mathbf{x}) = f(\mathbf{x}) + 2^{k-1}(\mathbf{u}\cdot\mathbf{x})$, and let $b_j^{(\mathbf{u})} = |\mathbf{x} \in \mathbb{V}_n : f_\mathbf{u}(\mathbf{x}) = j\}|$, $0 \le j \le 2^k - 1$. Then f is gbent if and only if for all $\mathbf{u} \in \mathbb{V}_n$ there exists an integer $\rho_\mathbf{u}$, $0 \le \rho_\mathbf{u} \le 2^{k-1} - 1$, such that*

$$b^{(\mathbf{u})}_{2^{k-1}+\rho_\mathbf{u}} = b^{(\mathbf{u})}_{\rho_\mathbf{u}} \pm 2^m \text{ and } b^{(\mathbf{u})}_{2^{k-1}+j} = b^{(\mathbf{u})}_j, \text{ for } 0 \le j \le 2^{k-1} - 1, j \ne \rho_\mathbf{u}.$$

Proof. First suppose that f is gbent. Then by Theorem 1, f is a regular gbent function. Hence

$$\mathcal{H}_f(\mathbf{u}) = \sum_{\mathbf{x}\in\mathbb{V}_n} \zeta^{f(\mathbf{x})}(-1)^{\mathbf{u}\cdot\mathbf{x}} = \sum_{\mathbf{x}\in\mathbb{V}_n} \zeta^{f(\mathbf{x})+2^{k-1}(\mathbf{u}\cdot\mathbf{x})} = \mathcal{H}_{f_\mathbf{u}}(0) = \sum_{j=0}^{2^k-1} b_j^{(\mathbf{u})}\zeta^j = 2^m\zeta^r$$

for some $0 \le r \le 2^k - 1$. With $\rho_\mathbf{u} = r$ if $0 \le r \le 2^{k-1} - 1$, and $\rho_\mathbf{u} = r - 2^{k-1}$ otherwise, the claim follows from Lemma 1.

The converse statement is verified in a straightforward manner.

We now can present connections between gbent functions and their components for the general case of gbent functions in $\mathcal{GB}_n^{2^k}$, $k > 1$. This generalizes the corresponding results for $k = 2$ and $k = 3$ in [8] and in [9].

Theorem 2. *Let n be even, and let $f(\mathbf{x})$ be a gbent function in $\mathcal{GB}_n^{2^k}$, $k > 1$, (uniquely) given as*

$$f(\mathbf{x}) = a_1(\mathbf{x}) + 2a_2(\mathbf{x}) + \cdots + 2^{k-2}a_{k-1}(\mathbf{x}) + 2^{k-1}a_k(\mathbf{x}),$$

$a_i \in \mathcal{B}_n$, $1 \le i \le k$. Then all Boolean functions of the form

$$g_\mathbf{c}(\mathbf{x}) = c_1a_1(\mathbf{x}) \oplus c_2a_2(\mathbf{x}) \oplus \cdots \oplus c_{k-1}a_{k-1}(\mathbf{x}) \oplus a_k(\mathbf{x}),$$

$\mathbf{c} = (c_1, c_2, \ldots, c_{k-1}) \in \mathbb{F}_2^{k-1}$, are bent functions.

Proof. As in Proposition 1, for the gbent function f we denote by $f_\mathbf{u}$ the function $f_\mathbf{u}(\mathbf{x}) = a_1(\mathbf{x}) + \cdots + 2^{k-2}a_{k-1}(\mathbf{x}) + 2^{k-1}(a_k(\mathbf{x}) + \mathbf{u}\cdot\mathbf{x})$ in $\mathcal{GB}_n^{2^k}$. Again, the integer $b_r^{(\mathbf{u})}$, $0 \le r \le 2^k - 1$, is defined as $b_r^{(\mathbf{u})} = |\{\mathbf{x} \in \mathbb{V}_n : f_\mathbf{u}(\mathbf{x}) = r\}|$. By Proposition 1, $b^{(\mathbf{u})}_{r+2^{k-1}} = b_r^{(\mathbf{u})}$ for all $0 \le r \le 2^{k-1} - 1$, except for one element $\rho_\mathbf{u} \in \{0, \ldots, 2^{k-1} - 1\}$ depending on $\mathbf{u}$, for which $b^{(\mathbf{u})}_{\rho_\mathbf{u}+2^{k-1}} = b^{(\mathbf{u})}_{\rho_\mathbf{u}} \pm 2^{n/2}$.

Since it is somewhat easier to follow, we first show the bentness of $a_k(\mathbf{x}) = g_0(\mathbf{x})$. In the second step we show the general case.

For $r \neq \rho_{\mathbf{u}}$, $0 \leq r \leq 2^{k-1} - 1$, consider all $\mathbf{x} \in \mathbb{V}_n$ for which $a_1(\mathbf{x}) + \cdots + 2^{k-2}a_{k-1}(\mathbf{x}) = r$. Since $b^{(\mathbf{u})}_{r+2^{k-1}} = b^{(\mathbf{u})}_r$, for exactly half of these $\mathbf{x}$ we have $a_k(\mathbf{x}) + \mathbf{u} \cdot \mathbf{x} = 0$ (note that the number of these $\mathbf{x}$ must be even). Among all $\mathbf{x} \in \mathbb{V}_n$ for which $a_1(\mathbf{x}) + \cdots + 2^{k-2}a_{k-1}(\mathbf{x}) = \rho_{\mathbf{u}}$, there are $b^{(\mathbf{u})}_{\rho_u}$ for which $a_k(\mathbf{x}) + \mathbf{u} \cdot \mathbf{x} = 0$, and there are $b^{(\mathbf{u})}_{\rho_{\mathbf{u}}+2^{k-1}} = b^{(\mathbf{u})}_{\rho_u} \pm 2^{n/2}$ for which $a_k(\mathbf{x}) + \mathbf{u} \cdot \mathbf{x} = 1$. Hence for the Walsh-Hadamard transform of a_k we get

$$\mathcal{W}_{a_k}(\mathbf{u}) = \sum_{\mathbf{x} \in \mathbb{V}_n} (-1)^{a_k(\mathbf{x}) \oplus \mathbf{u} \cdot \mathbf{x}} = \pm 2^{n/2},$$

which shows that a_k is bent.

To show that $g_{\mathbf{c}}$ is bent for every $\mathbf{c} \in \mathbb{F}_2^{k-1}$, we write $f_{\mathbf{u}}(\mathbf{x})$, $\mathbf{u} \in \mathbb{V}_n$, as

$$\begin{aligned} f_{\mathbf{u}}(\mathbf{x}) &= c_1 a_1(\mathbf{x}) + \cdots + c_{k-1} 2^{k-2} a_{k-1}(\mathbf{x}) \;+\; \bar{c}_1 a_1(\mathbf{x}) + \cdots + \bar{c}_{k-1} 2^{k-2} a_{k-1}(\mathbf{x}) \\ &\quad + 2^{k-1}(a_k(\mathbf{x}) + \mathbf{u} \cdot \mathbf{x}) := h(\mathbf{x}) + \bar{h}(\mathbf{x}) + 2^{k-1}(a_k(\mathbf{x}) + \mathbf{u} \cdot \mathbf{x}), \end{aligned}$$

where $\bar{c} = c \oplus 1$. Note that every $0 \leq r \leq 2^{k-1} - 1$ in the value set of $a_1(x) + \cdots + 2^{k-2}a_{k-2}(\mathbf{x})$ has then a unique representation as $h(\mathbf{x}) + \bar{h}(\mathbf{x})$. Consider $\mathbf{x}$ for which $h(\mathbf{x}) + \bar{h}(\mathbf{x}) = r + s \neq \rho_{\mathbf{u}}$. Again from $b^{(\mathbf{u})}_{r+2^{k-1}} = b^{(\mathbf{u})}_r$ we infer that for half of those $\mathbf{x}$ we have $a_k(\mathbf{x}) \oplus \mathbf{u} \cdot \mathbf{x} = 0$. Hence also

$$g_{\mathbf{c}}(\mathbf{x}) \oplus \mathbf{u} \cdot \mathbf{x} = c_1 a_1(\mathbf{x}) \oplus \cdots \oplus c_{k-1} a_{k-1}(\mathbf{x}) \oplus a_k(\mathbf{x}) \oplus \mathbf{u} \cdot \mathbf{x} = 0$$

for exactly half of those $\mathbf{x}$. (Observe that $h(\mathbf{x}_1) = h(\mathbf{x}_2) = r$ implies $c_1 a_1(\mathbf{x}_1) \oplus \cdots \oplus c_{k-1} a_{k-1}(\mathbf{x}_1) = c_1 a_1(\mathbf{x}_2) \oplus \cdots \oplus c_{k-1} a_{k-1}(\mathbf{x}_2)$.) Similarly as above, among all $\mathbf{x} \in \mathbb{V}_n$ for which $h(\mathbf{x}) + \bar{h}(\mathbf{x}) = \rho_{\mathbf{u}}$, there are $b^{(\mathbf{u})}_{\rho_u}$ for which $a_k(\mathbf{x}) \oplus \mathbf{u} \cdot \mathbf{x} = 0$, and there are $b^{(\mathbf{u})}_{\rho_{\mathbf{u}}+2^{k-1}} = b^{(\mathbf{u})}_{\rho_u} \pm 2^{n/2}$ for which $a_k(\mathbf{x}) \oplus \mathbf{u} \cdot \mathbf{x} = 1$. From this we conclude that $|\{\mathbf{x} \in \mathbb{V}_n : h(\mathbf{x}) + \bar{h}(\mathbf{x}) = \rho_u \text{ and } f_{\mathbf{u}}(\mathbf{x}) = 1\}| - |\{\mathbf{x} \in \mathbb{V}_n : h(\mathbf{x}) + \bar{h}(\mathbf{x}) = \rho_u \text{ and } f_{\mathbf{u}}(\mathbf{x}) = 0\}| = \pm 2^{n/2}$. Therefore

$$\mathcal{W}_{g_{\mathbf{c}}}(\mathbf{u}) = \sum_{\mathbf{x} \in \mathbb{V}_n} (-1)^{g_{\mathbf{c}}(\mathbf{x}) + \mathbf{u} \cdot \mathbf{x}} = \pm 2^{n/2},$$

and $g_{\mathbf{c}}$ is bent.

We remark that the necessary conditions in Theorem 2 are not sufficient when $k > 2$. The additional conditions on the Walsh spectra for $k = 3$ given in [9, Theorem 19] and for $k = 4$ given in our Theorem 7 are required, as one can easily confirm with examples employing vectorial Maiorana-McFarland bent functions.

The next result on the decomposition of a gbent function in $\mathcal{GB}_n^{2^k}$ into two gbent functions in $\mathcal{GB}_n^{2^{k-1}}$ reveals an inductive approach to the study of gbent functions in $\mathcal{GB}_n^{2^k}$. Note that for $k = 2$ we recover the result in [9] on the decomposition of a gbent function in $\mathcal{GB}_n^4$ into two bent functions.

Theorem 3. *Let $f \in \mathcal{GB}_n^{2^k}$ with $f(\mathbf{x}) = g(\mathbf{x}) + 2h(\mathbf{x}), g \in \mathcal{B}_n, h \in \mathcal{GB}_n^{2^{k-1}}$. If n is even, then the following statements are equivalent.*

(i) f is gbent in $\mathcal{GB}_n^{2^k}$;
(ii) h and $h+2^{k-2}g$ are both gbent in $\mathcal{GB}_n^{2^{k-1}}$ with $\mathcal{H}_{h+2^{k-2}g}(\mathbf{u}) = \pm\mathcal{H}_h(\mathbf{u})$, for all $\mathbf{u} \in \mathbb{V}_n$.

If n is odd, then (ii) implies (i).

Proof. We first show that for n even, h and $h+2^{k-2}g$ are gbent in $\mathcal{GB}_n^{2^{k-1}}$ if f is gbent in $\mathcal{GB}_n^{2^k}$. In a second step, we show that if h and $h+2^{k-2}g$ are both gbent in $\mathcal{GB}_n^{2^{k-1}}$, then f is gbent in $\mathcal{GB}_n^{2^k}$ if and only if $\mathcal{H}_h^{(2^{k-1})}(\mathbf{u}) = \pm\mathcal{H}_{h+2^{k-2}g}^{(2^{k-1})}(\mathbf{u})$, for all $\mathbf{u} \in \mathbb{V}_n$. This will conclude the proof for both, n even and n odd.

Let $\mathbf{u} \in \mathbb{V}_n$, and for $e \in \{0,1\}$ and $r \in \{0,\ldots,2^{k-1}-1\}$, let

$$S^{(\mathbf{u})}(e,r) = \{\mathbf{x} \in \mathbb{V}_n \ : \ g(\mathbf{x}) = e \text{ and } h(\mathbf{x}) + 2^{k-2}(\mathbf{u}\cdot\mathbf{x}) = r\}.$$

With the notations of Proposition 1, we have $f_{\mathbf{u}}(\mathbf{x}) = f(\mathbf{x}) + 2^{k-1}(\mathbf{u}\cdot\mathbf{x}) = g(\mathbf{x}) + 2(h(\mathbf{x}) + 2^{k-2}(\mathbf{u}\cdot\mathbf{x}))$, and $|S^{(\mathbf{u})}(e,r)| = b_{e+2r}^{(\mathbf{u})}$. If f is gbent, by Proposition 1, there exist $\epsilon \in \{0,1\}$ and $0 \le \rho_{\mathbf{u}} \le 2^{k-2}-1$, for which $|S^{(\mathbf{u})}(\epsilon, \rho_{\mathbf{u}} + 2^{k-2})| = |S^{(\mathbf{u})}(\epsilon, \rho_{\mathbf{u}})| \pm 2^{n/2}$. For $(e,r) \neq (\epsilon, \rho_{\mathbf{u}})$, we have $|S^{(\mathbf{u})}(e, r+2^{k-2})| = |S^{(\mathbf{u})}(e,r)|$. Observing that $\{\mathbf{x} \in \mathbb{V}_n \ : \ h(\mathbf{x}) + 2^{k-2}(\mathbf{u}\cdot\mathbf{x}) = r\} = S^{(u)}(0,r) \cup S^{(u)}(1,r)$, we obtain

$$\mathcal{H}_h^{(2^{k-1})}(\mathbf{u}) = \sum_{\mathbf{x}\in\mathbb{V}_n} \zeta_{2^{k-1}}^{h(\mathbf{x})}(-1)^{\mathbf{u}\cdot\mathbf{x}} = \sum_{\mathbf{x}\in\mathbb{V}_n} \zeta_{2^{k-1}}^{h(\mathbf{x})+2^{k-2}(\mathbf{u}\cdot\mathbf{x})} = \pm\zeta_{2^{k-1}}^{\rho_{\mathbf{u}}} 2^{n/2}.$$

Consequently, h is gbent in $\mathcal{GB}_n^{2^{k-1}}$. For $h + 2^{k-2}g \in \mathcal{GB}_n^{2^{k-1}}$ we have

$$\begin{aligned}\mathcal{H}_{h+2^{k-2}g}^{(2^{k-1})}(\mathbf{u}) &= \sum_{\mathbf{x}\in\mathbb{V}_n} \zeta_{2^{k-1}}^{h(\mathbf{x})+2^{k-2}(\mathbf{u}\cdot\mathbf{x})+2^{k-2}gx} = \sum_{\substack{e\in\mathbb{F}_2\\ r\in\mathbb{Z}_{2^{k-1}}}} \sum_{x\in S^{(u)}(e,r)} \zeta_{2^{k-1}}^{r+2^{k-2}e} \\ &= \sum_{\substack{e\in\mathbb{F}_2\\ r\in\mathbb{Z}_{2^{k-1}}}} |S^{(u)}(e,r)|\zeta_{2^{k-1}}^{r+2^{k-2}e} = \pm\zeta_{2^{k-1}}^{\rho_{\mathbf{u}}+2^{k-2}\epsilon} 2^{n/2},\end{aligned}$$

which implies that also $h + 2^{k-2}g$ is gbent in $\mathcal{GB}_n^{2^{k-1}}$.

To show the condition $\mathcal{H}_h^{(2^{k-1})}(\mathbf{u}) = \pm\mathcal{H}_{h+2^{k-2}g}^{(2^{k-1})}(\mathbf{u})$, we first observe that

$$\begin{aligned}2\mathcal{H}_f^{(2^k)}(\mathbf{u}) &= 2\sum_{\mathbf{x}\in\mathbb{F}_2^n} \zeta_{2^k}^{g(\mathbf{x})}\zeta_{2^{k-1}}^{h(\mathbf{x})}(-1)^{\mathbf{u}\cdot\mathbf{x}} \\ &= \sum_{\mathbf{x}\in\mathbb{F}_2^n} \left(1 + (-1)^{g(\mathbf{x})} + (1-(-1)^{g(\mathbf{x})})\zeta_{2^k}\right)\zeta_{2^{k-1}}^{h(\mathbf{x})}(-1)^{\mathbf{u}\cdot\mathbf{x}} \\ &= (1+\zeta_{2^k})\mathcal{H}_h^{(2^{k-1})}(\mathbf{u}) + (1-\zeta_{2^k})\mathcal{H}_{h+2^{k-2}g}^{(2^{k-1})}(\mathbf{u}). \qquad (1)\end{aligned}$$

Writing $\zeta_{2^k} = x + yi$, $\mathcal{H}_h^{(2^{k-1})}(\mathbf{u}) = a + bi$ and $\mathcal{H}_{h+2^{k-2}g}^{(2^{k-1})}(\mathbf{u}) = c + di$, from Eq. (1), taking the complex norm, squaring and rearranging terms (recall that $|\zeta_{2^k}|^2 = x^2 + y^2 = 1$), we get

$$
\begin{aligned}
2|\mathcal{H}_f^{(2^k)}(\mathbf{u})|^2 &= \frac{1}{2}(a^2+b^2)(1+2x+x^2+y^2) + \frac{1}{2}(c^2+d^2)(1-2x+x^2+y^2) \\
&\quad - (ac+bd)(\mathbf{x}^2+y^2-1) + 2(ad-bc)y \\
&= |\mathcal{H}_h^{(2^{k-1})}(\mathbf{u})|^2(1+x) + |\mathcal{H}_{h+2^{k-2}g}^{(2^{k-1})}(\mathbf{u})|^2(1-x) \\
&\quad + 2y\,\Im\left(\overline{\mathcal{H}_h^{(2^{k-1})}(\mathbf{u})}\mathcal{H}_{h+2^{k-2}g}^{(2^{k-1})}(\mathbf{u})\right). \qquad (2)
\end{aligned}
$$

If h, $h+2^{k-2}g$ are gbent, i.e. $|\mathcal{H}_h^{(2^{k-1})}(\mathbf{u})|^2 = |\mathcal{H}_{h+2^{k-2}g}^{(2^{k-1})}(\mathbf{u})|^2 = 2^n$ for all $\mathbf{u} \in \mathbb{V}_n$, then we immediately see that $|\mathcal{H}_f^{(2^k)}(\mathbf{u})|^2 = 2^n$ for all $\mathbf{u} \in \mathbb{V}_n$, and hence f is gbent if and only if $\Im\left(\overline{\mathcal{H}_h^{(2^{k-1})}(\mathbf{u})}\mathcal{H}_{h+2^{k-2}g}^{(2^{k-1})}(\mathbf{u})\right) = 0$, for all $\mathbf{u} \in \mathbb{V}_n$.

We now argue that the condition $\Im\left(\overline{\mathcal{H}_h^{(2^{k-1})}(\mathbf{u})}\mathcal{H}_{h+2^{k-2}g}^{(2^{k-1})}(\mathbf{u})\right) = 0$ is equivalent to $\mathcal{H}_{h+2^{k-2}g}(\mathbf{u}) = \pm\mathcal{H}_h(\mathbf{u})$. For easy writing, let f_0, f_1 be the gbent functions in the indices above. By the regularity of gbent functions (when n is even or $k \geq 3$), $\mathcal{H}_{f_0}(\mathbf{u}) = 2^{n/2}\zeta_{2^{k-1}}^i$, $\mathcal{H}_{f_1}(\mathbf{u}) = 2^{n/2}\zeta_{2^{k-1}}^j$ for some integers $0 \leq i, j \leq 2^k - 1$. Hence $\overline{\mathcal{H}_{f_0}(\mathbf{u})}\,\mathcal{H}_{f_1}(\mathbf{u})$ is real if and only if $\zeta_{2^k}^{j-i} = \pm 1$, i.e. $i = j$ or $i = j + 2^{k-1}$ (modulo 2^k). Equivalently, $\mathcal{H}_{f_1}(\mathbf{u}) = \pm\mathcal{H}_{f_0}(\mathbf{u})$. If n is odd and $k = 2$, then $\mathcal{H}_f(\mathbf{u}) = 2^{n/2}\zeta_8^i$, $i \in \{1, 3, 5, 7\}$, and the same argument works.

We close this section with some remarks on relations between gbent functions in $\mathcal{GB}_n^{2^k}$, n even, and relative difference sets. First note that the characters of $\mathbb{V}_n \times \mathbb{Z}_{2^k}$ are $\chi_{\mathbf{u},a}(\mathbf{x}, z) = \zeta_{2^k}^{az}(-1)^{\langle \mathbf{u},\mathbf{x}\rangle}$, $\mathbf{u} \in \mathbb{V}_n$, $a \in \mathbb{Z}_{2^k}$. Recall that if $|\chi_{\mathbf{u},a}(D)| = 2^{n/2}$ for all nonzero $a \in \mathbb{Z}_{2^k}$ and all $\mathbf{u} \in \mathbb{V}_n$, then the graph $D = \{(\mathbf{x}, f(\mathbf{x})) : \mathbf{x} \in \mathbb{V}_n\}$ of f forms a relative difference set in $\mathbb{V}_n \times \mathbb{Z}_{2^k}$ (see for instance Sect. 2.4. in [10]). Equivalently, if af is gbent for all nonzero a, then D is a relative difference set. As easily seen, it is sufficient that $2^t f$ is gbent for all $0 \leq t \leq k-1$. Using Theorem 2, it is not hard to show that $F(\mathbf{x}) = (a_0(\mathbf{x}), \ldots, a_{k-1}(\mathbf{x}))$ is then a vectorial bent function, hence also a relative difference set in an elementary abelian group. Such gbent functions, which seem quite rare, may be of particular interest for future research. For an example of a class of such gbent functions obtained from partial spreads we refer to [5].

3 Complete Characterization of Generalized Bent and Semibent Functions in $\mathcal{GB}_n^{16}$

We write $f \in \mathcal{GB}_n^{16}$ as

$$
\begin{aligned}
f(\mathbf{x}) &= a_1(\mathbf{x}) + 2a_2(\mathbf{x}) + 2^2 a_3(\mathbf{x}) + 2^3 a_4(\mathbf{x}) \\
&= b_1(\mathbf{x}) + 2^2 b_2(\mathbf{x}) = a_1(\mathbf{x}) + 2d(\mathbf{x}),
\end{aligned}
$$

where $a_i(\mathbf{x}) \in \mathcal{B}_n$, $i = 1, 2, 3, 4$, $b_1(\mathbf{x}) = a_1(\mathbf{x}) + 2a_2(\mathbf{x})$, $b_2(\mathbf{x}) = a_3(\mathbf{x}) + 2a_4(\mathbf{x})$ are in $\mathcal{GB}_n^4$, and $d(\mathbf{x}) = a_2(\mathbf{x}) + 2a_3(\mathbf{x}) + 2^2 a_4(\mathbf{x}) \in \mathcal{GB}_n^8$.

Our objective is to show necessary *and* sufficient conditions on the components $a_1, a_2, a_3, a_4, b_1, b_2, d$ for the gbentness of f. For the conditions on a_1 and d for the gbentness of $a_1(\mathbf{x}) + 2d(\mathbf{x})$ when n is even, we can apply Theorem 3: $f(\mathbf{x}) = a_1(\mathbf{x}) + 2d(\mathbf{x})$ is gbent if and only if d and $d + 4a_1$ are gbent in $\mathcal{GB}_n^8$ and $\mathcal{H}_d^{(8)}(\mathbf{u}) = \pm\mathcal{H}_{d+4a_1}^{(8)}(\mathbf{u})$ for all $\mathbf{u} \in \mathbb{V}_n$.

We start with a complete characterization of gbent functions in $\mathcal{GB}_n^{16}$ in terms of a_1, a_2, a_3, a_4. By this we extend results in [8,9] on gbent functions in $\mathcal{GB}_n^4$ and $\mathcal{GB}_n^8$. We then also will characterize gsemibent functions $f \in \mathcal{GB}_n^{16}$ in terms of a_1, a_2, a_3, a_4.

Theorem 4. *Suppose that $f(\mathbf{x}) = a_1(\mathbf{x}) + 2a_2(\mathbf{x}) + 2^2 a_3(\mathbf{x}) + 2^3 a_4(\mathbf{x})$, $a_i \in \mathcal{B}_n$, $1 \le i \le 4$. Then f is gbent in $\mathcal{GB}_n^{16}$ if and only if the conditions (i) (if n is even), or (ii) (if n is odd) hold:*

(i) For all $c_i \in \mathbb{F}_2$, $i = 1, 2, 3$, the Boolean function $c_1a_1 \oplus c_2a_2 \oplus c_3a_3 \oplus a_4$ is bent, and for all $\mathbf{u} \in \mathbb{V}_n$ we have

$$\begin{aligned}
&\mathcal{W}_{a_4}(\mathbf{u})\mathcal{W}_{a_2\oplus a_4}(\mathbf{u}) = \mathcal{W}_{a_3\oplus a_4}(\mathbf{u})\mathcal{W}_{a_2\oplus a_3\oplus a_4}(\mathbf{u}) \\
&= \mathcal{W}_{a_1\oplus a_4}(\mathbf{u})\mathcal{W}_{a_1\oplus a_2\oplus a_4}(\mathbf{u}) = \mathcal{W}_{a_1\oplus a_3\oplus a_4}(\mathbf{u})\mathcal{W}_{a_1\oplus a_2\oplus a_3\oplus a_4}(\mathbf{u}), \text{ and} \\
&\mathcal{W}_{a_4}(\mathbf{u})\mathcal{W}_{a_3\oplus a_4}(\mathbf{u}) = \mathcal{W}_{a_1\oplus a_4}(\mathbf{u})\mathcal{W}_{a_1\oplus a_3\oplus a_4}(\mathbf{u}).
\end{aligned}$$

(ii) For all $c_i \in \mathbb{F}_2$, $i = 1, 2, 3$, the Boolean function $c_1a_1 \oplus c_2a_2 \oplus c_3a_3 \oplus a_4$ is semibent, and for all $\mathbf{u} \in \mathbb{V}_n$ we either have

$$\begin{aligned}
&\mathcal{W}_{a_4}(\mathbf{u})\mathcal{W}_{a_2\oplus a_4}(\mathbf{u}) = \mathcal{W}_{a_1\oplus a_4}(\mathbf{u})\mathcal{W}_{a_1\oplus a_2\oplus a_4}(\mathbf{u}) = \pm 2^{n+1} \text{ and} \\
&\mathcal{W}_{a_3\oplus a_4}(\mathbf{u}) = \mathcal{W}_{a_2\oplus a_3\oplus a_4}(\mathbf{u}) = \mathcal{W}_{a_1\oplus a_3\oplus a_4}(\mathbf{u}) = \mathcal{W}_{a_1\oplus a_2\oplus a_3\oplus a_4}(\mathbf{u}) = 0,
\end{aligned}$$

or

$$\begin{aligned}
&\mathcal{W}_{a_2\oplus a_4}(\mathbf{u}) = \mathcal{W}_{a_4}(\mathbf{u}) = \mathcal{W}_{a_1\oplus a_4}(\mathbf{u}) = \mathcal{W}_{a_1\oplus a_2\oplus a_4}(\mathbf{u}) = 0 \text{ and} \\
&\mathcal{W}_{a_3\oplus a_4}(\mathbf{u})\mathcal{W}_{a_2\oplus a_3\oplus a_4}(\mathbf{u}) = \mathcal{W}_{a_1\oplus a_3\oplus a_4}(\mathbf{u})\mathcal{W}_{a_1\oplus a_2\oplus a_3\oplus a_4}(\mathbf{u}) = \pm 2^{n+1}.
\end{aligned}$$

Our proof for Theorem 4 is quite technical and in parts computer-assisted. Hence we omit is here and present it in the appendix.

The result on the semibentness of functions in $\mathcal{GB}_n^{16}$ is obtained with the same approach. For the proof we again refer to the appendix.

Theorem 5. *Let $f \in \mathcal{GB}_n^{16}$ be given as $f(\mathbf{x}) = a_1(\mathbf{x}) + 2a_2(\mathbf{x}) + 2^2 a_3(\mathbf{x}) + 2^3 a_4(\mathbf{x})$, $a_i \in \mathcal{B}_n$, $1 \le i \le 4$. Then f is gsemibent when n is odd, and generalized 2-plateaued when n is even, if and only if the Boolean function $c_1a_1 \oplus c_2a_2 \oplus c_3a_3 \oplus a_4$ is semibent for all $c_i \in \mathbb{F}_2$, $i = 1, 2, 3$, such that for all $\mathbf{u} \in \mathbb{V}_n$ their Walsh-Hadamard transforms are either all zero, or they satisfy*

$$\begin{aligned}
&\mathcal{W}_{a_4}(\mathbf{u})\mathcal{W}_{a_2+a_4}(\mathbf{u}) = \mathcal{W}_{a_3+a_4}(\mathbf{u})\mathcal{W}_{a_2+a_3+a_4}(\mathbf{u}) \\
&= \mathcal{W}_{a_1+a_4}(\mathbf{u})\mathcal{W}_{a_1+a_2+a_4}(\mathbf{u}) = \mathcal{W}_{a_1+a_3+a_4}(\mathbf{u})\mathcal{W}_{a_1+a_2+a_3+a_4}(\mathbf{u}), \text{ and} \\
&\mathcal{W}_{a_4}(\mathbf{u})\mathcal{W}_{a_3+a_4}(\mathbf{u}) = \mathcal{W}_{a_1+a_4}(\mathbf{u})\mathcal{W}_{a_1+a_3+a_4}(\mathbf{u}).
\end{aligned}$$

In the light of Theorem 4, one may expect that with a similar approach one can also show relations between gbentness in $\mathcal{GB}_n^{16}$ and in $\mathcal{GB}_n^4$. We here only state the theorem. For the proof we refer to our eprint [4].

Theorem 6. *Let $f \in \mathcal{GB}_n^{16}$ with $f(\mathbf{x}) = a_1(\mathbf{x}) + 2a_2(\mathbf{x}) + 2^2a_3(\mathbf{x}) + 2^3a_4(\mathbf{x}) = b_1(\mathbf{x}) + 2^2b_2(\mathbf{x})$, where $b_1 = a_1 + 2a_2, b_2 = a_3 + 2a_4 \in \mathcal{GB}_n^4$. The function f is gbent in $\mathcal{GB}_n^{16}$ if and only if $b_2, b_1+b_2, 2b_1+b_2, 3b_1+b_2$ are gbent in $\mathcal{GB}_n^4$ with their generalized Walsh-Hadamard transforms satisfying the following conditions, (i) for n even, respectively, (ii) for n odd, for all $\mathbf{u} \in \mathbb{V}_n$:*

(i) $2^{-n/2}(\mathcal{H}_{3b_1+b_2}(\mathbf{u}), \mathcal{H}_{b_1+b_2}(\mathbf{u}), \mathcal{H}_{2b_1+b_2}(\mathbf{u}), \mathcal{H}_{b_2}(\mathbf{u}))$ belongs to one of $(\epsilon, \epsilon, \epsilon, \epsilon)$, $(\epsilon, \epsilon, -\epsilon, -\epsilon)$, $(\epsilon, -\epsilon, \epsilon i, -\epsilon i)$, $(\epsilon - \epsilon, -\epsilon i, \epsilon i)$, $(\epsilon i, \epsilon i, \epsilon i, \epsilon i)$, $(\epsilon i, \epsilon i, -\epsilon i, -\epsilon i)$, $(\epsilon i, -\epsilon i, \epsilon, -\epsilon)$, $(-\epsilon i, \epsilon i, -\epsilon, \epsilon)$, where $\epsilon \in \{\pm 1\}$.

(ii) $2^{-(n-1)/2}(\mathcal{H}_{3b_1+b_2}(\mathbf{u}), \mathcal{H}_{b_1+b_2}(\mathbf{u}), \mathcal{H}_{2b_1+b_2}(\mathbf{u}), \mathcal{H}_{b_2}(\mathbf{u}))$ belongs to one of $(\epsilon + \mu i, \epsilon + \mu i, \epsilon + \mu i, \epsilon + \mu i)$, $(\epsilon + \mu i, \epsilon + \mu i, -\epsilon - \mu i, -\epsilon - \mu i)$, $(\epsilon + \mu i, -\epsilon - \mu i, \epsilon - \mu i, -\epsilon + \mu i)$, $(\epsilon + \mu i, -\epsilon - \mu i, -\epsilon + \mu i, \epsilon - \mu i)$, for $\epsilon, \mu \in \{\pm 1\}$.

We close this section with some results on the Gray image $\psi(f)$ of gbent functions f in $\mathcal{GB}_n^8$ and $\mathcal{GB}_n^{16}$, extending the corresponding results in [9].

Lemma 2. *Let $n, k \geq 2$ be positive integers and $\psi : \mathbb{V}_{n+k-1} \to \mathbb{F}_2$ be defined by $\psi(\mathbf{x}, y_1, y_2, \ldots, y_{k-1}) = a_k(\mathbf{x}) \oplus \bigoplus_{i=1}^{k-1} y_i a_i(\mathbf{x})$, where $a_i \in \mathcal{B}_n$, $1 \leq i \leq k$. Denote by $\mathbf{a}(\mathbf{x})$ the vectorial Boolean function $\mathbf{a}(\mathbf{x}) = (a_1(\mathbf{x}), \ldots, a_{k-1}(\mathbf{x}))$ and let $\mathbf{u} \in \mathbb{V}_n$ and $\mathbf{v} = (v_1, \ldots, v_{k-1}) \in \mathbb{V}_{k-1}$. The Walsh-Hadamard transform of ψ at $(\mathbf{u}, \mathbf{v})$ is then*

$$\mathcal{W}_\psi(\mathbf{u}, v_1, \ldots, v_{k-1}) = \sum_{\alpha \in \mathbb{V}_{k-1}} (-1)^{\alpha \cdot \mathbf{v}} \mathcal{W}_{a_k \oplus \alpha \cdot \mathbf{a}}(\mathbf{u}).$$

Proof. We will show our claim by induction on k. For $k = 2$ we have

$$\begin{aligned}\mathcal{W}_\psi(\mathbf{u}, v_1) &= \sum_{\substack{\mathbf{x} \in \mathbb{V}_n \\ y_1 \in \mathbb{F}_2}} (-1)^{y_1 a_1(\mathbf{x}) \oplus a_2(\mathbf{x})} (-1)^{v_1 y_1 \oplus \mathbf{u} \cdot \mathbf{x}} \\ &= \sum_{\mathbf{x} \in \mathbb{V}_n} (-1)^{a_2(\mathbf{x})} (-1)^{\mathbf{u} \cdot \mathbf{x}} + \sum_{\mathbf{x} \in \mathbb{V}_n} (-1)^{a_1(\mathbf{x}) \oplus a_2(\mathbf{x})} (-1)^{v_1 \oplus \mathbf{u} \cdot \mathbf{x}} \\ &= \mathcal{W}_{a_2}(\mathbf{u}) + (-1)^{v_1} \mathcal{W}_{a_1 \oplus a_2}(\mathbf{u}).\end{aligned}$$

Now let

$$\psi(\mathbf{x}, y_1, \ldots, y_k) = \psi_1(\mathbf{x}, y_1, \ldots, y_{k-1}) \oplus y_k a_k(\mathbf{x}), \text{ where}$$
$$\psi_1(\mathbf{x}, y_1, \ldots, y_{k-1}) = a_{k+1}(\mathbf{x}) \oplus \bigoplus_{i=1}^{k-1} y_i a_i(\mathbf{x}).$$

Then $\mathcal{W}_\psi(\mathbf{u}, \mathbf{v}, v_k) = \mathcal{W}_{\psi_1}(\mathbf{u}, \mathbf{v}) + (-1)^{v_k} \mathcal{W}_{\psi_1 \oplus a_{k+1}}(\mathbf{u}, \mathbf{v})$, which implies our claim by the induction assumption.

Theorem 7. *We have:*

(i) Let $f(\mathbf{x}) = a_1(\mathbf{x}) + 2a_2(\mathbf{x}) + 2^2a_3(\mathbf{x}) \in \mathcal{GB}_n^8$ *be gbent. Then its Gray image* $\psi(f)$ *is semibent in* $\mathcal{B}_{n+2}$*.*

(ii) Let $f = a_1(\mathbf{x}) + 2a_2(\mathbf{x}) + 2^2a_3(\mathbf{x}) + 2^3a_4(\mathbf{x}) \in \mathcal{GB}_n^{16}$ *be gbent. Then* $\psi(f)$ *is semibent in* $\mathcal{B}_{n+3}$ *if* n *is odd, and 3-plateaued in* $\mathcal{B}_{n+3}$ *if* n *is even.*

Proof. (*i*) By Lemma 2, for $\psi(f)(\mathbf{x}, y_1, y_2) = a_1(\mathbf{x})y_1 + a_2(\mathbf{x})y_2 + a_3(\mathbf{x})$,

$$\begin{aligned}\mathcal{W}_{\psi(f)}(\mathbf{u}, v_1, v_2) =&\mathcal{W}_{a_3}(\mathbf{u}) + (-1)^{v_1}\mathcal{W}_{a_3\oplus a_1}(\mathbf{u})\\ &+ (-1)^{v_2}\mathcal{W}_{a_3\oplus a_2}(\mathbf{u}) + (-1)^{v_1+v_2}\mathcal{W}_{a_3\oplus a_2\oplus a_1}(\mathbf{u}).\end{aligned} \tag{3}$$

Assume first that n is even. Since f is gbent, by [9, Theorem 19], for the bent components we have $\mathcal{W}_{a_3}(\mathbf{u})\mathcal{W}_{a_1\oplus a_2\oplus a_3}(\mathbf{u}) = \mathcal{W}_{a_1\oplus a_3}(\mathbf{u})\mathcal{W}_{a_2\oplus a_3}(\mathbf{u})$, for all $\mathbf{u} \in \mathbb{V}_n$. Take $\mathcal{W}_{a_3}(\mathbf{u}) = \mu_1(\mathbf{u})2^{n/2}, \mathcal{W}_{a_3\oplus a_1}(\mathbf{u}) = \mu_2(\mathbf{u})2^{n/2}, \mathcal{W}_{a_3\oplus a_2}(\mathbf{u}) = \mu_3(\mathbf{u})2^{n/2}, \mathcal{W}_{a_3\oplus a_2\oplus a_1}(\mathbf{u}) = \mu_4(\mathbf{u})2^{n/2}$, for some $\mu_i \in \{-1, 1\}, 1 \le i \le 4$. Thus, $\mu_1(\mathbf{u})\mu_4(\mathbf{u}) = \mu_2(\mathbf{u})\mu_3(\mathbf{u})$. Using these in Eq. (3), we obtain

$$2^{-n/2}\mathcal{W}_{\psi(f)}(\mathbf{u}, v_1, v_2) = \mu_1(\mathbf{u}) + (-1)^{v_1}\mu_2(\mathbf{u}) + (-1)^{v_2}\mu_3(\mathbf{u}) + (-1)^{v_1\oplus v_2}\mu_4(\mathbf{u}).$$

For $(\mu_1(\mathbf{u}), \mu_2(\mathbf{u}), \mu_3(\mathbf{u}), \mu_4(\mathbf{u}))$ with values in the set

$$\begin{aligned}&(-1,-1,-1,-1), (1,1,-1,-1), (-1,-1,1,1), (-1,1,-1,1),\\ &(1,-1,-1,1), (-1,1,1,-1), (1,-1,1,-1), (1,1,1,1),\end{aligned}$$

$2^{-n/2}\mathcal{W}_{\psi(f)}(\mathbf{u}, v_1, v_2)$ takes one of the following values

$$\begin{aligned}&(-1)^{v_1\oplus v_2\oplus 1} + (-1)^{v_1\oplus 1} + (-1)^{v_2\oplus 1} - 1, (-1)^{v_1\oplus v_2} + (-1)^{v_1\oplus 1} + (-1)^{v_2} - 1,\\ &(-1)^{v_1\oplus v_2} + (-1)^{v_1} + (-1)^{v_2\oplus 1} - 1, (-1)^{v_1\oplus v_2\oplus 1} + (-1)^{v_1} + (-1)^{v_2} - 1,\\ &(-1)^{v_1\oplus v_2} + (-1)^{v_1\oplus 1} + (-1)^{v_2\oplus 1} + 1, (-1)^{v_1\oplus v_2\oplus 1} + (-1)^{v_1\oplus 1} + (-1)^{v_2} + 1,\\ &(-1)^{v_1\oplus v_2\oplus 1} + (-1)^{v_1} + (-1)^{v_2\oplus 1} + 1, (-1)^{v_1\oplus v_2} + (-1)^{v_1} + (-1)^{v_2} + 1.\end{aligned}$$

Therefore, $\mathcal{W}_{\psi(f)}$ attains the values $0, \pm 2^{(n+4)/2}$, thus $\psi(f)$ is semibent.

We now consider the case of odd n. Then, by [9, Theorem 19], $a_3, a_1 \oplus a_3, a_2 \oplus a_3, a_1 \oplus a_2 \oplus a_3$ are all semibent and, $\mathcal{W}_{a_3}(\mathbf{u}) = \mathcal{W}_{a_1\oplus a_3}(\mathbf{u}) = 0$ and $|\mathcal{W}_{a_1\oplus a_2\oplus a_3}(\mathbf{u})| = |\mathcal{W}_{a_2\oplus a_3}(\mathbf{u})| = 2^{(n+1)/2}$, or $|\mathcal{W}_{a_3}(\mathbf{u})| = |\mathcal{W}_{a_1\oplus a_3}(\mathbf{u})| = 2^{(n+1)/2}$ and $\mathcal{W}_{a_1\oplus a_2\oplus a_3}(\mathbf{u}) = \mathcal{W}_{a_2\oplus a_3}(\mathbf{u}) = 0$.

Case 1. Let $\mathcal{W}_{a_3}(\mathbf{u}) = \mathcal{W}_{a_1\oplus a_3}(\mathbf{u}) = 0$, $\mathcal{W}_{a_1\oplus a_2\oplus a_3}(\mathbf{u}) = \epsilon_1(\mathbf{u})2^{(n+1)/2}$, $\mathcal{W}_{a_2\oplus a_3}(\mathbf{u}) = \epsilon_2(\mathbf{u})2^{(n+1)/2}$, with $\epsilon_1, \epsilon_2 \in \{-1, 1\}$. With (3),

$$\mathcal{W}_{\psi(f)}(\mathbf{u}, v_1, v_2) = (-1)^{v_2}2^{(n+1)/2}\left(\epsilon_1(\mathbf{u}) + (-1)^{v_1}\epsilon_2(\mathbf{u})\right),$$

from which we infer that $\mathcal{W}_{\psi(f)}(\mathbf{u}, v_1, v_2) \in \{0, \pm 2^{(n+3)/2}\}$, for all combinations of $\epsilon_i(\mathbf{u})$ and v_i, $i = 1, 2$. Therefore $\psi(f)$ is semibent.

Case 2. Let $\mathcal{W}_{a_3}(\mathbf{u}) = \epsilon_1(\mathbf{u})2^{(n+1)/2}$, $\mathcal{W}_{a_1\oplus a_3}(\mathbf{u}) = \epsilon_2(\mathbf{u})2^{(n+1)/2}$, $\mathcal{W}_{a_1\oplus a_2\oplus a_3}(\mathbf{u}) = \mathcal{W}_{a_2\oplus a_3}(\mathbf{u}) = 0$, with $\epsilon_1, \epsilon_2 \in \{-1, 1\}$. As before, from (3) we obtain

$$\mathcal{W}_{\psi(f)}(\mathbf{u}, v_1, v_2) = 2^{(n+1)/2}\left(\epsilon_1(\mathbf{u}) + (-1)^{v_1}\epsilon_2(\mathbf{u})\right),$$

from which we infer that $\mathcal{W}_{\psi(f)}(\mathbf{u}, v_1, v_2) \in \{0, \pm 2^{(n+3)/2}\}$ and therefore $\psi(f)$ is semibent.

(*ii*) By Lemma 2, for $\psi(f)(\mathbf{x}, y_1, y_2, y_3) = a_4(\mathbf{x}) \bigoplus_{i=1}^{3} y_i a_i(\mathbf{x})$,

$$\begin{aligned}\mathcal{W}_{\psi(f)}(\mathbf{u}, v_1, v_2, v_3) &= \mathcal{W}_{a_4}(\mathbf{u}) + (-1)^{v_1}\mathcal{W}_{a_4\oplus a_1}(\mathbf{u})\\ &+ (-1)^{v_2}\mathcal{W}_{a_4\oplus a_2}(\mathbf{u}) + (-1)^{v_3}\mathcal{W}_{a_4\oplus a_3}(\mathbf{u})\\ &+ (-1)^{v_1\oplus v_2}\mathcal{W}_{a_4\oplus a_2\oplus a_1}(\mathbf{u}) + (-1)^{v_1\oplus v_3}\mathcal{W}_{a_4\oplus a_3\oplus a_1}(\mathbf{u})\\ &+ (-1)^{v_2\oplus v_3}\mathcal{W}_{a_4\oplus a_3\oplus a_2}(\mathbf{u}) + (-1)^{v_1\oplus v_2\oplus v_3}\mathcal{W}_{a_4\oplus a_3\oplus a_2\oplus a_1}(\mathbf{u}).\end{aligned}$$

By going through the 32 cases of Theorem 4 for the Walsh-Hadamard transforms in the expression above (16 for n even and 16 for n odd), we obtain that the Walsh-Hadamard spectrum is $\{0, \pm 2^{3+n/2}\}$ (for n even) and $\{0, \pm 2^{2+(n+1)/2}\}$ (for n odd), hence the claim.

Acknowledgements. Work by P.S. started during a very enjoyable visit at RICAM. Both the second and third named author thank the institution for the excellent working conditions. The second author is supported by the Austrian Science Fund (FWF) Project no. M 1767-N26.

A Appendix

Proof of Theorem 4: Let $\mathbf{u} \in \mathbb{V}_n$. Eq. (2) for $k = 4$ implies

$$\begin{aligned}16\sqrt{2}|\mathcal{H}_f^{(16)}(\mathbf{u})|^2 &= (2 + \sqrt{2+\sqrt{2}})4\sqrt{2}|\mathcal{H}_d^{(8)}(\mathbf{u})|^2 + (2 - \sqrt{2+\sqrt{2}})4\sqrt{2}|\mathcal{H}_{d+4a_1}^{(8)}(\mathbf{u})|^2\\ &+ 8\sqrt{4-2\sqrt{2}}\,\Im\left(\mathcal{H}_d^{(8)}(\mathbf{u})\mathcal{H}_{d+4a_1}^{(8)}(\mathbf{u})\right).\end{aligned} \tag{4}$$

We denote by A, C, D, W the Walsh-Hadamard transforms $\mathcal{W}_{a_4}(\mathbf{u})$, $\mathcal{W}_{a_2\oplus a_4}(\mathbf{u})$, $\mathcal{W}_{a_3\oplus a_4}(\mathbf{u})$, $\mathcal{W}_{a_2\oplus a_3\oplus a_4}(\mathbf{u})$ (in that order). We denote by B, X, Y, Z the Walsh-Hadamard transforms $\mathcal{W}_{a_1\oplus a_4}(\mathbf{u})$, $\mathcal{W}_{a_1\oplus a_2\oplus a_4}(\mathbf{u})$, $\mathcal{W}_{a_1\oplus a_3\oplus a_4}(\mathbf{u})$, $\mathcal{W}_{a_1\oplus a_2\oplus a_3\oplus a_4}(\mathbf{u})$ (in that order). By [9, Lemma 17], we know that the generalized Walsh-Hadamard transform of any function in $\mathcal{GB}_n^8$, say d and $d + 4a_1$ with $d = a_2 + 2a_3 + 2^2 a_4$, is of the form

$$4\mathcal{H}_d^{(8)}(\mathbf{u}) = \alpha_0 A + \alpha_1 C + \alpha_2 D + \alpha_3 W, \quad 4\mathcal{H}_{d+4a_1}^{(8)}(\mathbf{u}) = \alpha_0 B + \alpha_1 X + \alpha_2 Y + \alpha_3 Z,$$

where $\alpha_0 = 1 + (1+\sqrt{2})i$, $\alpha_1 = 1 + (1-\sqrt{2})i$, $\alpha_2 = 1 + \sqrt{2} - i$, $\alpha_3 = 1 - \sqrt{2} - i$, and moreover that (see also [9, Corollary 18]),

$$\begin{aligned}4\sqrt{2}|\mathcal{H}_d^{(8)}(\mathbf{u})|^2 &= A^2 - C^2 + 2CD + D^2 - 2AW - W^2 + \sqrt{2}(A^2 + C^2 + D^2 + W^2)\\ 4\sqrt{2}|\mathcal{H}_{d+4a_1}^{(8)}(\mathbf{u})|^2 &= B^2 - X^2 + 2XY + Y^2 - 2BZ - Z^2 + \sqrt{2}(B^2 + X^2 + Y^2 + Z^2).\end{aligned} \tag{5}$$

Furthermore, with straightforward computations we get

$$\begin{aligned}&8\sqrt{4-2\sqrt{2}}\,\Im\left(\mathcal{H}_b^{(8)}(\mathbf{u})\mathcal{H}_{b+4a_1}^{(8)}(\mathbf{u})\right)\\&=2\sqrt{2-\sqrt{2}}\,(BC+BD-AX-WX-AY+WY+CZ-DZ)\\&\quad+2\sqrt{4-2\sqrt{2}}\,(BD+WX-AY-CZ).\end{aligned}\tag{6}$$

Using (5) and (6) in Eq. (4) we obtain

$$\begin{aligned}&16\sqrt{2}|\mathcal{H}_f^{(16)}(\mathbf{u})|^2\\&=2(A^2+B^2-C^2+2CD+D^2-2AW-W^2-X^2+2XY+Y^2-2BZ-Z^2)\\&\quad+2\sqrt{2}(A^2+B^2+C^2+D^2+W^2+X^2+Y^2+Z^2)\\&\quad+\sqrt{2-\sqrt{2}}\,(A^2-B^2+2BC+C^2+D^2+W^2-2AX-4WX-X^2\\&\qquad+2WY-Y^2+4CZ-2DZ-Z^2)\\&\quad+2\sqrt{2+\sqrt{2}}\,(A^2-B^2+BD+CD+D^2-AW+WX-AY-XY\\&\qquad-Y^2+BZ-CZ).\end{aligned}\tag{7}$$

If f is gbent in $\mathcal{GB}_n^{16}$, i.e., $|\mathcal{H}_f^{(16)}(\mathbf{u})|^2=2^n$, by the linear independence of $\{1,\sqrt{2},\sqrt{2-\sqrt{2}},\sqrt{2+\sqrt{2}}\}$ (as easily shown, the set forms a basis of $\mathbb{Q}(\sqrt{2},\sqrt{2-\sqrt{2}})$), we arrive at the following system of equations with solutions in $\mathbb{Z}$,

$$\begin{aligned}&A^2+B^2+C^2+D^2+W^2+X^2+Y^2+Z^2=2^{n+3}\\&A^2+B^2-C^2+2CD+D^2-2AW-W^2-X^2+2XY+Y^2-2BZ-Z^2=0\\&A^2-B^2+2BC+C^2+D^2+W^2-2AX-4WX-X^2\\&\qquad\qquad+2WY-Y^2+4CZ-2DZ-Z^2=0\\&A^2-B^2+BD+CD+D^2-AW+WX-AY-XY-Y^2+BZ-CZ=0.\end{aligned}\tag{8}$$

Let 2^t be the largest power of 2 which divides all, A,B,C,D,X,Y,Z and W, i.e., $A=2^tA_1$, etc., with at least one of the $A_1,B_1,\ldots$ being odd. First, if n is even and $t>\frac{n}{2}$, then $t=\frac{n}{2}+1$ only. Dividing by 2^{2t}, the first equation of (8) becomes $A_1^2+B_1^2+C_1^2+D_1^2+W_1^2+X_1^2+Y_1^2+Z_1^2=2$, which gives the solution $(\pm1,\pm1,0,0,0,0,0,0)$ (and permutations of these values). However, a simple computation reveals that none of these possibilities also satisfies the last three equations of (8). If n is odd and $t>\frac{n+1}{2}$, then $t=\frac{n+3}{2}$, but this implies that only one value out of $A,B,\ldots$ is nonzero. Again, that is impossible to satisfy the last three equations of (8). Assume now that $t<\frac{n}{2}$. The first equation of (8) becomes $A_1^2+B_1^2+C_1^2+D_1^2+W_1^2+X_1^2+Y_1^2+Z_1^2=2^{n+3-2t}$, which is divisible by 2^5 (when n is even, since $t\le\frac{n-2}{2}$), respectively 2^4 (when n is odd, since $t\le\frac{n-1}{2}$). If n is even, this can only happen if $A_1,B_1,\ldots,$ are all even, that is, $\equiv 0,2,4,6 \pmod 8$, but that contradicts our assumption that t is the largest power of 2 dividing $A,B,\ldots$. If n is odd and $t\le\frac{n-3}{2}$, the previous argument would work, and if $t=\frac{n-1}{2}$, then $A_1^2+B_1^2+C_1^2+D_1^2+W_1^2+X_1^2+Y_1^2+Z_1^2=16$. By considering every residues for $A_1, B_1, \ldots$, modulo 4 and imposing the condition that the 2nd, 3rd, 4th equations of our system also must be 0 modulo 16, we only get possibilities $(0,0,2,2,0,0,2,2)$, $(0,2,0,2,0,2,0,2)$, $(0,2,2,0,0,2,2,0)$, $(2,0,0,2,2,0,0,2)$, $(2,0,2,0,2,0,2,0)$, $(2,2,0,0,2,2,0,0)$ for $(A_1, B_1, \ldots)$ modulo 4, but that implies that all $A_1, B_1, \ldots$ are even, contradicting our assumption on t. This shows that the only possibility is $2^t = 2^{n/2}$ if n is even, and $2^t = 2^{(n+1)/2}$ if n is odd.

Thus, one needs to find integer solutions for the equation $A_1^2+B_1^2+C_1^2+D_1^2+\mathcal{W}_1^2+X_1^2+Y_1^2+Z_1^2 = 8$, for n even, or $A_1^2+B_1^2+C_1^2+D_1^2+\mathcal{W}_1^2+X_1^2+Y_1^2+Z_1^2 = 4$ for n odd, which also satisfy the last three equations in (8). Mathematica renders the following: if n is even, then $2^{-\frac{n}{2}}(A, C, D, W, B, X, Y, Z)$ (note the order) is one of

$$\begin{aligned}
&\pm(-1,-1,-1,-1,-1,-1,-1,-1), && \pm(-1,1,-1,1,-1,1,-1,1),\\
&\pm(-1,-1,-1,-1,1,1,1,1), && \pm(-1,1,-1,1,1,-1,1,-1),\\
&\pm(1,-1,-1,1,-1,1,1,-1), && \pm(1,1,-1,-1,-1,-1,1,1),\\
&\pm(1,-1,-1,1,1,-1,-1,1), && \pm(1,1,-1,-1,1,1,-1,-1),
\end{aligned} \tag{9}$$

and, if n is odd, then $2^{-\frac{n+1}{2}}(A, C, D, W, B, X, Y, Z)$ is one of

$$\begin{aligned}
&\pm(-1,1,0,0,-1,1,0,0), && \pm(-1,1,0,0,1,-1,0,0),\\
&\pm(0,0,-1,1,0,0,-1,1), && \pm(0,0,-1,1,0,0,1,-1),\\
&\pm(0,0,1,1,0,0,-1,-1), && \pm(0,0,1,1,0,0,1,1),\\
&\pm(1,1,0,0,-1,-1,0,0), && \pm(1,1,0,0,1,1,0,0).
\end{aligned}$$

We see that in both cases, if f is gbent, then the conditions of the theorem are satisfied. The converse follows with straightforward calculations. □

Proof of Theorem 5*:* Assume that f is gsemibent in $\mathcal{GB}_n^{16}$ when n is odd, respectively generalized 2-plateaued when n is even. Then $|\mathcal{H}_f^{(16)}(\mathbf{u})| \in \{0, \pm 2^{(n+1)/2}\}$ for n odd, respectively, $|\mathcal{H}_f^{(16)}(\mathbf{u})| \in \{0, \pm 2^{(n+2)/2}\}$ for n even. Using the notations of Theorem 4, from Eq. (7), we immediately get $A = B = C = D = X = Y = W = Z = 0$ if $\mathcal{H}_f^{(16)}(\mathbf{u}) = 0$. If $|\mathcal{H}_f^{(16)}(\mathbf{u})| = 2^{(n+1)/2}$ (for n odd), respectively, $|\mathcal{H}_f^{(16)}(\mathbf{u})| = 2^{(n+2)/2}$ (for n even), then (7) again yields the system of Eq. (8) with the one difference that in the first equation the power of 2 on the right side is 2^{n+4}, respectively, 2^{n+5}. With the same argument as in the proof of Theorem 4 we see that for such $\mathbf{u}$, $2^{-\frac{n+1}{2}}(A, C, D, W, B, X, Y, Z)$ (for n odd), respectively, $2^{-\frac{n+2}{2}}(A, C, D, W, B, X, Y, Z)$ (for n even) can only take the values from Eq. (9).

Straightforwardly, one confirms that the converse is also true, and the theorem is shown. □

References

1. Carlet, C.: $\mathbb{Z}_{2^k}$-linear codes. IEEE Trans. Inf. Theory **44**(4), 1543–1547 (1998)
2. Hodžić, S., Pasalic, E.: Generalized bent functions-some general construction methods and related necessary and sufficient conditions. Crypt. Commun. **7**, 469–483 (2015)
3. Liu, H., Feng, K., Feng, R.: Nonexistence of generalized bent functions from $\mathbb{Z}_2^n$ to $\mathbb{Z}_m$. Des. Codes Crypt. **82**, 647–662 (2017)
4. Martinsen, T., Meidl, W., Stănică, P.: Generalized bent functions and their Gray images. http://arxiv.org/pdf/1511.01438
5. Martinsen, T., Meidl, W., Stănică, P.: Partial spread and vectorial generalized bent functions. Des.Codes Crypt. (2017, to appear)
6. Kumar, P.V., Scholtz, R.A., Welch, L.R.: Generalized bent functions and their properties. J. Comb. Theory - Ser. A **40**, 90–107 (1985)
7. Schmidt, K.U.: Quaternary constant-amplitude codes for multicode CDMA. IEEE Trans. Inform. Theory **55**(4), 1824–1832 (2009)
8. Solé, P., Tokareva, N.: Connections between quaternary and binary bent functions. Prikl. Diskr. Mat. **1**, 16–18 (2009). http://eprint.iacr.org/2009/544.pdf
9. Stănică, P., Martinsen, T., Gangopadhyay, S., Singh, B.K.: Bent and generalized bent Boolean functions. Des. Codes Crypt. **69**, 77–94 (2013)
10. Tan, Y., Pott, A., Feng, T.: Strongly regular graphs associated with ternary bent functions. J. Comb. Theory - Ser. A **117**, 668–682 (2010)

Cryptography

Enhanced Digital Signature Using RNS Digit Exponent Representation

Thomas Plantard[1] and Jean-Marc Robert[2,3(✉)]

[1] CCISR, SCIT, University of Wollongong, Wollongong, Australia
[2] Team DALI, Université de Perpignan Via Domitia, Perpignan, France
jean-marc.robert@univ-perp.fr
[3] LIRMM, UMR 5506, Université Montpellier and CNRS, Montpellier, France

Abstract. Digital Signature Algorithm (DSA) involves modular exponentiation, of a public and known base by a random one-time exponent. In order to speed-up this operation, well-known methods take advantage of the memorization of base powers. However, due to the cost of the memory, to its small size and to the latency of access, previous research sought for minimization of the storage. In this paper, taking into account the modern processor features and the growing size of the cache memory, we improve the storage/efficiency trade-off, by using a RNS Digit exponent representation. We then propose algorithms for modular exponentiation. The storage is lower for equivalent complexities for modular exponentiation computation. The implementation performances show significant memory saving, up to 3 times for the largest NIST standardized key sizes compared to state of the art approaches.

Keywords: RNS · Digital signature · Modular exponentiation · Memory storage · Efficient software implementation

1 Introduction

In the DSS (Digital Signature Standard), DSA (Digital Signature Algorithm) is a popular authentication protocol. According to the NIST standard (see [3]), the public parameters are p, q and g. The parameter g is a generator of the multiplicative group $\mathbb{Z}/p\mathbb{Z}$ of order q, which is a prime of size corresponding to the required security level. Therefore, p is a prime chosen such that q divides $p-1$. The recommended security levels in the standard are 80–256 bits, corresponding to 160–512 bit sizes for the prime q. When a server needs to sign a batch of documents or authentications, the main operations are modular exponentiations $g^k \bmod p$ (one per signature), where k is a one time random parameter. Taking advantage of the fixed public parameter g is a natural way to speed-up the signature protocol, by storing well chosen powers of g. The main known methods of the state of the art are the one presented by Gordon in [6], which stores the $g^{R^i} \bmod p$ values, and also the *Fixed-base Comb*, which is presented by Lim and Lee in [10]. While improving the complexity, and therefore, lowering the

S. Duquesne and S. Petkova-Nikova (Eds.): WAIFI 2016, LNCS 10064, pp. 177–192, 2016.
DOI: 10.1007/978-3-319-55227-9_13

computation time, these methods require some storage. The trade-off between the efficiency and the storage amount is the comparison criteria between these different approaches.

All the protocols derived from the DSA can use these different approaches, since they all need an exponentiation of a known base by a random exponent. Blind signature and E-voting are examples of protocols using fixed-base modular exponentiation [11]. Moreover, El-Gamal encryption and signature also use a public generator g to the power of a randomly chosen exponent (see [4]). However, the decryption uses the result of this operation, and the idea does not apply in this case.

On the arithmetic side, the Residue Number System (RNS), based on the Chinese Remainder Theorem, is a classical way to speed-up and/or parallelize arithmetic computations and was first presented by Svoboda in [14] and by Garner in [5]. One can find a complete classical presentation of the RNS in Knuth [9].

Contributions: In this paper, we propose to use the memorization of base powers with numeration scales in radix $R = m_0 \cdot m_1$ and the RNS representation of each digit using the base $\mathcal{B} = \{m_0, m_1\}$. We study the recoding algorithm and apply it to the exponent in modular exponentiation. We propose a modular exponentiation algorithm using this recoding of the exponent and memorization. We called this algorithm the m_0m_1 exponentiation method. We studied the corresponding complexities and storage amounts, and compared the results with the *Fixed-base Comb* and Radix-R methods. We showed that our m_0m_1 exponentiation method has better storage/complexity trade-off that the aforementioned methods, for the NIST recommended field sizes and a large range of storage amount. We then made software implementations of our algorithms and performed tests in order to validate the storage/timing trade-off. The speed-up comparison shows the benefits. This approach provides also a fair flexibility in terms of required storage amount: one can choose the storage amount according to the device resources available and compatible to the global computation load of the system.

This paper is organized as follows: Sect. 2, we review the main classical fixed-base exponentiation algorithms, taking advantage of storage and give their complexities and storage requirements; Sect. 3, we present our approach for the m_0m_1 recoding; Sect. 4, we then show the application on modular exponentiation; Sect. 5 shows the implementation strategies and results in terms of performances we got in this work; finally Sect. 6, we give some concluding remarks and perspectives.

2 State of the Art Review

In this section, we review the state of the art classical approaches for fixed-base modular exponentiation.

When a server needs to sign a document or a message, the computation consists of several operations, the main one being a modular exponentiation $g^k \bmod p$, with k being a one-time random exponent. This computation can use the classical Square-and-Multiply algorithm (see Algorithm 1). In terms of complexity, given the exponent length t (that is, the size of the prime q), the number of modular squaring is $t-1$ and the number of modular multiplications to be computed is $t/2$ on average, half of this length, for a randomly chosen exponent. There is no storage in this case.

The method presented by Gordon in [6] first suggests to store the t successive squarings of g (that is the sequence of g^{2^i}). In terms of complexity, given the exponent length t (again, the size of the prime q), one has now no squarings and the number of modular multiplications to be computed is half of this length on average as in the previous case, for a randomly chosen exponent. The storage amount is t values in $\mathbb{Z}/p\mathbb{Z}$ as mentioned above. Gordon in [6] mentions the generalization of this idea into a radix R method, which consists of the memorization of the values $g^{i \cdot R^j}$, with $i \in [1, ..., R-1]$ and $0 \leq j < \ell$ where ℓ is the radix R length of the exponent, which we denote by $w = \log_2(R)$ ($\ell = \lceil t/w \rceil$). In this case, the complexity is $\ell - 1$ modular multiplications, for a storage amount of $\ell \cdot (R-1)$ values in $\mathbb{Z}/p\mathbb{Z}$. In the sequel, we will call this approach the Radix-R Exponentiation Method (see Algorithm 2).

Algorithm 1. Left-to-Right Square-and-Multiply Modular Exponentiation

Require: $k = (k_{t-1}, \ldots, k_0)$, the DSA modulus p, g a generator of $\mathbb{Z}/p\mathbb{Z}$ of order q.
Ensure: $X = g^k \bmod p$

$X \leftarrow 1$
for i from $t-1$ downto 0 **do**
  $X \leftarrow X^2 \bmod p$
  if $k_i = 1$ **then**
    $X \leftarrow X \cdot g \bmod p$
  end if
end for
return (X)

Algorithm 2. Radix-R Exponentiation Method

Require: $k = (k_{\ell-1}, \ldots, k_0)_R$, the DSA modulus p, g a generator of $\mathbb{Z}/p\mathbb{Z}$ of order q.
Ensure: $X = g^k \bmod p$

Precomputation. Store $G_{i,j} \leftarrow g^{i \cdot R^j}$, with $i \in [1, ..., R-1]$ and $0 \leq j < \ell$.
$X \leftarrow 1$
for i from $\ell - 1$ downto 0 **do**
  $X \leftarrow X \cdot G_{k_i,i} \bmod p$
end for
return (X)

Algorithm 3. *Fixed-base Comb* Exponentiation Method

Require: $k = (k_{t-1}, \ldots, k_1, k_0)_2$, the DSA modulus p, g a generator of $\mathbb{Z}/p\mathbb{Z}$ of order q, window width w, $d = \lceil t/w \rceil$.
Ensure: $X = g^k \bmod p$

Precomputation. Compute and store $g^{[a_{w-1}, \ldots, a_0]} \bmod p, \quad \forall (a_{w-1}, \ldots, a_0) \in \mathbb{Z}_2^w$.
By padding k on the left by 0's if necessary, write $k = K^{w-1} \| \ldots \| K^1 \| K^0$, where each K^j is a bit string of length d. Let K_i^j denote the ith bit of K^j.
$X \leftarrow 1$
for i from $d-1$ downto 0 **do**
  $X \leftarrow X^2 \bmod p$
  $X \leftarrow X \cdot g^{[K_i^{w-1}, \ldots, K_i^1, K_i^0]} \bmod p$
end for
return (X)

Another classical method is the so called *Fixed-base Comb* method. In [8], Hankerson *et al.* present this method proposed by Lim and Lee in [10]. The window width w is the number of comb-teeth, and $d = \lceil t/w \rceil$ is the distance in bits between two teeth. This method is shown in Algorithm 3, in which we denote $[a_{w-1}, \cdots, a_1, a_0] = a_{w-1} 2^{(w-1)d} + \ldots + a_2 2^{2d} + a_1 2^d + a_0$. The complexity of this approach is $d-1$ modular squarings and d multiplications, for a storage amount of $2^w - 1$ values in $\mathbb{Z}/p\mathbb{Z}$. One drawback of this method is the lack of flexibility for the storage amount, which increases exponentially with respect to the window width w.

Table 1, we give the complexities and the storage amounts of all these approaches.

Table 1. Complexities and storage amounts of state of the art methods, average case, binary exponent length t. #MM denotes the number of modular multiplications, #MS the number of modular squarings.

	#MM	#MS	Storage (# values $\in \mathbb{Z}/p\mathbb{Z}$)
Square-and-multiply, Algorithm 1	$t/2$	$t-1$	-
Radix-R method, Algorithm 2	$\lceil t/w \rceil$	-	$\lceil t/w \rceil \cdot (R-1)$
Fixed-base Comb, Algorithm 3	$d = \lceil t/w \rceil$	$d-1$	$2^w - 1$

3 $m_0 m_1$ Recoding

In this section, we present our approach for the $m_0 m_1$ recoding. Our goal is to use this representation in a modular exponentiation computation. The RNS digit representation with two moduli splits the digits in two parts. The first part will be used to select the precomputed values and the second part for final computation of the modular exponentiation, with the best possible trade-off.

3.1 Algorithm

We first remind the RNS representation with RNS base $\mathcal{B} = \{m_0, m_1\}$ of two moduli. Let $R = m_0 \cdot m_1$ and $x \in \mathbb{Z}$ such that $0 \leq x < R$. Let us also assume m_0 is prime, since this allows us to invert all integers $<m_0$ modulo m_0, and we choose $m_1 < m_0$. In the sequel, we denote $|x|_m = x \bmod m$.

One represents x with the residues

$$\begin{cases} x^{(0)} = |x|_{m_0} \\ x^{(1)} = |x|_{m_1} \end{cases}$$

and x can be retrieved using the Chinese Remainder Theorem as follows:

$$x = \left| x^{(0)} \cdot m_1 \cdot |m_1^{-1}|_{m_0} + x^{(1)} \cdot m_0 \cdot |m_0^{-1}|_{m_1} \right|_R .$$

We now present our recoding approach. Our idea here is to use an exponent k recoding in radix $R = m_0 \cdot m_1$. We represent every radix-R digits in RNS with RNS base $\mathcal{B} = \{m_0, m_1\}$. Let k_i be the digits of k in radix-R, and let us denote $(k_i^{(0)}, k_i^{(1)})$ their RNS representations in base $\mathcal{B}$. Thus, one has:

$$k = \sum_{i=0}^{\ell-1} k_i R^i, \text{ with } \ell = \lceil t/\log_2(R) \rceil,$$
$$\text{and } \begin{cases} k_i^{(0)} = |k_i|_{m_0}, \\ k_i^{(1)} = |k_i|_{m_1}. \end{cases}$$

Let us denote (when $k_i^{(1)} \neq 0$)

$$\begin{aligned} m'_0 &= m_1 \cdot |m_1^{-1}|_{m_0}, \\ m'_1 &= m_0 \cdot |m_0^{-1}|_{m_1}, \\ k'_i &= |k_i^{(0)} \cdot (k_i^{(1)})^{-1}|_{m_0}. \end{aligned}$$

One keeps $\kappa_i \leftarrow (k'_i, k_i^{(1)})$ as a representation of k_i and this leads to $k_i = \left| k_i^{(1)} |k'_i \cdot m'_0 + m'_1|_R \right|_R$. We handle the modular reduction mod R as follows:

$$k_i = k_i^{(1)} |k'_i \cdot m'_0 + m'_1|_R - \lfloor k_i^{(1)} \cdot |k'_i \cdot m'_0 + m'_1|_R / R \rfloor \cdot R.$$

Let us denote $C = \lfloor k_i^{(1)} \cdot (k'_i \cdot m'_0 + m'_1)/R \rfloor$. By noticing that $0 \leq C < m_1$, we now consider C as a carry that one can subtract to k_{i+1}. We then compute

$$\text{if } k_{i+1} \geq C \text{ then } k_{i+1} \leftarrow k_{i+1} - C, C \leftarrow 0, \text{else } k_{i+1} \leftarrow k_{i+1} + R - C, C \leftarrow 1,$$

and one gets $k_{i+1} \geq 0$.

When $k_i^{(1)} = 0$, we handle this by slightly rewriting κ_i as follows: $\kappa_i = (|k_i^{(0)} + 1|_{m_0} \cdot m'_0 - m'_0)$, thus keeping $\kappa_i \leftarrow (|k_i^{(0)} + 1|_{m_0}, 0)$ as a representation of k_i in this case. In addition, one notices that the carry C is not modified here (it is either 0 or 1 and has been previously settled).

The sequence of the $\kappa_i \leftarrow (k'_i, k_i^{(1)})$ is the $m_0 m_1$ recoding of k we can use to compute a modular exponentiation.

One notices it might be necessary to process a last carry C, with a final correction. The recoding algorithm is shown in Algorithm 4.

Algorithm 4. m_0m_1 Recoding

Require: $\{m_0, m_1\}$ RNS base with $R = m_0 \cdot m_1$, $k = \sum_{i=0}^{\ell-1} k_i R^i$.
Ensure: $\{\kappa_i, 0 \le i < \ell, (C)\}$, m_0m_1 recoding of scalar k.

1: $C \leftarrow 0$
2: **for** i from 0 to $\ell - 1$ **do**
3: $\quad k_i \leftarrow k_i - C, C \leftarrow 0$
4: $\quad$ **if** $k_i < 0$ **then**
5: $\quad\quad k_i \leftarrow k_i + R$, $C \leftarrow 1$
6: $\quad$ **end if**
7: $\quad k_i^{(0)} = |k_i|_{m_0}, k_i^{(1)} = |k_i|_{m_1}$.
8: $\quad$ **if** $k_i^{(1)} = 0$ **then**
9: $\quad\quad \kappa_i \leftarrow (|k_i^{(0)} + 1|_{m_0}, 0)$
10: $\quad$ **else**
11: $\quad\quad k_i' \leftarrow |k_i^{(0)} \cdot (k_i^{(1)})^{-1}|_{m_0}$
12: $\quad\quad C \leftarrow C + \lfloor k_i^{(1)} \cdot |k_i' \cdot m_0' + m_1'|_R / R \rfloor$
13: $\quad\quad \kappa_i \leftarrow (k_i', k_i^{(1)})$
14: $\quad$ **end if**
15: **end for**
16: **return** $\{\kappa_i, 0 \le i < \ell, (-C)\}$

3.2 Example

We present here an example of m_0m_1 recoding with an exponent size t of 20 bits ($0 < k < 2^{20}$), and $\mathcal{B} = \{11, 8\}$ (i.e. $m_0 = 11, m_1 = 8$). Thus, in this case, one has the radix $R = m_0 \cdot m_1 = 88$, $\ell = \lceil 20/\log_2(88) \rceil = 4$, and therefore

$$m_0' = 8 \cdot |8^{-1}|_{11} = 56,$$
$$m_1' = 11 \cdot |11^{-1}|_8 = 33.$$

Let us take $k = 936192_{10}$, the random exponent. By rewriting k in radix-R, one has

$$k = 48 + 78 \cdot 88 + 32 \cdot 88^2 + 1 \cdot 88^3.$$

We now use Algorithm 4, which consists of a `for` loop, steps 2 to 15.

- In the first iteration ($i = 0$), one has $k_0 = 48$.
  - One has $C \leftarrow 0$ and one skips the `if`-test steps 4 to 6 since $k_0 \ge 0$.
  - Step 7, one computes the RNS representation in base $\mathcal{B}$ of $k_0 = 48$:

$$k_0^{(0)} = |k_0|_{11} = 4, k_0^{(1)} = |k_0|_8 = 0.$$

  - Steps 6 and 7, since $k_0^{(1)} = 0$, one sets

$$\kappa_0 \leftarrow (|k_0^{(0)} + 1|_{11}, 0) = (5, 0)$$

- In the second iteration ($i = 1$), one has $k_1 = 78$.
  - One has $C \leftarrow 0$ and one skips the `if`-test steps 4 to 6 since $k_1 \ge 0$.

- Step 7, one computes the RNS representation in base $\mathcal{B}$ of $k_1 = 78$:

$$k_1^{(0)} = |k_1|_{11} = 1, k_1^{(1)} = |k_1|_8 = 6.$$

- Steps 11 and 12, since $k_1^{(1)} \neq 0$, one has

$$\begin{array}{ll} |(k_1^{(1)})^{-1}|_{11} & \leftarrow 2 \\ k_1' = |k_1^{(0)} \cdot (k_1^{(1)})^{-1}|_{11} & \leftarrow 2 \\ C \leftarrow \lfloor (k_1^{(1)} \cdot |k_1' \cdot 56 + 33|_{88})/88 \rfloor & \leftarrow 3 \end{array}$$

and finally

$$\kappa_1 \leftarrow (2, 6)$$

- In the third iteration ($i = 2$), one has now $k_2 \leftarrow k_2 - C = 29$.
  - The RNS representation in base $\mathcal{B}$ of k_2 is $k_2^{(0)} = 7, k_2^{(1)} = 5$.
  - The computation steps 11–12 gives $C \leftarrow 2$, and

$$\kappa_2 \leftarrow (8, 5).$$

Without providing all the details, one finally gives the values returned by the algorithm:

$$\kappa = ((5, 0), (2, 6), (8, 5), (3, 7)), \text{ and } C = -2.$$

4 $m_0 m_1$ Modular Exponentiation

4.1 Algorithm

We now present the use of our recoding in the modular exponentiation. One wants to compute

$$\begin{aligned} g^k \bmod p &= g^{\sum_{i=0}^{\ell-1} k_i \cdot R^i} \bmod p \\ &= g^{\sum_{i=0}^{\ell-1} \kappa_i \cdot R^i} \cdot g^{C \cdot R^\ell} \bmod p \\ &= g^{C \cdot R^\ell} \cdot \prod_{i=0}^{\ell-1} g^{\kappa_i \cdot R^i} \bmod p \end{aligned}$$

with

$$g^{\kappa_i \cdot R^i} \bmod p = g^{\kappa_1^{(1)} \cdot R^i \cdot |\kappa_i' \cdot m_0' + m_1'|_R} \bmod p, \text{ when } \kappa_1^{(1)} \neq 0$$

and

$$g^{\kappa_i \cdot R^i} \bmod p = g^{R^i \cdot |\kappa_i' \cdot m_0' + m_1'|_R} \cdot g^{-R^i \cdot |m_0' + m_1'|_R} \bmod p, \text{ when } \kappa_1^{(1)} = 0.$$

In order to compute the fixed-base modular exponentiation $g^k \bmod p$, with p prime, one stores the following values:

$$G_{i,j} = g^{R^i \cdot |j \cdot m_0' + m_1'|_R} \bmod p, \text{ with } 0 \leq i \leq \ell - 1, 0 \leq j < m_0$$

$$\text{and } G_{\ell,1} = g^{R^\ell \cdot |m_0' + m_1'|_R} \bmod p.$$

The field inversion is very costly over $\mathbb{Z}/p\mathbb{Z}$, therefore, one also stores the following inverses:

$$G_{i,-1} = g^{-|m'_0+m'_1|_R \cdot R^i} \bmod p \text{ avec } 0 \leq i \leq \ell$$

One uses one value K_j per possible values of $1 \leq \kappa_i^{(1)} < m_1$, that is m_1 points. Thus, one now has

$$K_j = \left(\prod_{\text{for all } \kappa_i^{(1)}=j} G_{i,\kappa_i^{(0)}} \right) \times \left(G_{\ell,sign(C)1}\right)_{|C|=j} \bmod p$$

and

$$K_0 = \prod_{\text{for all } \kappa_i^{(1)}=0} G_{i,\kappa_i^{(0)}} \times G_{i,-1} \bmod p.$$

This leads to

$$g^k \quad \bmod p = K_0 \times \prod_{j=1}^{m_1} K_j^j.$$

Every single individual modular exponentiation K_j^j is performed with a square-and-multiply approach, which is more efficient than performing $j-1$ modular multiplications, even for small m_1.

One may notice that the amount of storage is now $(m_0+1) \times \ell + 1$ points. This approach is depicted in Algorithm 5.

4.2 Example

We now go back to our previous example in Sect. 3.2 page 6. One considers again the same values and parameters:

- an exponent size t of 20 bits ($0 < k < 2^{20}$), and $\mathcal{B} = \{11, 8\}$ (i.e. $m_0 = 11, m_1 = 8$);
- radix $R = m_0 \cdot m_1 = 88$ ($\ell = 4$);
- one has $k = 936192_{10}$;
- and we use the $m_0 m_1$ recoding previously computed:

$$\kappa = ((5,0),(2,6),(8,5),(3,7)), \text{ and } C = -2.$$

We present the computation of $g^k \bmod p$ using Algorithm 5. In terms of storage, one computes the values

$$G_{i,j} = g^{R^i \cdot |j \cdot m'_0 + m'_1|_R} \quad \bmod p \text{ with } 0 \leq i \leq \ell - 1.$$

One has the following values of $|j \cdot m'_0 + m'_1|_R$ for $0 \leq j < 11$:

$$\{33, 1, 57, 25, 81, 49, 17, 73, 41, 9, 65\}$$

Algorithm 5. Fixed-base m_0m_1 method modular exponentiation

Require: $\{m_0, m_1\}$ RNS base with $R = m_0m_1, k = \sum_{i=0}^{\ell-1} k_iR^i$ and $\kappa = \{\kappa_i, 0 \leq i < \ell, (C)\}$ the m_0m_1 recoding of k, p, the DSA modulus, $g \in \mathbb{Z}/p\mathbb{Z}$, public generator of order q.
Ensure: $X = g^k \mod p$
1: *Precomputation.* Store $G_{i,j} \leftarrow g^{R^i \cdot |j \cdot m'_0 + m'_1|_R}$ with $0 \leq i < \ell - 1, 0 \leq j < m_0, G_{\ell,1} \leftarrow g^{R^\ell \cdot |m'_0+m'_1|_R}, G_{i,-1} \leftarrow g^{-R^i \cdot |m'_0+m'_1|_R}, 0 \leq i \leq \ell$
2: $A \leftarrow 1, K_j \leftarrow 1$ for $0 \leq j < m_1$
3: **for** i from 0 to $\ell - 1$ **do**
4: **if** $\kappa_i^{(1)} = 0$ **then**
5: $K_0 \leftarrow K_0 \times G_{i,\kappa_i^{(0)}} \times G_{i,-1}$
6: **else**
7: $K_{\kappa_i^{(1)}} \leftarrow K_{\kappa_i^{(1)}} \times G_{i,\kappa_i^{(0)}}$
8: **end if**
9: **end for**
10: $K_{|C|} \leftarrow K_{|C|} \times G_{\ell,sign(C)1}$
11: $W \leftarrow$ size of m_1 in bits
12: **for** i from W downto 0 **do**
13: $A \leftarrow A^2$
14: **for** j from $m_1 - 1$ downto 1 **do**
15: **if** bit i of j is non zero **then**
16: $A \leftarrow A \times K_j$
17: **end if**
18: **end for**
19: **end for**
20: **return** $(A \times K_0)$

In our case, with the chosen parameters, this brings us to store the following values in $\mathbb{Z}/p\mathbb{Z}$:

$$G_i = \{g^{88^i \cdot 33}, g^{88^i}, g^{88^i \cdot 57}, g^{88^i \cdot 25}, g^{88^i \cdot 81}, g^{88^i \cdot 49}, g^{88^i \cdot 17}, g^{88^i \cdot 73}, g^{88^i \cdot 41}, g^{88^i \cdot 9}, g^{88^i \cdot 65}\}.$$

We now use κ in Algorithm 5.

- the first steps are a `for` loop (steps 3 to 9):
  - in the first iteration, one has $\kappa_0^{(1)} = 0$ (and $\kappa_0^{(0)} = 5$), and this gives
  $$K_0 \leftarrow G_{0,\kappa_0^{(0)}} \times G_{0,-1} = g^{49} \times g^{-1} = g^{48}.$$
  - in the second iteration, one has $\kappa_1^{(1)} = 6$ (and $\kappa_1^{(0)} = 2$), and this gives
  $$K_6 \leftarrow G_{1,\kappa_1^{(0)}} = g^{88 \cdot 57} = g^{5016}.$$
  - in the third iteration, one has $\kappa_2^{(1)} = 5$ (and $\kappa_2^{(0)} = 8$), and this gives
  $$K_5 \leftarrow G_{1,\kappa_2^{(0)}} = g^{88^2 \cdot 41} = g^{317504}.$$

- in the fourth and last iteration, one has $\kappa_2^{(1)} = 7$ (and $\kappa_2^{(0)} = 3$), and this gives

$$K_7 \leftarrow G_{1,\kappa_2^{(0)}} = g^{88^3 \cdot 25} = g^{17036800}.$$

- the last carry $C = -2$ is now processed (step 10):

$$K_2 \leftarrow G_{4,-1} = g^{88^4 \cdot (-1)} = g^{-59969536}.$$

- the reconstruction in the second `for` loop (steps 12 to 16) provides the final result by computing

$$\begin{aligned} g^k \bmod p &= K_0 \times \textstyle\prod_{j=1}^{m_1} K_j^j \bmod p \\ &= g^{48+2\cdot(-59969536)+5\cdot 317504+6\cdot 5016+7\cdot 17036800} \bmod p \\ &= g^{936192} \bmod p, \end{aligned}$$

which is the desired result.

4.3 Complexity

The complexity of Algorithm 5 is evaluated step by step in Table 2 for the average case. The number of field multiplications (MM) is evaluated as follows:

- the MMs in step 5 are performed only in case of $K_0 \neq 1$, instead, it is only an instantiation of K_0;
- the MMs in step 7 are performed only in case of $K_{\kappa_i^{(1)}} \neq 1$, instead, it is only an instantiation of $K_{\kappa_i^{(1)}}$;
- the same applies for step 10;

This saves on average m_1 MMs, and this is taken into account in the Total line in Table 2 (it explains the $-m_1$ term). The number of operations in the final reconstruction is evaluated as follows:

Table 2. m_0m_1 modular exponentiation complexity and storage, average case.

Complexity		
Step	Operation	Complexity
$\ell/m_1 \times$ step 5	$K_0 \leftarrow K_0 \times G_{i,\lvert\kappa_i^{(0)}+1\rvert_{m_0}} \times G_{i,-1}$	2 MM
$\ell\frac{m_1-1}{m_1} \times$ step 7	$K_{\kappa_i^{(1)}} \leftarrow K_{\kappa_i^{(1)}} \times G_{i,\kappa_i^{(0)}}$	1 MM
$1 \times$ step 10	$K_{\lvert C\rvert} \leftarrow K_{\lvert C\rvert} \times G_{\ell,1}^{sign(C)}$	1 MM
$(W-1) \times$ step 13	$A \leftarrow A^2$	1 MS
$(\mathcal{H}-1) \times$ steps 15	$A \leftarrow A \times K_j$	1 MM
$1 \times$ step 18	$(A \times K_0)$	1 MM
Total	$(\ell\frac{m_1+1}{m_1} - m_1 + \mathcal{H})$ MM + $(W-1)$ MS	
Total storage	$(m_0+1) \times \ell + m_1 + 2$ elements of $\mathbb{Z}/p\mathbb{Z}$	

- the modular squaring in step 13 is performed only in case of $A \neq 1$;
- the MMs in step 15 and 18 are performed only in case of $K_j \neq 1$;

For the sake of simplicity, we denote by $\mathcal{H}$ the sum of the j Hamming weights for each j from $m_1 - 1$ downto 1 (`foreach` loop step 14). The value of $\mathcal{H}$ is as follows for the different values of m_1:

m_1	2	3	4	5	6	7	8	9
$\mathcal{H}$	1	2	4	5	7	9	12	13

We now discuss the complexity comparison of the considered methods (*Fixed-base Comb* Algorithm 3, Radix-R Algorithm 2 and m_0m_1 Algorithm 3). Since the parameters are very different between these three methods, a formal comparison is difficult. Therefore, we present a comparison based on numerical application, for NIST recommended sizes. In the sequel of this section, we then provide complexity evaluations in terms of field multiplications MM, under the assumption of squaring MS = 0.86 MM, which is the average value of our implementations for the NIST DSA recommended field sizes.

Figure 1 gives the general behavior of the three algorithms in terms of storage with respect to the complexity. One can see that the *Fixed-base Comb* method is the best for small storage amount. Our m_0m_1 approach is better for larger amount of storage, however, the Radix-R method is the best when the storage is increasing. In the figure, the field size is the largest of the ones recommended in the NIST standards (see [13]). Thus, the storage amount for such size is very big. Nevertheless, the behavior is roughly the same for smaller sizes, although the

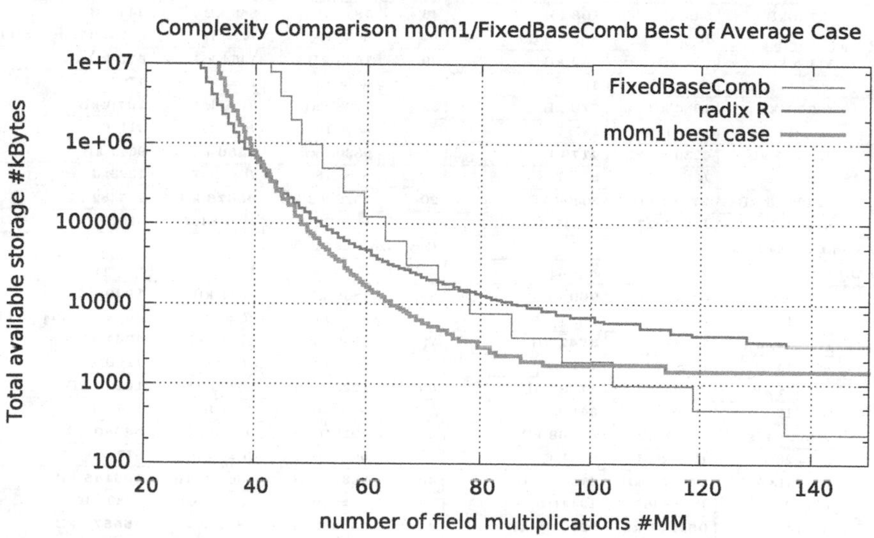

Fig. 1. Complexity comparison, Fixed base modular exponentiation NIST DSA, key size 512 bits (field size 15360 bits).

benefit of our approach is lower. The NIST provides recommended key sizes and corresponding field size (respectively the size of the primes q and p, see NIST SP800-57 [13]). This standardized sizes are as follows:

Key size (bits)	160	224	256	384	512
Field size (bits)	1024	2048	3072	7680	15360

For these sizes, Table 3 shows the complexity comparison between the *Fixed-base Comb* Algorithm 3, the Radix-R Method Algorithm 2 and our m_0m_1-Recoding approach Algorithm 5. For an equivalent number of MMs, we provide the minimum amount of storage.

We now provide a few comments about this table.

- For all sizes, we do not provide the results for small amount of storage (values of $w < 8$). For such storage, the *Fixed-base Comb* method is the best. One may notice that the Radix-R approach needs the greatest storage at this complexity level.
- For intermediate values of complexity, our proposed m_0m_1 approach shows the best storage/complexity trade-off. However, the benefits are greater for the larger key sizes.

Table 3. Storage amount comparison, *Fixed-base Comb* method and m_0m_1 modular exponentiation fixed-base, average case, NIST recommended exponent sizes.

Key size $t = 224$ bits				Key size $t = 256$ bits			
$\#MM$	*Fixed-base C.*	Radix-R	m_0m_1	$\#MM$	*Fixed-base C.*	Radix-R	m_0m_1
45	127.5 kB	345 kB	**108** kB	46	383 kB	845 kB	**241** kB
	$w = 9$	$R = 31$	$m_0 = 11$; $m_1 = 9$		$w = 10$	$R = 47$	$m_0 = 17$; $m_1 = 11$
37	511.5 kB	594 kB	**242** kB	39	1535 kB	1454 kB	**579** kB
	$w = 11$	$R = 61$	31; 7		$w = 12$	$R = 97$	47; 7
30	4095.5 kB	1386 kB	**770** kB	32	12287 kB	3179 kB	**2070** kB
	$w = 14$	$R = 179$	127; 7		$w = 15$	$R = 257$	211; 6
24	32767.5 kB	4230 kB	**4173** kB	26	98303 kB	**9486** kB	9642 kB
	$w = 17$	$R = 677$	877; 7		$w = 18$	$R = 937$	1223; 6
19	524287.5 kB	**27084** kB	50409 kB	20	1572863 kB	**66676** kB	225482 kB
	$w = 21$	$R = 5417$	13441; 5		$w = 22$	$R = 8467$	37579; 5
Key size $t = 384$ bits				Key size $t = 512$ bits			
$\#MM$	*Fixed-base C.*	Radix-R	m_0m_1	$\#MM$	*Fixed-base C.*	Radix-R	m_0m_1
63	1918 kB	4081 kB	**969** kB	86	3836 kB	9841 kB	**1940** kB
	$w = 11$	$R = 67$	$m_0 = 19$; $m_1 = 11$		$w = 11$	$R = 59$	$m_0 = 13$; $m_1 = 11$
50	15358 kB	10087 kB	**3742** kB	73	15356 kB	17855 kB	**4747** kB
	$w = 14$	$R = 191$	101; 11		$w = 13$	$R = 127$	41; 10
41	122878 kB	26655 kB	**17284** kB	60	122876 kB	46775 kB	**16224** kB
	$w = 17$	$R = 677$	541; 6		$w = 16$	$R = 409$	179; 11
35	983038 kB	80357 kB	**64768** kB	52	491516 kB	93110 kB	**54680** kB
	$w = 20$	$R = 2381$	2381; 6		$w = 18$	$R = 937$	677; 7
30	7864318 kB	**246070** kB	315053 kB	48	983036 kB	156091 kB	**106185** kB
	$w = 23$	$R = 8467$	13441; 5		$w = 19$	$R = 1699$	1489; 10
26	62914558 kB	**951217** kB	3256278 kB	41	7864316 kB	489112 kB	**355573** kB
	$w = 26$	$R = 37579$	165397; 5		$w = 22$	$R = 6211$	5417; 7
24	503316478 kB	**1750756** kB	- kB	35	62914556 kB	**2048419** kB	2113890 kB
	$w = 29$	$R = 74699$	–		$w = 25$	$R = 30347$	37579; 7

- $t = 224$, the best gain of our m_0m_1 approach is for $\#MM \approx 24$, with a storage 5 to 8 times smaller than the storage required for the *Fixed-base Comb* method, respectively for $\#MM = 30$ and $\#MM = 24$, and 35% less than the one of the Radix-R method. However, below $\#MM \approx 24$, the Radix-R approach is better.
- $t = 256$, the best gain of our m_0m_1 approach is for $\#MM \approx 32$, with a storage about 6 times smaller than the storage required for the *Fixed-base Comb* method, and 44% less than the one of the Radix-R method. Again, with decreasing values of $\#MM$ (below 26), the Radix-R approach is better.
- $t = 384$, the best gain of our m_0m_1 approach is for $\#MM \approx 35$, with a storage about 15 times smaller than the storage required for the *Fixed-base Comb* method, and 19% less than the one of the Radix-R method. Again, with decreasing values of $\#MM$ (below 33), the Radix-R approach is better.
- $t = 512$, the best gain of our m_0m_1 approach is for $\#MM \approx 41$, with a storage about 22 times smaller than the storage required for the *Fixed-base Comb* method, and 27% less than the one of the Radix-R method. Again, with decreasing values of $\#MM$ (below 38), the Radix-R approach is better.

However, one may notice that the bigger memory storage sizes exceed the common values of Random Access Memory, and also the maximum allowed for the `malloc` function of the standard C library for memory allocation. Nevertheless, the storage savings proposed by our method and the Radix-R ones allow to keep the level under the limit for lower complexities.

As a conclusion, our m_0m_1 approach shows lower storage amount for intermediate values of storage, whatever the standardized key size.

5 Implementation Results

5.1 Implementation Strategies

We review hereafter the main implementation strategies and test process. This applies for the three considered exponentiation algorithms. The algorithms were coded in C, compiled with `gcc 4.8.3` and run on the same platform.

Multiprecision Multiplication and Squaring: We used the low level functions performing multi-precision multiplication and squaring of the `GMP` library as building blocks of our codes (`GMP 6.0.0`, see GMP library [1]). According to the GMP documentation, the classical schoolbook algorithm is used for small sizes, and Karatsuba and Toom-Cook sub quadratic methods for size ≥2048 bits.

Modular Reduction: This operation implements the Montgomery representation and modular reduction method, which avoid multi-precision division in the computation of the modular reduction. This approach has been presented by Montgomery in [12]. More specifically, we used the block Montgomery algorithm suggested by Bosselaers *et al.* in [2]. In this algorithm, the multi-precision operations combine full size operand with one word operand and are also available in the GMP library [1]. Although the complexity is the same, the implementation is more computer friendly.

m_0m_1 Recoding: The conversion in radix-R needs multi-precision divisions. These operations are implemented using the GMP library [1]. The size of these operations is decreasing along the algorithm, and this is managed through GMP. The other operations are classical long integer operations. Steps 9 and 21 in Algorithm 4, an inversion modulo m_0 is required $(|(k_i^{(1)})^{-1}|_{m_0})$. This operation is performed using the Extended Euclidean Algorithm, over long integer data. For the considered exponent sizes, the cost of the recoding is negligible. This is explained by the small size of the exponent in comparison with the size of the data processed during the modular exponentiation (see the key sizes given page 11). The timings given in the next Section include this recoding.

Test Processing: The tests involve a few hundred dataset, which consists of random exponent inputs and an exponentiation base with the precomputed val-

Table 4. Synthesis of implementation results, clock cycles and storage (kB). Test performed on an Intel XEON E5-2650 (Ivy bridge), gcc 4.8.3, CENTOS 7.0.1406.

Modular exponentiation			
State of the art methods			
Fixed-base Comb	Radix R	m_0, m_1 rec	Ratio
$\#CC$ storage	$\#CC$ storage	$\#CC$ storage	m_0, m_1/ Best S.o.A.
Key size 224 bits, field size 2048 bits (level of security: 112 bits)			
221108 1023.5 kB ($w = 12$)	227838 829 kB ($R = 91$)	**219864** **580 kB** ($m_0 = 89, m_1 = 6$)	×0.994 ×0.700
210074 2047.5 kB ($w = 13$)	206888 1324 kB ($R = 163$)	**207072** **766 kB** ($m_0 = 127, m_1 = 7$)	×0.985 ×0.579
149690 65535 kB ($w = 18$)	147877 **7289kB** ($R = 1223$)	**146156** 21599 kB ($m_0 = 5417, m_1 = 6$)	×0.988 ×2.96
Key size 256 bits, field size 3072 bits (level of security: 128 bits)			
524539 1535 kB ($w = 12$)	502981 1411 kB ($R = 91$)	**501466** **897 kB** ($m_0 = 79, m_1 = 6$)	×0.997 ×0.636
449397 6143 kB ($w = 14$)	445871 2251 kB ($R = 163$)	**446444** **2056 kB** ($m_0 = 211, m_1 = 6$)	×1.001 ×0.913
356892 98303 kB ($w = 18$)	354640 **6414 kB** ($R = 571$)	354071 12843 kB ($m_0 = 1721, m_1 = 7$)	×0.998 ×2.002
Key size 384 bits, field size 7680 bits (level of security: 192 bits)			
4442590 1918 kB ($w = 11$)	4492191 3430 kB ($R = 53$)	**4409584** **1134 kB** ($m_0 = 23, m_1 = 10$)	×0.993 ×0.467
3554339 15358 kB ($w = 14$)	**3524896** 8290 kB ($R = 163$)	3551437 **4164 kB** ($m_0 = 113, m_1 = 10$)	×1.008 ×0.502
2736341 245758 kB ($w = 18$)	**2543480** 45221 kB ($R = 1223$)	2743399 **29961 kB** ($m_0 = 1031, m_1 = 7$)	×1.079 ×0.662
Key size 512 bits, field size 15360 bits (level of security: 256 bits)			
18632429 15536 kB ($w = 13$)	19260731 13765 kB ($R = 91$)	**18550238** **4745 kB** ($m_0 = 41, m_1 = 10$)	×0.996 ×0.345
14848261 122876 kB ($w = 16$)	15401002 34418 kB ($R = 163$)	**14813453** **22109 kB** ($m_0 = 257, m_1 = 11$)	×0.998 ×0.642
12477816 983036 kB ($w = 19$)	**12193232** 119061 kB ($R = 1223$)	12499600 **102820 kB** ($m_0 = 1381, m_1 = 7$)	×1.025 ×0.863

ues stored. We compute 2000 times the corresponding exponentiation for each dataset and keep the minimum number of clock cycles. This avoids the cold-cache effect and system issues. The timings are obtained by averaging the timings of all dataset.

5.2 Tests Results and Comparison

The three considered exponentiation algorithms were coded in C, compiled with `gcc 4.8.3` and run on the following platform: the CPU is an Intel XEON® E5-2650 (Ivy bridge), and the operating system is CENTOS 7.0.1406. On this platform, the Random Access Memory is 12.6 GBytes. One notices that the performance results include the recoding in the radix-R and m_0, m_1 cases. The implementation results confirm the complexity evaluation, for key sizes of 224, 256, 384, and 512 bits. However, the better results are for 384 and 512 bits.

In Table 4, we provide the most significant results. The gains shown are roughly in the same order of magnitude as the one of the complexity evaluation. In particular, for the largest key size (512 bits), the storage of our m_0, m_1 approach is nearly ten times less than the one required with the *Fixed-base Comb* method, and nearly 14% less than the one required for the Radix-R method, for the same computation timing, about $12.5 \times 10^6 \#CC$.

6 Conclusion and Future Work

In this paper, we have presented a new method for fixed-base exponentiation using a radix-R conversion with RNS representation of every radix-R digits, using an RNS base with two moduli $\mathcal{B} = \{m_0, m_1\}$. We have designed a recoding algorithm, which computes our $m_0 m_1$ representation of the exponent, and we have used it in a modular exponentiation algorithm which provides memory storage savings or improve the performance in terms of clock cycles per modular exponentiation, while offering a total flexibility in terms of storage amount. We have provided a complexity evaluation, which shows that our approach improves significantly the complexity/storage trade-off. We have then implemented this approach in order to check the performance benefits. We have compared our approach with two other classical approaches, the *Fixed-base Comb* and the Radix-R, and have confirmed the complexity results, showing the better storage/performance trade-off of our approach.

Two issues remain opened:

- Side-channel analysis is also a major threat, even in case of software implementation. For example, Gueron in [7] mentions the cache attack. In the present paper, we did not take this threat into account in the memorization process. However, by using a storage pattern spreading the data in memory, we could ensure the resistance against cache-attacks in the same way as the one used by Gueron without penalty. This needs to be implemented in all algorithms for fair comparison.

- Our approach can be applied to Elliptic Curve Cryptography, particularly to the ECDSA signature protocol. In this case, one needs to compute Elliptic Curve Scalar Multiplication. However, the relatively cheap doubling of point operation in comparison with point addition for the NIST recommended curves makes the benefits of our approach not as good as the one in the modular exponentiation case. Therefore, this approach needs to be implemented in relevant curves. For example, the twisted Edwards curve is an example of curve with relatively equivalent doubling and addition in terms of complexity.

References

1. The GNU Multiple Precision Arithmetic Library (GMP). http://gmplib.org/
2. Bosselaers, A., Govaerts, R., Vandewalle, J.: Comparison of three modular reduction functions. In: Stinson, D.R. (ed.) CRYPTO 1993. LNCS, vol. 773, pp. 175–186. Springer, Heidelberg (1994). doi:10.1007/3-540-48329-2_16
3. Acting Secretary Cameron Kerry and USST/Director Patrick Gallagher: Digital Signature Standard (DSS). In: Federal Information Processing Standards Publications, FIPS PUB 186-4. NIST (2013)
4. ElGamal, T.: A public key cryptosystem and a signature scheme based on discrete logarithms. In: Blakley, G.R., Chaum, D. (eds.) CRYPTO 1984. LNCS, vol. 196, pp. 10–18. Springer, Heidelberg (1985). doi:10.1007/3-540-39568-7_2
5. Garner, H.L.: The residue number system. In: Proceedings of the Western Joint Computer Conference, pp. 146–153 (1959)
6. Gordon, D.M.: A survey of fast exponentiation methods. J. Algorithms **27**(1), 129–146 (1998)
7. Gueron, S.: Efficient software implementations of modular exponentiation. J. Cryptogr. Eng. **2**(1), 31–43 (2012)
8. Hankerson, D., Hernandez, J., Menezes, A.: Software implementation of elliptic curve cryptography over binary fields. In: Koç, Ç.K., Paar, C. (eds.) CHES 2000. LNCS, vol. 1965, pp. 1–24. Springer, Heidelberg (2000). doi:10.1007/3-540-44499-8_1
9. Knuth, D.E.: The Art of Computer Programming, Volume II: Seminumerical Algorithms, 3rd edn. Addison-Wesley, Boston (1998)
10. Lim, C.H., Lee, P.J.: More flexible exponentiation with precomputation. In: Desmedt, Y.G. (ed.) CRYPTO 1994. LNCS, vol. 839, pp. 95–107. Springer, Heidelberg (1994). doi:10.1007/3-540-48658-5_11
11. López-García, L., Dominguez Perez, L.J., Francisco Rodríguez-Henríquez, F.: A pairing-based blind signature e-voting scheme. Comput. J. **57**(10), 1460–1471 (2014)
12. Montgomery, P.: Modular multiplication without trial division. Math. Comput. **44**(170), 519–521 (1985)
13. U.S.D.C. Rebecca Blank and USST/Director Patrick Gallagher: Recommendation for key management. In: Computer Security, Part 1, Rev 3. NIST Special Publication 800-7, pp. 62–64. NIST (2012)
14. Svoboda, A.: The numerical system of residual classes in mathematical machines. In: IFIP Congress, pp. 419–421 (1959)

Efficient Finite Field Multiplication for Isogeny Based Post Quantum Cryptography

Angshuman Karmakar[1(✉)], Sujoy Sinha Roy[1], Frederik Vercauteren[1,2], and Ingrid Verbauwhede[1]

[1] KU Leuven ESAT/COSIC and iMinds, Kasteelpark Arenberg 10 bus 2452, 3001 Leuven-Heverlee, Belgium
{Angshuman.Karmakar,sujoy.sinharoy,Ingrid.Verbauwhede}@esat.kuleuven.be, frederik.vercauteren@gmail.com

[2] Open Security Research, Fangda 704, Kejinan-12th, Nanshan, 518000 Shenzhen, China

Abstract. Isogeny based post-quantum cryptography is one of the most recent addition to the family of quantum resistant cryptosystems. In this paper we propose an efficient modular multiplication algorithm for primes of the form $p = 2 \cdot 2^a 3^b - 1$ with b even, typically used in such cryptosystem. Our modular multiplication algorithm exploits the special structure present in such primes. We compare the efficiency of our technique with Barrett reduction and Montgomery multiplication. Our C implementation shows that our algorithm is approximately 3 times faster than the normal Barrett reduction.

Keywords: Modular multiplication · Isogeny · Post-quantum cryptography

1 Introduction

The rapid development in the field of quantum computing has increased the possibility of practical quantum computer arriving within a few decades [17]. Using a powerful quantum computer, Shor's [2] algorithm can factor integers and can compute discrete logarithm in polynomial time. This has rendered cryptosystems such as RSA and those using elliptic curve cryptography highly vulnerable.

Due to these developments, research in post-quantum cryptography has seen a flurry of activity that resulted in many novel post-quantum cryptosystems. Though the cryptosystems based on learning with errors or LWE has gained the most interest, there exist other cryptosystems such as the McEliece cryptosystem [18], cryptosystems based on isogeny between elliptic curves [1,13], the multivariate cryptosystem [19] etc. Many cryptographic schemes based on these primitives have been proposed which are analogous to their classical counterparts and hopefully will replace them in the near future.

A cryptosystem based on the computation of isogenies between elliptic curves was first proposed by Anton Stolbunov [13]. The security of this cryptosystem was based on the hardness of computing isogenies between ordinary elliptic

S. Duquesne and S. Petkova-Nikova (Eds.): WAIFI 2016, LNCS 10064, pp. 193–207, 2016.
DOI: 10.1007/978-3-319-55227-9_14

curves. The best known classical algorithm to solve this problem has exponential [11] complexity. But the work of Childs et al. [8] has shown that this problem has sub-exponential complexity on a quantum computer. Also their system was slow for practical purposes.

The isogeny based post quantum cryptosystem proposed by De Feo et al. [1] uses supersingular elliptic curves instead of ordinary elliptic curves. The authors in [1] have argued that the problem of computing isogenies between supersingular elliptic curves is quantum secure. They have also shown that their cryptosystem is many times faster than the previous system and offers post-quantum security for practical parameter sizes.

2 Motivation

The isogeny based post-quantum cryptosystem proposed by De Feo et al. [1] is based on the difficulty of computing isogenies between supersingular elliptic curves. Computing isogenies and applying them to the points of elliptic curves ultimately boils down to arithmetic operations in a finite field over which the supersingular curve is defined. In isogeny based cryptography the prime p is of the form $p = f \cdot 2^a 3^b - 1$ where f is a small number. Such a special structure of the prime is essential for the scheme. Like many other cryptosystems, isogeny based cryptosystem rely heavily on modular multiplication.

Montgomery multiplication [3] and Barrett reduction [7] are two ingenious methods to replace computationally costly divisions used in modular reduction with additional multiplications, additions, bit shifts etc. These methods tackle the costly modular multiplication quite efficiently and they can be applied for any general prime. So they are unable to exploit any special structure of the prime for even faster reduction.

Mersenne primes [5] and Pseudo-Mersenne primes [6] offer very fast reduction due to their special structure. Also the NIST-curves [20] which are used in elliptic curve cryptography frequently use fields over generalized Mersenne primes [4] for the advantage of extremely fast modular reduction. Even though the primes we discuss cannot be categorized as a Mersenne prime, generalized Mersenne prime or Pseudo-Mersenne prime, the possibility of exploiting the special structure of the prime for an efficient modular multiplication calculation is highly intriguing. The parameters a and b for the prime $p = 2 \cdot 2^a \cdot 3^b - 1$ in the isogeny based post-quantum protocol are chosen in such a way that $\log_2(2^a) \approx \log_2(3^b)$. For example, the 771-bit prime $p = 2 \cdot 2^{386} 3^{242} - 1$ is used in [1] for 128-bit security.

Our Contribution. In this work we describe a fast modular multiplication algorithm for the primes used in isogeny based post-quantum cryptosystems. Our algorithm is inspired by the Barrett reduction [7] and leverages special structures of the primes used in such cryptosystems. While there are several techniques for performing efficient arithmetic in fields whose characteristic is a Mersenne prime or a Pseudo-Mersenne prime [4], we are not aware of any techniques that could accelerate modular arithmetic in finite fields of characteristic $p = f \cdot 2^a 3^b - 1$ where f is a small number. In this paper we propose an efficient algorithm to perform fast modular arithmetic with primes of the form $p = 2 \cdot 2^a 3^b - 1$ with b even. Besides the new algorithm, we list a number of such primes for different security levels. These primes are listed in Appendix C.

3 Mathematical Background

In this section we will briefly describe the isogeny based key exchange protocol and then focus on efficient modular multiplication techniques. For a detailed description of isogeny based key exchange interested readers may follow [1].

3.1 Isogenies of Elliptic Curves

An isogeny $\phi : E_1 \rightarrow E_2$ is a basepoint preserving, i.e. $\phi(\mathcal{O}) \rightarrow \mathcal{O}$, morphism between two elliptic curves E_1 and E_2 defined over $\mathbb{F}_q$ (Sect. III.4 in [12]). Two elliptic curves are said to be *isogenous* if there exists an isogeny between them. This is an equivalence relation and symmetry is given by the existence of a *dual isogeny*. As mentioned in [1], an isogeny class is an equivalence class under the above equivalence relation. Inside the same isogeny class the curves are either all supersingular or all ordinary curves. The post-quantum key exchange scheme by De Feo et al. in [1] uses supersingular curves.

In this key-exchange scheme the public parameters are a supersingular curve E_0 defined over a field $\mathbb{F}_{p^2}$ with $p = f \cdot 2^a 3^b \pm 1$, and bases $\{P_a, Q_a\}$ and $\{P_b, Q_b\}$ which generate the *torsion groups* $E_0[2^a]$ and $E_0[3^b]$ respectively. Alice chooses $m_a, n_a \in_R \mathbb{Z}/2^a\mathbb{Z}$ and computes the isogeny $\phi_a : E_0 \rightarrow E_a$, $E_a = E_0/\langle [m_a]P_a + [n_a]Q_a \rangle$. Alice also computes $\phi_a(P_b)$ and $\phi_a(Q_b)$ under this isogeny and sends E_a, $\phi_a(P_b)$, and $\phi_a(Q_b)$ to Bob. Similarly Bob chooses $m_b, n_b \in_R \mathbb{Z}/3^b\mathbb{Z}$ computes the isogeny $\phi_b : E_0 \rightarrow E_b$, $E_b = E_0/\langle [m_b]P_b + [n_b]Q_b \rangle$ and sends E_b, $\phi_a(P_b)$, and $\phi_a(Q_b)$ to Alice. After this Alice calculates the isogeny $\phi'_a : E_a \rightarrow E_{ab}$, $E_{ab} = E_a/\langle [m_a]\phi_b(P_a) + [n_a]\phi_b(Q_a) \rangle$ and similarly Bob calculates $\phi'_b : E_b \rightarrow E_{ba}$. Bob and Alice then use their common *j-invariant* $j(E_{ab}) = j(E_{ba})$ as their shared key.

The difficulty of the key-exchange scheme is based on the hardness of computing isogenies between supersingular elliptic curves. The authors in [1] have argued that the complexity of the best known algorithm [16] for solving this problem is $\sqrt[4]{p}$ using classical computers and $\sqrt[6]{p}$ using a quantum computer, where p is the characteristic of the field over which the curves are defined (more details in Sect. 5 and 6 of [1]). The authors have described post quantum protocols for zero knowledge proof, key-exchange and public key cryptosystem in their paper [1]. Hash functions [15] and digital signature schemes [14] based on the isogenies have also been proposed.

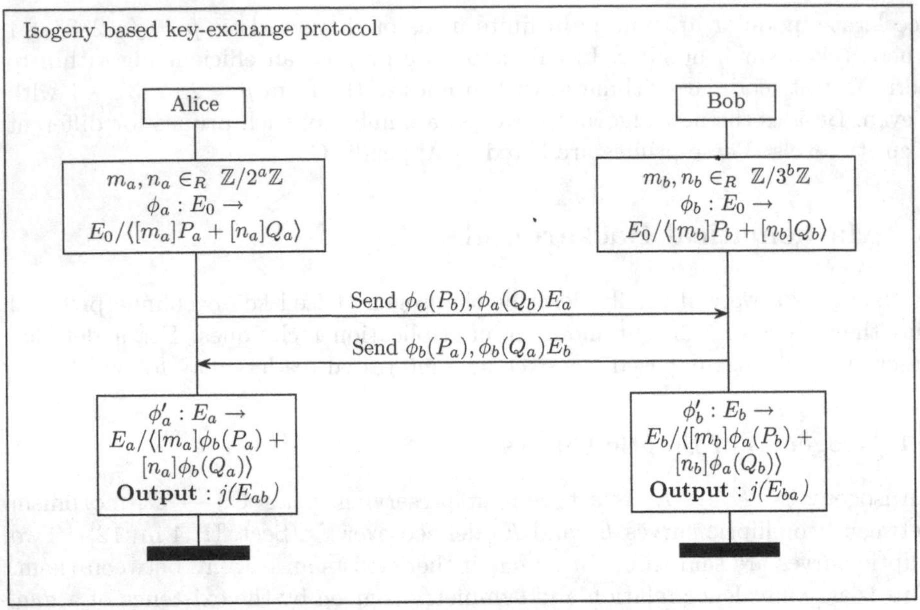

3.2 Efficient Modular Arithmetic

In this section we describe two famous algorithms for efficient modular reductions: the Barrett reduction, and the Montgomery reduction.

Barrett Reduction: Euclid's division lemma tells us that for any two positive integers a and b there exist q and r such that $a = q \cdot b + r$, $r \in [0, b-1]$. Here of course, $a = r \pmod{b}$, but finding such q and r requires division of a by b. There exist fast methods for division by small constants [9], but in general for practical cryptographic settings, division is a computationally costly operation.

For constant divisors, Barrett's reduction is a clever trick. It estimates $1/b$ to substitute division by a few multiplications and bit shifts. The $1/b$ in Barrett reduction is approximated as,

$$1/b = \frac{(2^k)/b}{b \cdot 2^k/b} = \frac{(2^k)/b}{2^k} \approx \frac{x}{2^k}$$

Usually the value of x is taken as $x = \lfloor 2^k/b \rfloor$ where the parameter k depends on a. The error e of the approximation of $1/b$ is $e = 1/b - x/2^k$. Hence, the error in approximating the quotient q is ae. As $q \in \mathbb{Z}^+$, for a correct result we require that the error in approximating q is smaller than 1. This condition is satisfied when $k = \log_2(a)$. The Barrett reduction algorithm is shown in Algorithm 1.

Input: Two numbers a and b, parameter k, $x = \left\lfloor \frac{2^k}{b} \right\rfloor$
Output: $a \pmod b$
$q \leftarrow (a \times x) >> k;$
$r \leftarrow a - q \times b;$
if $r \geq b$ **then**
  $r \leftarrow r - b$
end
return r

Algorithm 1. Barrett's Reduction Algorithm

Montgomery Multiplication: Montgomery multiplication [3] is another technique used to remove the necessity of performing modular reduction after each multiplication of the field elements. To use Montgomery's technique we need a number r co-prime to the modulus p or equivalently $r \cdot r' + p \cdot p' = 1$. The values r' and p' can be calculated by the extended Euclidean algorithm [10]. Montgomery multiplication first converts the operands a and b to the Montgomery domain as $a_M = a \cdot r \pmod p$, $b_M = b \cdot r \pmod p$, the multiplication algorithm described in Algorithm 2 ensures that the product also stays in the Montgomery domain as $a_M \times b_M = c_M \pmod p = a \cdot b \cdot r \pmod p$. Also the result of addition and subtraction between operands in the Montgomery domain stays in the Montgomery domain. As the conversion to and from the Montgomery domain is a costly procedure, this technique is useful where we need many multiplications, additions or subtractions in close succession.

Input: Two numbers $a_M = a \cdot r \pmod p$ and $b_M = b \cdot r \pmod p$
Output: $c_M = a \cdot b \cdot r \pmod p$
$t \leftarrow a_M \cdot b_M;$
$c_M \leftarrow (t + (t \cdot p' \pmod r)) \cdot p)/r;$
if $c_M \geq p$ **then**
  $c_M \leftarrow c_M - p$
end
return c_M

Algorithm 2. Montogomery Multiplication

As mentioned before, the above two methods do not utilize the special structure of the primes for faster modular multiplication. In the next section we are going to describe our modular multiplication algorithm which exploits the special structure of the prime for efficient modular multiplication.

4 New Modular Multiplication Algorithm

In our method, the representation of field elements plays an important role in the efficiency of the method. We represent a field element, let's say $A \in \mathbb{F}_p$, where $p = 2 \cdot 2^a 3^b - 1$, as

$$A = a_1 \cdot 2^a 3^b + a_2 \cdot 2^{a/2} 3^{b/2} + a_3, \quad a_1 \in [0, 1], \ a_2, a_3 \in [0, 2^{a/2} 3^{b/2}) \tag{1}$$

In the above representation we have assumed that a is even. However it is not mandatory. If a is odd we can write $p = 4 \cdot 2^{a-1} 3^b - 1$. This change in the value of the cofactor (from 2 to 4) does not affect the performance of the algorithm. During the course of our modular multiplication algorithm the only significance of the value of the cofactor is to determine the value of the coefficient a_1, where we need to divide some numbers by the cofactor. As division by 4 is almost as easy as division by 2 in binary representation, the change of value of the cofactor from 2 to 4 has little impact on the performance. In case a is odd the range of a_1 will change to $[0, 3]$. Here we want to note that we could have written $p = 2^{a+1} 3^b - 1$ instead of $p = 4 \cdot 2^{a-1} 3^b - 1$ with the cofactor equal to one, both of these representations of p have no major impact on the performance and can be switched between one another trivially by simple mathematical manipulations. Using the same argument as above we need b to be even else it will impact the performance significantly, as division by 6 or 12 is not as easy as division by 2 or 4.

We note that this conversion from normal integer representation to this special representation and vice versa is a costly procedure. But we explain at the end of this section that this conversion and reconversion are one-time procedures that we need to perform at the beginning and the end of the key-exchange algorithm.

4.1 Multiplication Algorithm

Let's suppose we have two numbers $A, B \in \mathbb{F}_p$ as represented in Eq. (1). After multiplying them we get the result C as per the equation shown below:

$$\begin{aligned} C = a_1 b_1 \cdot 2^{2a} 3^{2b} + (a_1 b_2 + a_2 b_1) 2^{3a/2} 3^{3b/2} + (a_1 b_3 + a_2 b_2 + a_3 b_1) 2^a 3^b \\ + (a_2 b_3 + a_3 b_2) 2^{a/2} 3^{b/2} + a_3 b_3. \end{aligned} \tag{2}$$

Since the prime p is of the form $2 \cdot 2^a 3^b - 1$, we can replace $2^a 3^b$ in Eq. (2) by $2^{-1} (mod\ p)$. Hence $a_1 b_1 \cdot 2^{2a} 3^{2b}$ gets replaced by 0 or $2^{-2} \pmod{p}$ as $a_1, b_1 \in \{0, 1\}$ and $a_1 b_1 \in \{0, 1\}$. Note that for a fixed prime we can precompute the value of $2^{-2} \pmod{p}$ and use that for the above replacement in Eq. (2).

We can replace $(a_1 b_3 + a_2 b_2 + a_3 b_1) 2^a 3^b$ as follows. If $(a_1 b_3 + a_2 b_2 + a_3 b_1)$ is even, we can write $(a_1 b_3 + a_2 b_2 + a_3 b_1) 2^a 3^b = (a_1 b_3 + a_2 b_2 + a_3 b_1)/2 \pmod{p}$. Otherwise we can write $(a_1 b_3 + a_2 b_2 + a_3 b_1) 2^a 3^b = ((a_1 b_3 + a_2 b_2 + a_3 b_1 - 1)/2) \pmod{p} + (a_1 b_3 + a_2 b_2 + a_3 b_1) \pmod{2} \cdot 2^a 3^b$. Considering both the even and odd cases we can write the following equation:

$$\begin{aligned} & (a_1 b_3 + a_2 b_2 + a_3 b_1) 2^a 3^b \\ \implies & \Big(\lfloor (a_1 b_3 + a_2 b_2 + a_3 b_1)/2 \rfloor \Big) + \Big((a_1 b_3 + a_2 b_2 + a_3 b_1) \mod 2 \Big) 2^a 3^b. \end{aligned}$$

Similarly,

$$\left(a_1b_2 + a_2b_1\right) \cdot 2^{3a/2}3^{3b/2}$$
$$\implies \left(\lfloor (a_1b_2 + a_2b_1)/2 \rfloor\right) \cdot 2^{a/2}3^{b/2} + \left((a_1b_2 + a_2b_1) \mod 2\right) \cdot 2^{a/2-1}3^{b/2}.$$

Rewriting Eq. (2) by replacing the coefficients we get the following equation:

$$A \times B = \Big(\underbrace{2^{-2} \pmod p)}_{\text{replacing} 2^{2a}3^{2b}} a_1b_1 + a_3b_3 + ((a_1b_2 + a_2b_1) \pmod 2))2^{a/2-1}3^{b/2} +$$
$$\underbrace{\lfloor (a_1b_3 + a_2b_2 + a_3b_1)/2) \rfloor}_{\text{replacing}(a_1b_3+a_2b_2+a_3b_1)2^a3^b}\Big) + \Big(\underbrace{\lfloor (a_1b_2 + a_2b_1)/2 \rfloor}_{\substack{\text{replacing}\\(a_1b_2+a_2b_1)2^{3a/2}3^{3b/2}}} + (a_2b_3 + a_3b_2)\Big)2^{a/2}3^{b/2}$$
$$+ \Big((a_1b_3 + a_2b_2 + a_3b_1) \pmod 2)\Big)2^a3^b.$$

The algorithm is described in Algorithm 4. To compute the above expression we have to perform four smaller multiplications: a_2b_2, a_2b_3, a_3b_2 a_3b_3, as the other terms which are multiplied with $a_1, b_1 \in \{0, 1\}$.

Now we have the product as $A \times B = C = C_1 \cdot 2^a3^b + C_2 \cdot 2^{a/2}3^{b/2} + C_3$, but in this expression the coefficients C_2 and C_3 lie in the range $[0, 2^a3^b)$, which is not consistent with our representation where C_2 and C_3 should lie in the range $[0, 2^{a/2}3^{b/2})$. Hence we need to split them further so that they fit according to our representation scheme. This splitting involves divisions of the coefficients C_i for $i = 2$ and 3 by $2^{a/2}3^{b/2}$. In the next section we are going to explain how we can do this division efficiently.

4.2 Efficient Division

Our purpose is to divide a number $C_i \in [0, 2^a3^b)$ by $2^{a/2}3^{b/2}$ and calculate the quotient q and remainder r in an efficient way. We note that division by two is a simple right shift operation. Hence we perform the division by $2^{a/2}3^{b/2}$ using the steps shown below.

1. Extract the $a/2$ least significant bits of C_i and store them in a variable r_1.
2. Right shift C_i by $a/2$ bits to obtain C_i'.
3. Divide C_i' by $3^{b/2}$ to get the quotient q and the remainder r_2.

Hence we have $C_i = q \cdot 2^{a/2}3^{b/2} + (r_2 \cdot 2^{a/2} + r_1) = q \cdot 2^{a/2}3^{b/2} + r$.

The division operation by $3^{b/2}$ in Step 3 is not as easy as the division by $2^{a/2}$. However since b is a fixed integer, the division can be performed using multiplications similar to the Barrett reduction technique [7] as described in Algorithm1 in Sect. 3.

Obtaining the quotients and remainders after dividing C_2 and C_3 by $2^{a/2}3^{b/2}$, it is trivial to write C in the desired representation of a finite field element.

Input: 2 numbers $Q \in [0, 2^a 3^b)$ and $P = 2^{a/2} 3^{b/2}$ and $\log_2 Q \approx 2 \cdot \log_2 P$. $P' = P/2^{a/2}$ precomputed $x = 2^k/P'$, k is as described in Sect. 3.2

Output: q and r such that $Q = q \cdot P + r$

1 $t \leftarrow \lfloor Q/2^{a/2} \rfloor, s = Q \pmod{2^{a/2}}$;
2 $q \leftarrow t \times x >> k$;
3 $r \leftarrow t - P' \times q$;
4 $r \leftarrow r \times 2^{a/2} + s$;
5 **if** $r > P$ **then**
6     $r \leftarrow r - P$;
7     $q \leftarrow q + 1$
8 **end**
9 **return** q, r

Algorithm 3. Our Division Algorithm

In the next part of this section we will compare the cost of our modular reduction technique with the original Barrett reduction technique. Note that the parameters a and b in the prime $p = 2 \cdot 2^a \cdot 3^b - 1$ are chosen in such a way that $\log_2(2^a) \approx \log_2(3^b)$. For convenience let us take $\log_2(2^a) \approx \log_2(3^b) \approx N$. So the prime is of size $2N$ bits.

Comparison with Barrett Reduction: In the Barrett reduction technique in Algorithm 1 the result of an integer multiplication that is of size $\leq 4N$ bits is reduced by a prime of size $2N$ bits. For correct computation k is of size $4N$ bits. In this scenario we have to perform one $4N \times 2N$ bit multiplication to compute the quotient (line 1 in Algorithm 1) and one $2N \times 2N$ bit multiplication to compute the remainder (line 2 in Algorithm 1). Thus using a quadratic complexity multiplier, the Barrett reduction technique has a cost of $12N^2$. In our modular reduction technique we perform divisions of two numbers C_2 and C_3 of size $\leq 2N$ by an N bit number $2^{a/2}3^{b/2}$. Since division by a power of two is almost free, the cost of each division reduces to the cost of dividing a number of size $\leq 3N/2$ bits by a $N/2$ bit number. To perform the divisions correctly we need to fix the value of k to $3N/2$. Hence each of the two division operations perform a $3N/2 \times N$ bit multiplication and an $N \times N/2$ bit multiplication (lines 2 and 3 in Algorithm 3). Thus using a quadratic complexity multiplier, our reduction technique has a cost of $4N^2$.

Comparison with Montogomery Multiplication: In this section we provide a comparison of the computational cost of Montgomery multiplication with our technique. As defined in Sect. 4.2 our prime is of size $2 \cdot N$ bits. For executing a single round of Montgomery multiplication we need two $2N \times 2N$ bit multiplications. And a relatively easier multiplication of $t \cdot p' \pmod{r}$ where only the last $2 \cdot N$ bits of the result are required. In our case we need only four $N \times N$ bit multiplications for the first part of our algorithm and two $3N/2 \times N$ bit and $N \times N/2$ bit multiplications for the final reduction.

Input: $A, B \in \mathbb{F}_p$, $A = a_1 \cdot 2^a 3^b + a_2 \cdot 2^{(a/2)} 3^{(b/2)} + a_3$ and
$B = b_1 \cdot 2^a 3^b + b_2 \cdot 2^{(a/2)} 3^{(b/2)} + b_3$; $2^{-2} \pmod p$ precalculated

Output: $C = A \times B \pmod p$,$C = C_1 \cdot 2^a 3^b + C_2 \cdot 2^{(a/2)} 3^{(b/2)} + C_3$

1 $C_1 = 0; C_2 = 0; C_3 = 0;$
2 Multiply $a_2 \times b_2$, $a_2 \times b_3$, $a_3 \times b_2$, $a_3 \times b_3$; `//` $\in [0, 2^a 3^b)$
3 Multiply $a_1 \times b_1$, $a_1 \times b_2$, $a_1 \times b_3$, $b_1 \times a_2$ $b_1 \times a_3$; `//` $\in [0, 2^{a/2} 3^{b/2})$
4 $C_3 \leftarrow a_1 b_1 \cdot 2^{-2} \pmod p + a_3 b_3$;
5 $C_2 \leftarrow a_2 b_3 + a_3 b_2$;
6 $t \leftarrow (a_1 b_2 + a_2 b_1)$; `// replacing` $(a_1 b_2 + a_2 b_1) 2^{3a/2} 3^{3b/2}$
7 **if** *isEven*(t) **then**
8 | $C_2 \leftarrow C_2 + t/2$
9 **else**
10 | $t \leftarrow t - 1$;
11 | $C_2 \leftarrow C_2 + t/2$;
12 | $C_3 \leftarrow C_3 + 2^{a/2-1} 3^{b/2}$
13 **end**
14 $t \leftarrow (a_1 b_3 + a_2 b_2 + a_3 b_1)$; `// replacing` $(a_1 b_3 + a_2 b_2 + a_3 b_1) 2^a 3^b$
15 **if** *isEven*(t) **then**
16 | $C_3 \leftarrow C_3 + t/2$;
17 | $C_1 \leftarrow 0$
18 **else**
19 | $t \leftarrow t - 1$;
20 | $C_3 \leftarrow C_3 + t/2$;
21 | $C_1 \leftarrow 1$
22 **end**
`/* End of first part` $C = C_1 2^a 3^b + C_2 2^{a/2} 3^{b/2} + C_3$`, reduce` $C_2, C_3 = O(2^a 3^b)$ `further by Barrett division */`
23 $q, r \leftarrow$ *BarrettDivision*(C_3);
24 $C_3 \leftarrow r$;
25 $C_2 \leftarrow C_2 + q$;
26 $q, r \leftarrow$ *BarrettDivision*(C_2);
27 $C_2 \leftarrow r$;
28 $C_1 \leftarrow C_1 + q$;
29 **if** *isEven*(C_1) **then**
30 | $C_3 \leftarrow C_3 + C_1/2$;
31 | $C_1 \leftarrow 0$
32 **else**
33 | $C_3 \leftarrow C_3 + (C_1 - 1)/2$;
34 | $C_1 \leftarrow 1$
35 **end**
36 if C_3 overflows i.e. $C_3 > 2^{a/2} 3^{b/2}$, then $C_3 \leftarrow C_3 - 2^{a/2} 3^{b/2}$, $C_2 \leftarrow C_2 + 1$ if C_2 also overflows $C_1 \leftarrow C_1 + 1$ and repeat steps 29 to 35, this situation occurs rarely and also then we have to perform this step at most once;
37 **return** $(C_1 \cdot 2^a 3^b + C_2 \cdot 2^{a/2} 3^{b/2} + C_3)$

Algorithm 4. Multiplication Algorithm

Here we want to mention that the two *Barrett Divisions* performed in the reduction stage (23 and 26) of Algorithm 4 can be run in parallel, effectively reducing the computing time by half (Fig. 1).

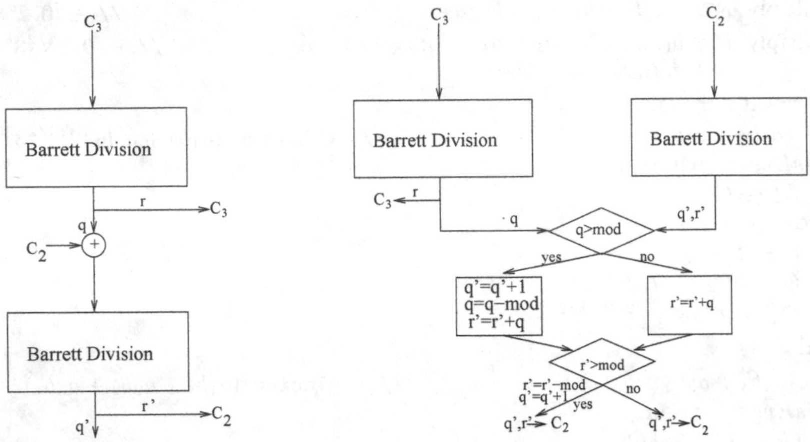

(a) Serial Execution (b) Parallel Execution

Fig. 1. Serial and Parallel execution for the reduction part of Algorithm 4

5 Software Implementation

For a comparison of the effective speedup of our algorithm we implemented our algorithm using C in a 32 bit multi-precision format for a security level of 128 bits. We also implemented a normal Barrett reduction using the same multi-precision format. The cost of multiplication when multiplying two input numbers in both of the algorithms is expected to remain the same. Therefore we used normal schoolbook multiplication. Upon running multiple instances of both the algorithms on a computer with CentOS on a core i5 CPU and averaging the running times we obtain the results as given in Table 1. As we can see from the table, we achieve an approximate 62% speed-up in reduction and 43% speed-up for modular multiplication (multiplication + reduction) with our method against the normal Barrett reduction. This result is consistent with our prediction in Sect. 4.2.

Table 1. Comparison of our algorithm with normal Barrett reduction algorithm

Operation	Running time (μ s)
Barrett reduction	50.547
Normal multiplication	67.097
Our reduction	19.565
Our multiplication	38.490

6 Hardware Implementation

To check the performance of the new modular multiplication scheme, we have designed a hardware architecture that performs modular multiplications following Algorithm 4. The arithmetic unit of the architecture consists of a combinational multiplier of input size N/2 and an addition/subtraction circuit of input size N. The operands are stored as arrays of N/2 bit words in a register file that contains 52 registers in total and of which 16 registers were used to store the pre-computed values as required by the Algorithm 4. Since the proposed algorithm performs arithmetic operations on two operands, we kept two output ports and one input port in the register file. During a multiplication, the multiplier performs multiplications of words and the adder helps to accumulate the result in the accumulator register *ACC*. For performing multi-precision additions and subtractions, only the lower half (i.e. the N/2 bits) of the addition/subtraction circuit is used. The control signals are generated by a hierarchy of finite state machines for multi-precision addition, subtraction, shifting and multiplication. On the top of the hierarchy, a finite state machine executes the operations required for the modular multiplication operation (Fig. 2).

We have compiled the hardware architecture using the Xilinx ISE 14.4 tool targetting the Virtex 6 FPGA (xc6vcx240t-2ff784). For this evaluation, we chose the field generated by the prime $2 \cdot 2^{386} 3^{242} - 1$ (hence N is 385 bits). After

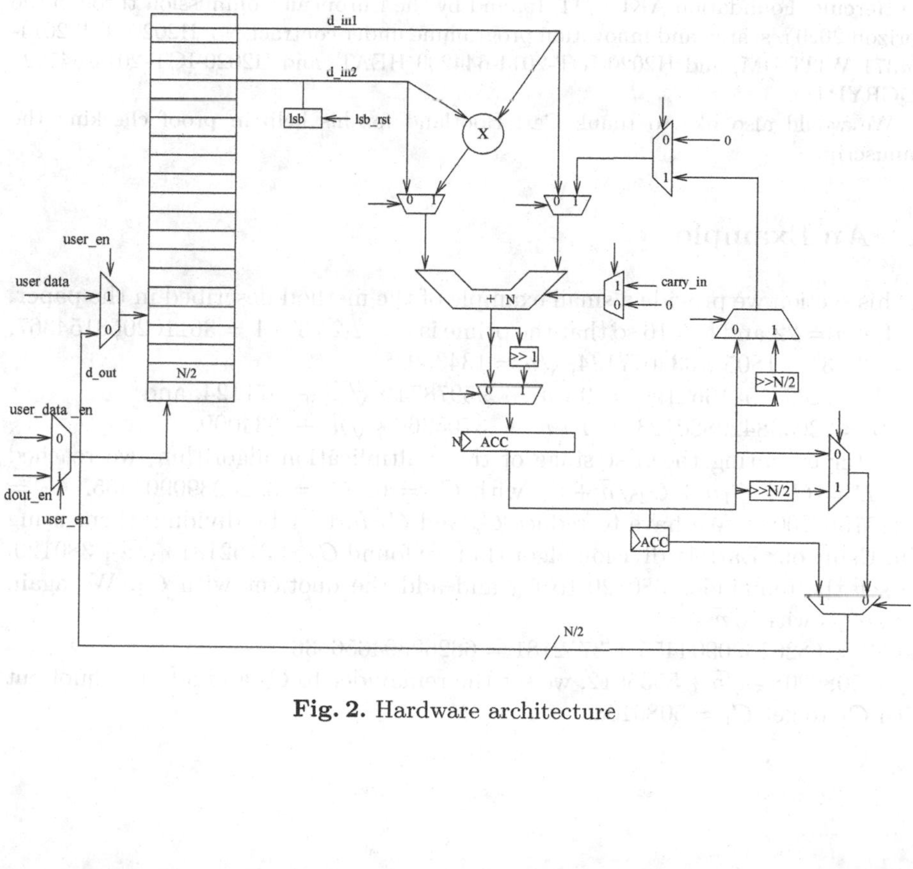

Fig. 2. Hardware architecture

place and route operation, the architecture consumes 11, 924 registers and 12, 790 look-up-tables, accounting to 3% and 8% of the resources available in the FPGA. The operating frequency of the architecture is 31 MHz. One modular multiplication (integer multiplication + modular reduction) takes 236 cycles and hence 7.61 μs.

7 Conclusion

We presented a fast modular multiplication algorithm that exploits the special structure of primes of the form $p = 2 \cdot 2^a 3^b - 1$, used in isogeny based post-quantum cryptography. To our knowledge there is no other algorithm that exploits the structure of such primes for fast reduction. We have shown that our algorithm is more efficient than Montgomery multiplication and Barrett reduction. We believe that with our algorithm will significantly decrease the time required to calculate isogenies between supersingular elliptic curves, which will strengthen the potential of isogeny based post-quantum cryptography as a practical post-quantum cryptosystem.

Acknowlegments. A. Karmakar and S. Sinha Roy were supported by Erasmus Mundus PhD Scholarship. This work was supported in part by the Research Council KU Leuven: C16/15/058. In addition, this work was supported in part by iMinds, the Flemish Government, FWO G.0550.12N, G.00130.13N and FWO G.0876.14N, by the Hercules Foundation AKUL/11/19, and by the European Commission through the Horizon 2020 research and innovation programme under contract No. H2020-ICT-2014-644371 WITDOM, and H2020-ICT-2014-644209 HEAT, and H2020-ICT-2014-645622 PQCRYPTO.

We would also like to thank Carl Bootland for his help in proof checking the manuscript.

A An Example

In this section we provide a small example of the method described in the paper.

Let $a = 22$,and $b = 16$ so that the prime is $p = 2{\cdot}2^a{\cdot}3^b{-}1 = 361102068154367$, $n = 2^a \cdot 3^b = 180551034077184$, $\sqrt{n} = 13436928$

$A = 128965951662196 \ = 0 * n \ + \ 9597874 * \sqrt{n} \ + \ 971124$, and

$B = 230338429880123 \ = 1 * n \ + \ 3705266 * \sqrt{n} \ + \ 334009$

After executing the first stage of the multiplication algorithm, we reached $A \times B = C = C_1 n + C_2\sqrt{n} + C_3$ with $C_1 = 0$, $C_2 = 68262390904455$, $C_3 = 50417786320088$. We have to reduce C_2 and C_3 further by dividing them using $\sqrt{n}$. Using our Barrett division algorithm we found $C_3 = 3752181 * \sqrt{n} + 380120$, we set the remainder 380120 to C_3 and add the quotient with C_2. We again divide C_2 with $\sqrt{n}$

$C_2 = 68262390904455 + 3752181 = 68262394656636$

$C_2 = 5080208 * \sqrt{n} + 5535612$, we set the remainder to C_2 and add the quotient with C_1 to get $C_1 = 5080208$.

As $C_1 \pmod 2) = 0$, we add $C_1/2 = 2540104$ to C_3 to get $C_3 = 380120 + 2540104 = 2920224$.
Here C_3 is smaller than $\sqrt{n}$ and there is no overflow. So we stop our algorithm here. Finally, we get the result as $C = 0 * n + 5535612 * \sqrt{n} + 2920224 = 74381622800160$, which is indeed $A \times B \pmod p$.

B Application in Isogeny Based Post-quantum Key Exchange Protocol

The isogeny based post-quantum protocol, described in Sect. 3 works by computing and applying isogenies over supersingular elliptic curve groups. These operations are fundamentally field arithmetic operations over the field $\mathbb{F}_{p^2}$, where the curve is defined.

Here we want to mention that modular addition and subtraction is also easy in our representation. Let's say we want to add two numbers $A, B \in \mathbb{F}_p$ to get the sum $C = (a_1 + b_1) \cdot n + (a_2 + b_2) \cdot \sqrt{n} + (a_3 + b_3) = C_1 \cdot n + C_2 \cdot \sqrt{n} + c_3$ for convenience we have assumed $n = 2^a 3^b$. Here again, similar to multiplication algorithm, C_1, C_2 and C_3 may not be consistent with our representation as given in Eq. (1). To make C consistent with our representation we follow steps 23 to 36 of Algorithm 4. But here we do not have to use the division Algorithm 3, only a subtraction by $2^{a/2}3^{b/2}$ will suffice. For subtraction we first negate a number $B \in \mathbb{F}_p$ as $-B = p - b = (1 - b_1) \cdot n + (\sqrt{n} - 1 - b_1) \cdot \sqrt{n} + (\sqrt{n} - 1 - b_3)$ followed by an addition.

To apply our method to the isogeny based key exchange algorithm as mentioned in Sect. 3.1, we changed the representation of the public parameters in the beginning of the algorithm and executed the algorithm. In the last step we changed the representation back to the original form and matched both Alice and Bob's *j-invariant*.

To further test the correctness of the algorithm we ran an instance of the unmodified algorithm with same parameter set and numbers m and n. We verified that both executions produce identical results.

C List of Primes

In this section we list values for a and b for security level of around 256 bit and 512 bit. We found these values by a simple brute force search using a C implementation. As mentioned before the prime is $p = 2 \cdot 2^a 3^b + k$, with the value of $\log_2(3^b)$ close to a. The primality has been tested using *GMP* [21] and *PARI/GP* [22]. Also we should mention that this list is not exhaustive (Tables 2 and 3).

Table 2. Table for primes with around 256 bit PQ security

#	a	b	k	#	a	b	k	#	a	b	k
1	738	514	+1	10	760	490	−1	19	814	538	−1
2	741	510	+1	11	764	484	−1	20	819	552	+1
3	747	468	+1	12	768	518	+1	21	826	528	+1
4	748	468	−1	13	772	478	−1	22	826	538	−1
5	750	482	−1	14	774	476	−1	23	829	458	+1
6	750	490	+1	15	778	484	+1	24	830	512	+1
7	752	542	−1	16	784	496	+1	25	832	470	+1
8	756	468	−1	17	792	480	+1	26	834	488	−1
9	758	514	−1	18	798	526	+1				

Table 3. Table for primes with around 512 bit PQ security

#	a	b	k	#	a	b	k
1	1538	946	+1	5	1556	958	+1
2	1541	982	+1	6	1569	966	+1
3	1550	1018	−1	7	1570	942	−1
4	1551	964	+1	8	1598	1034	+1

References

1. De Feo, L., Jao, D., Plût, J.: Towards quantum-resistant cryptosystems from supersingular elliptic curve isogenies. https://eprint.iacr.org/2011/506.pdf
2. Shor, P.W.: Polynomial-time algorithms for prime factorization and discrete logarithms on a quantum computer. SIAM J. Comput. **26**(5), 1484–1509 (1997). arXiv:quant-ph/9508027v2
3. Montgomery, P.: Modular multiplication without trial division **44**, 519–521 (1985) http://www.ams.org/journals/mcom/1985-44-170/S0025-5718-1985-0777282-X/home.html
4. Solinas, J.A.: Generalized Mersenne prime. In: van Tilborg, H.C.A., Jajodia, S. (eds.) Encyclopedia of Cryptography and Security, pp. 509–510. Springer US, New York City (2011)
5. Solinas, J.A.: Mersenne prime. In: van Tilborg, H.C.A. (ed.) Encyclopedia of Cryptography and Security, pp. 774–775. Springer, US, New York City (2011)
6. Solinas, J.A.: Pseudo-Mersenne prime. In: van Tilborg, H.C.A., Jajodia, S. (eds.) Encyclopedia of Cryptography and Security, pp. 992–992. Springer US, New York City (2011)
7. Barrett, P.: Implementing the rivest shamir and adleman public key encryption algorithm on a standard digital signal processor. In: Odlyzko, A.M. (ed.) CRYPTO 1986. LNCS, vol. 263, pp. 311–323. Springer, Heidelberg (1987). doi:10.1007/3-540-47721-7_24
8. Childs, A., Jao, D., Soukharev, V.: Constructing elliptic curve isogenies in quantum subexponential time (2010). http://arxiv.org/abs/1012.4019/

9. Dinechin, F., Didier, L.-S.: Table-based division by small integer constants. In: Choy, O.C.S., Cheung, R.C.C., Athanas, P., Sano, K. (eds.) ARC 2012. LNCS, vol. 7199, pp. 53–63. Springer, Heidelberg (2012). doi:10.1007/978-3-642-28365-9_5
10. Donald, K.: The Art of Computer Programming, vol. 2. Addison-Wesley, Boston. Chapter 4
11. Galbraith, S., Stolbunov, A.: Improved algorithm for the isogeny problem for ordinary elliptic curves. Appl. Algebra Eng. Commun. Comput. **24**(2), 107–137 (2013)
12. Silverman, J.H.: The Arithmetic of Elliptic Curves. Graduate Texts in Mathematics, vol. 106. Springer, New York (2009)
13. Stolbunov, A.: Constructing public-key cryptographic schemes based on class group action on a set of isogenous elliptic curves. Adv. Math. Commun. **4**(2), 215–235 (2010)
14. Jao, D., Soukharev, V.: Isogeny-based quantum-resistant undeniable signatures. In: Mosca, M. (ed.) PQCrypto 2014. LNCS, vol. 8772, pp. 160–179. Springer, Heidelberg (2014). doi:10.1007/978-3-319-11659-4_10
15. Charles, D.X., Lauter, K.E., Goren, E.Z.: Cryptographic hash functions from expander graphs. J. Cryptol. **22**(1), 93–113 (2009)
16. Tani, S.: Claw Finding Algorithms Using Quantum Walk, March 2008. http://arxiv.org/abs/0708.2584
17. Microsoft predicts practical quantum computers within 10 years. http://www.ibtimes.co.uk/microsoft-predicts-practical-quantum-computers-within-10-years-1524268
18. McEliece, R.J.: A public-key cryptosystem based on algebraic coding theory. DSN Prog. Rep. **44**, 114–116 (1978)
19. Ding, J., Schmidt, D.: Rainbow, a new multivariable polynomial signature scheme. In: Ioannidis, J., Keromytis, A., Yung, M. (eds.) ACNS 2005. LNCS, vol. 3531, pp. 164–175. Springer, Heidelberg (2005). doi:10.1007/11496137_12
20. Recommended elliptic curves for federal government use. http://csrc.nist.gov/groups/ST/toolkit/documents/dss/NISTReCur.pdf
21. The GNU multiple precision arithmetic library. https://gmplib.org/
22. The PARI/GP computer algebra system. http://pari.math.u-bordeaux.fr/

A Search Strategy to Optimize the Affine Variant Properties of S-Boxes

Stjepan Picek(✉), Bohan Yang, and Nele Mentens

KU Leuven ESAT/COSIC and IMinds,
Kasteelpark Arenberg 10, 3001 Leuven-Heverlee, Belgium
stjepan@computer.org

Abstract. Affine transformations are an often used tool in symmetric key cryptography. They are mostly known as a way of removing fixed points in S-boxes, as for instance in the AES S-box. In general, affine transformations do not have an influence on most cryptographic properties, since those properties are affine invariant; affine transformations only change the representation of the S-box. Because of that, there is not much research on what would be the best affine transformation in terms of usability in practical scenarios. With this research, we try to close that gap; we concentrate on several cryptographic properties and one implementation property that are variable under various affine transformations. To provide experimental validations, we concentrate on affine transformations in S-boxes of three sizes, namely, 4×4, 5×5, and 8×8. Our results indicate that it is possible to optimize one or more of the considered properties. Finally, although we experiment with only a handful of properties, our methodology is of a general nature and could be used for other cryptographic properties that are affine variant.

1 Introduction

In the process of designing a symmetric key cipher, one often uses some form of affine transformation. The easiest example is to consider the AES S-box, where an exclusive OR (XOR) operation with a constant was added after the matrix multiplication to remove a fixed point at the first position [1]. In fact, this can be regarded as a representative example of the use of affine transformations in cipher design. When Leander and Poschmann defined optimal 4-bit S-boxes, they presented them through canonical representatives, i.e. those that are first in the lexicographical order, where all other optimal S-boxes are obtained via affine transformations of those canonical representatives [2]. Furthermore, the authors of the PRINCE cipher gave eight class representatives of suitable S-boxes where one can choose any S-box that is affine equivalent to one of those

This work has been supported in part by Croatian Science Foundation under the project IP-2014-09-4882. In addition, this work was supported in part by the Research Council KU Leuven (C16/15/058) and IOF project EDA-DSE (HB/13/020).

S. Duquesne and S. Petkova-Nikova (Eds.): WAIFI 2016, LNCS 10064, pp. 208–223, 2016.
DOI: 10.1007/978-3-319-55227-9_15

eight S-boxes [3]. One more example is the Ascon cipher [4], for which the S-box is an affine transformation of the Keccak S-box [5].

The biggest advantage of using affine transformations lies in the fact that most of the cryptographic properties will not change, but only the representation of the S-box changes. Therefore, it may seem that affine transformations do not deserve a more thorough analysis. However, today we know of a number of properties that can change under various affine transformations (i.e. properties that are affine variant) such as the transparency order [6], the modified transparency order [7], the branch number [8], and the confusion coefficient [9]. Besides those cryptographic properties, it is straightforward to investigate that implementation properties such as power and area also change under affine transformations.

In this paper, we consider the selection of affine transformations in order to optimize certain properties of S-boxes while retaining other properties. Naturally, not all properties considered in this work are relevant for all designs. Therefore, we want to stress that our focus is on the methodology for finding a suitable affine transformation. One can say that, since we are experimenting with the 4×4 size, it is possible to conduct an exhaustive search to find the best possible affine transformation. Indeed, that would be possible, at least when the considered cryptographic properties allow a fast evaluation. Nevertheless, we extend our experiments and show that our approach is also working for the sizes 5×5 and 8×8. Therefore, we conduct affine transformations on the S-boxes used in PRESENT [10], Keccak [5], and AES [1].

Our Contributions. In this paper, we present two main contributions. The first one is a general methodology for generating affine transformations that improve several affine variant properties. Naturally, this technique has merits for all S-box sizes, but it is particularly useful for S-box sizes that are too large for exhaustive search (i.e. bigger than 4×4). The second contribution is a novel way of reducing the search space size. As presented here, some properties change only under certain transformations and our methodology can find exactly those transformations.

Outline. The rest of this paper is organized as follows. In Sect. 2, we discuss a number of relevant cryptographic properties for S-boxes. In Sect. 3, we enumerate related work on the definition of the affine equivalence of S-boxes. Section 4 presents our experimental setup, the methods we use and the results. Next, in Sect. 5, we discuss the obtained results and offer several possible future research avenues. Finally, in Sect. 6, we give a short conclusion.

2 Cryptographic Properties of S-Boxes

The addition modulo 2 (XOR) is denoted as "$\oplus$". The inner product of the vectors $\bar{a}$ and $\bar{b}$ is denoted as $\bar{a} \cdot \bar{b}$ and equals $\bar{a} \cdot \bar{b} = \oplus_{i=1}^{n} a_i b_i$. The Hamming weight HW of a vector $\bar{v}$, where $\bar{v} \in \mathbb{F}_2^n$, is the number of non-zero positions in the vector. An (n, m)-function is any mapping F from $\mathbb{F}_2^n$ to $\mathbb{F}_2^m$. In this paper, we are interested only in cases where $n = m$.

As already stated, affine transformations cannot change many cryptographic properties. Three properties that are affine invariant, yet highly significant for the rest of this paper are the nonlinearity, the bijectivity, and the δ-uniformity of an S-box.

An S-box is called bijective (balanced) if it takes every value of $\mathbb{F}_2^m$ the same number of times, namely 2^{n-m} [11].

The nonlinearity N_F of an S-box F is equal to the minimum nonlinearity of all non-zero linear combinations of its coordinate functions [11].

$$N_F = 2^{n-1} - \frac{1}{2} \max_{\bar{a} \in \mathbb{F}_2^n, \bar{v} \in \mathbb{F}_2^{m*}} |W_F(\bar{a}, \bar{v})|. \quad (1)$$

$W_F(\bar{a}, \bar{v})$ represents the Walsh-Hadamard transform of F:

$$W_F(\bar{a}, \bar{v}) = \sum_{\bar{x} \in \mathbb{F}_2^n} (-1)^{\bar{v} \cdot F(\bar{x}) \oplus \bar{a} \cdot \bar{x}}. \quad (2)$$

Let F be a function from $\mathbb{F}_2^n$ into $\mathbb{F}_2^n$ and $a, b \in \mathbb{F}_2^n$. We denote:

$$D(a, b) = |\{x \in \mathbb{F}_2^n : F(x + a) + F(x) = b\}|. \quad (3)$$

The entry at the position (a, b) corresponds to the cardinality of $D(a, b)$ and is denoted as $\delta(a, b)$. The δ-uniformity δ_F is then defined as [12,13]:

$$\delta_F = \max_{a \neq 0, b} \delta(a, b). \quad (4)$$

The results for the aforesaid invariant properties of the PRESENT, Keccak, and AES S-box are given in Table 1. We note that those S-boxes are bijective. For the PRESENT and AES case we have S-boxes that are the best possible (or believed to be in the case of AES and its nonlinearity value) with regards to those properties. However, when considering the Keccak S-box, both nonlinearity and δ-uniformity are not optimal (cf. with e.g. the Almost Bent (AB) function in the PRIMATEs S-box that has nonlinearity 12 and δ-uniformity 2 [14]). Furthermore, the Keccak S-box has two fixed points and a branch number equal to two, while its affine equivalent S-box as used in Ascon has zero fixed points and a branch number equal to 3 [4].

Table 1. The values of the considered affine invariant properties.

S-box	Size	N_F	δ_F
PRESENT	4×4	4	4
Keccak	5×5	8	8
AES	8×8	112	4

We use the aforesaid properties to establish informal equivalence classes considering affine variant properties. When considering the 4×4 S-box size, Leander and Poschmann defined optimal S-boxes as those being bijective, with maximal nonlinearity (equal to 4), and minimal δ-uniformity (again equal to 4). Therefore, all resulting S-boxes in our experiments (those that were obtained with affine transformations) are still optimal S-boxes and we do not report those properties. Naturally, to ensure there are no mistakes, in our design process we still check them and consider S-boxes valid only if those properties do not change. Similarly, for sizes 5×5 and 8×8 we do not report the values of the affine invariant properties, but we check them in our analysis.

We emphasize that we are only interested in the design of S-boxes and not in the design of a whole cipher, and therefore we do not presume our S-boxes can replace the ones that are currently used in existing ciphers. Rather, in the process of the design of new ciphers, we believe our methodology can play a role. Therefore, since we do not aim to replace the original S-boxes, we do not conduct any cryptanalysis on complete ciphers. We concentrate only on the affine equivalence notion. Two S-boxes S_1 and S_2 of dimension $n \times n$ are affine equivalent if the following equation holds [2]:

$$S_1(x) = B(S_2(A(x) \oplus a)) \oplus b, \tag{5}$$

where A and B are invertible $n \times n$ matrices in $GF(2)$ and $a, b \in \mathbb{F}_2^n$.

We are also interested in the following cryptographic properties that change under affine transformations: the number of fixed points, the modified transparency order (MT_F) property, and the branch number (b_F).

An S-box has no fixed points if the following equation holds [1]:

$$S(a) \oplus a \neq \mathbf{0}, \forall a. \tag{6}$$

Although fixed points are generally considered as not desired [1], there are still a number of block ciphers that use S-boxes with fixed points, e.g. Noekeon [15], Midori [16], etc.

After the design of the transparency order property [6], Chakraborty et al. found some errors in the definition and consequently they suggested the modified transparency order property that is defined as [7]:

$$MT_F = \max_{\bar{\beta} \in \mathbb{F}_2^m} (m - \frac{1}{2^{2n} - 2^n} \sum_{\bar{a} \in \mathbb{F}_2^{n*}} \sum_{j=1}^{m} |A_{F_j}(a) + \sum_{i=1, i \neq j}^{m} (-1)^{\beta_i \oplus \beta_j} C_{F_i, F_j}(a)|), \tag{7}$$

where $A_{F_j}(a)$ represents the autocorrelation function of F and $C_{F_i,F_j}(a)$ represents the crosscorrelation function. The crosscorrelation $C_{F_i,F_j}(a)$ between functions F_i and F_j equals:

$$C_{F_i,F_j}(a) = \sum_{x \in \{0,1\}^n} (-1)^{F_i(x) \oplus F_j(x \oplus a)}. \tag{8}$$

The modified transparency order property was intended to show the level of resilience of S-boxes against side-channel attacks (SCA). However, we emphasize that it has been shown that it cannot serve as a definitive measure for a better side-channel resistance [17]. Nevertheless, the property has some merits since S-boxes with smaller modified transparency order indeed possess somewhat better SCA resilience. Here, the smaller the value of the modified transparency order, the better the resilience against SCA.

The branch number b_F can be defined as [8]:

$$b_F = \min_{a,b \neq a} (HW(a \oplus b) + HW(S(a) \oplus S(b))). \quad (9)$$

A bijective S-box must have a branch number equal to at least two [18]. Note that this definition of the branch number differs from the definition given in [1] and is suitable for evaluating a single S-box. The branch number describes the diffusion capabilities of an S-box; the higher the value, the better it is.

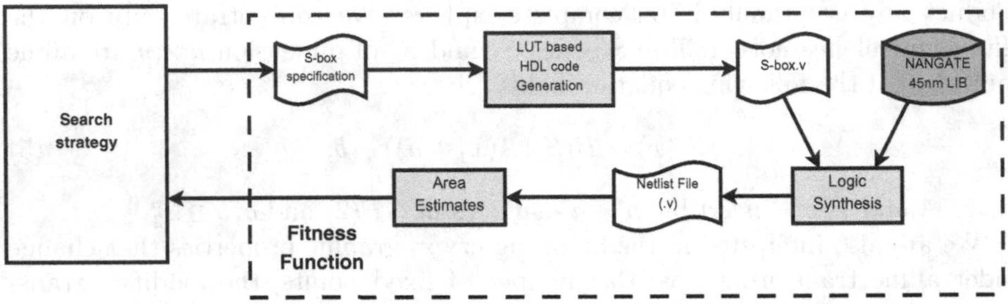

Fig. 1. Simulation setup for the generation/evaluation of S-boxes.

It is straightforward to experimentally verify that properties like area, power, and latency are affine variant, and here we consider area as a case study for affine variant implementation properties. The area cost of S-boxes is estimated by means of simulation before placement and routing. Figure 1 shows our simulation setup. In the first step, an affine transformation of an S-box is generated by the search strategy. The 4×4, 5×5, and 8×8 S-boxes are generated in the style of a lookup table (LUT). A Matlab (R2014b) script is then used to generate the HDL (Hardware Description Language) description of the S-box (Verilog file *S-box.v*) and to control the simulation flow and Synopsys Design Compiler (I-2013.12) to produce the gate-level netlist. All S-boxes are synthesized to the NANGATE 45 open cell library (PDKv1_3_v2010_12). The area consumption of the S-box is estimated by the Synopsys tool chain and represented with the unit GE, which stands for Gate Equivalent, i.e. the number of equivalent NAND gates in the specified technology.

3 Related Work

When designing the Rijndael cipher, Daemen and Rijmen constructed the S-box as a sequence of a function g and an invertible affine transformation f [1]. As the authors stated, the affine transformation has no impact on the nonlinearity property of the S-box, but enables the S-box to have a complex algebraic expression and no fixed or opposite fixed points [1]. They used an affine transformation that consists of a matrix multiplication followed by an XOR with a constant value. Osvik used a search algorithm with heuristics to find efficient instruction sequences for the Serpent S-boxes. In this work, the S-boxes are fixed and only the implementation changes [19]. Leander and Poschmann defined optimal 4-bit S-boxes as those having nonlinearity and δ-uniformity equal to 4 and being bijective [2]. Furthermore, they were able to find that there are only 16 optimal 4×4 S-boxes up to the affine equivalence. Saarinen conducted an exhaustive search of all 16! bijective 4×4 S-boxes [18]. The author investigated two types of equivalence, namely, the linear equivalence (LE) and the permutation-XOR equivalence (PE). The exhaustive search is conducted over all 4×4 S-boxes and they are classified into 142 090 700 different PE classes. Furthermore, the author defined Golden S-boxes and found that all Golden S-boxes belong to only four PE classes. Golden S-boxes are all S-boxes that have the following properties: S-boxes and their inverses satisfy a differential probability $p \leq 1/4$, a bias of linear approximation $\eta \leq 1/4$, and a branch number of 3. Further, all output bits have algebraic degree 3 and are dependent on all input bits in a nonlinear fashion. The Ph.D. thesis of de Cannière conducted the exhaustive analysis of all affine equivalence classes for 4×4 S-boxes [20]. In total, he found 302 affine equivalence classes. Biryukov et al. presented two algorithms for solving linear and affine equivalence problems for arbitrary S-boxes [21]. With this tool, the authors are able to find a number of new equivalent representations for a number of ciphers. Carlet et al. defined a more general version of an equivalence called the CCZ (Carlet, Charpin, Zinoviev) equivalence [22]. Two S-boxes F, G of size $n \times n$ are CCZ equivalent if there exists a linear permutation on $\mathbb{F}_2^n \times \mathbb{F}_2^n$ such that the graph of F is mapped to the graph of G. Ullrich et al. introduced an iterative deepening depth first search strategy to find the most efficient bitsliced implementations of S-boxes. The authors classify 4×4 S-boxes on the basis of some affine invariant properties and then find the most efficient S-box per class [8]. Berghoff et al. defined eight suitable classes to select an S-box when using the PRINCE cipher [3]. To be acceptable for PRINCE, the S-box needs to fulfill the following criteria: the maximal probability of a differential is 1/4 and there are exactly 15 such differentials, the maximal absolute bias of a linear approximation is 1/4 and there are exactly 30 of such approximations, and each of the 15 non-zero component functions has an algebraic degree of three. Picek et al. made a classification of 4×4 S-boxes where the best S-boxes are those that are optimal (i.e. belong to one of the 16 optimal classes), but also with an increased side-channel resistance (S-boxes with the lowest value of the transparency order property) [23]. We note that the transparency order property was shown to be flawed, which makes this classification relatively useless in practice. Sarkar et al. discussed how to choose an S-box in the (extended) affine equivalence class based on differential power analysis using a Hamming weight model [24]. Zhang et al. made a new classification of 4×4 S-boxes in which all S-boxes are classified under 183 categories out of which three are platinum categories (which are a subset of optimal S-boxes) [25]. The authors found that for the PRESENT, RECTANGLE [26], and SPONGENT [27] ciphers, one can get potentially better S-boxes by choosing from those three platinum categories. Canteaut and Roué investigated the effect of the S-box affine transformations on the maximal expected differential probability $MEDP$ and linear potential $MELP$ over two rounds of a substitution permutation network [28].

4 Experimental Setting and Results

In this section, we first briefly discuss the difficulty of finding the best possible affine transformation and then we present our search strategy. Next, we show the obtained results through two case studies. Finally, we discuss how our search strategy can be used to reduce the search space size when considering certain affine variant properties.

4.1 The Number of Affine Transformations

Here, we discuss the number of possible affine transformations. Recall from Eq. (5) that the matrices A and B need to be invertible in $GF(2)$. The number of $n \times n$ invertible binary matrices is the order of the General Linear Group over $GF(2)$ [29]:

$$GL(n) = \prod_{i=0}^{n-1} (2^n - 2^i). \qquad (10)$$

It is easy to calculate that for $n = 4$ there are in total 20 160 invertible matrices. However, since there are two matrices and additionally two constants $a, b \in \mathbb{F}_2^n$, the total number of combinations is $\approx 2^{36}$. Although this is a huge number, it is still within reasonable computation time if we consider certain properties that can be calculated efficiently (e.g. the nonlinearity property). However, if we consider implementation properties like power or area, then the time necessary to calculate the respective property for a single 4×4 S-box is in the order of magnitude of 10 s. Therefore, it is impossible to run an exhaustive search. Furthermore, for the 4×4 size, there are 16 optimal classes, which means that we need to run such a search 16 times. Already for the 5×5 size, there are 9 999 360 invertible matrices and therefore, the total number of combinations equals $\approx 2^{56}$. Based on the aforesaid, we see that an exhaustive search is often not a realistic option. Therefore, we need a faster way to obtain good results. To that end, we experiment with a heuristic search technique.

4.2 Heuristic Search Strategy

For heuristics, we use a genetic algorithm (GA); we utilize the simplest version of the algorithm we could design for this problem. Individuals (solutions) are encoded as a set of genotypes of bitstring values. Each genotype represents one matrix or a constant as given in Eq. (5). Each individual consists of four genotypes where the first two represent the matrices A and B and the third and the fourth genotypes represent the constants a and b. The first two genotypes can be considered as row vectors of size n^2, where the transformation to a matrix is done by splitting the vector in n rows of size n. We use the tournament selection mechanism in order to avoid the need to tune the crossover rate parameter. We work with the 3-tournament selection which is the option that offers the fastest convergence [30]. In that selection mechanism, three solutions are selected randomly and the worst one is discarded. Then, from the remaining two solutions, one offspring is created by the crossover operator. For variation operators, we use the Simple mutation and the One-point crossover [30]. In the One-point crossover, a random crossover point is selected and all bits before that point are taken from the first solution and the remaining bits are taken from the second solution. In the Simple mutation, a randomly selected bit is inverted with a probability p_m per individual (i.e. each individual will mutate with a probability equal to p_m where the mutation operator is executed only once on a given individual). We set the mutation probability to 0.8. In all experiments, we use a population size of 100 individuals. The algorithm starts with a random initial population, where we do not impose any criteria on the starting population (e.g. we do not require that the matrices A and B are invertible). As a termination criterion, we use the number of evaluations without improvement, which is in our case 50 generations.

Next, we give a description of one generation of our heuristics for the 4×4 S-box size. First, the algorithm creates 100 random individuals where every individual consists of 4 bitstring vectors. Two bitstring vectors represent the matrices A and B and have a length of 16, and two bitstring vectors represent the constants a and b of size 4. Then, then algorithm randomly selects 3 individuals and calculates their fitness (i.e. when setting the values of the matrices and the constants to Eq. (5) it calculates for instance the modified transparency order). Then, the worst individual is discarded and from the two better ones, an offspring is created with the One-point crossover. This process repeats N times to ensure that most of the population is evaluated and improved. Afterwards, the mutation is done on a number of individuals and finally, all individuals are evaluated again and their fitness value is updated. This procedure runs until the stopping criterion is met and at that moment, the evolution is finished.

Regarding the speed of the algorithm, on average one generation (100 individuals) needs around 1 s to evolve. In that estimation we include the cost of the evaluation of the cryptographic properties, but not the cost of the evaluation of the implementation properties. To calculate the area estimate of a single S-box, we require a processing time in the order of magnitude of 10 s, which means one generation lasts for around 1 000 s in total. We note that although here we work with GA, our methodology is not exclusive for that algorithm, but it could work with any other heuristics that supports the bitstring representation. Naturally, it is to be expected that in such case one could also need to change the fitness function and the stopping criterion. For further details about genetic algorithms, we refer readers to [30].

4.3 The Obtained Results

Next, we give the results for two cryptographic properties as the first case study and then for the implementation property as the second case study. In all tables we display in the *Original* column the value that denotes the initial value of the S-boxes we investigate and in the *New* column the value that denotes the value of the property after the affine transformation.

The Modified Transparency Order. When looking for S-boxes with a minimal modified transparency order value, we aim to minimize the following equation, since the smaller the value of MT_F the better:

$$objective = MT_F + fixed. \tag{11}$$

However, it has been shown that the minimal necessary affine transformation (minimal in a sense that it consists of the smallest number of terms) needed to change the modified transparency order property is of the form [17]:

$$S_2(x) = B(S_1(x)). \tag{12}$$

Therefore, we can simplify our potential solutions (affine transformations) such that they consist of only a single matrix (B), which is what we do in the next set of experiments. Note that now the search space is rather small for sizes 4×4 and 5×5. Since we also want to avoid fixed points, we include this goal in the objective function. The results obtained with the GA approach are given in Table 2. Column *Matrix B* presents examples of values one should use for the matrix B to obtain the reported MT_F values. There, one can see that our approach finds the minimal modified transparency value that is possible for the PRESENT class (G_1). Since there are no previous results for the Keccak S-box (or any S-box of size 5×5) and the modified transparency order, we report the difference between the original S-box and the transformed one where we see that the difference is significant. Furthermore, for the 8×8 size, our approach yields a marginally better value than the previously known one that equals 6.89 [17]. Although the difference is negligible, it assures us that our technique is a viable choice. This is especially apparent since the modified transparency order is computationally expensive (when compared to other cryptographic properties) so by following our approach one requires less time compared to random search or running heuristics with the goal of finding new S-boxes.

Branch Number Results. Here, our objective function equals:

$$objective = b_F + (2^n - fixed), \tag{13}$$

where the goal is to maximize the b_F value. Note that here we subtract the number of fixed points from the theoretical maximal number of fixed points since we require to increase the branch number, but without adding fixed points. When considering the branch number property, we do not know what is the smallest affine transformation we need to use to change that property. Therefore, we work with Eq. (5) and report our results in Table 3.

Table 2. Results for MT_F, Eq. (12).

S-box	MT_F		Matrix B
	Original	New	
PRESENT	2.467	1.9	[4, 2, 7, 8]
Keccak	3.871	2.645	[4, 25, 1, 16, 6]
AES	6.916	6.88	[35, 242, 8, 80, 64, 184, 138, 52]

Table 3. Results for GA, Eq. (13).

S-box	b_F	
	Original	New
PRESENT	3	3
Keccak	2	3
AES	2	2

For the PRESENT S-box, we could not improve the original value of the branch number property (since 3 is also the maximal possible value), but we notice the property is quite sensitive to changes and it is easy to degrade the value to 2.

Area Results. In order to find S-boxes that have minimal area, we use the following simple objective function, with the goal to minimize the value:

$$objective = area + fixed. \tag{14}$$

The results for the area are given in Table 4. Note that again here we do not allow that our transformed S-boxes have fixed points. Here, we omit the AES S-box case since we believe it is unrealistic to expect that S-box of such a size would be implemented in a lookup table fashion. The results are given in gate equivalence (GE) unit.

Table 4. Results for GA, Eq. (14).

S-box	Area [GE]	
	Original	New
PRESENT	26	13.3
Keccak	17	20.33

To put those results into a better perspective, we give area results for lookup table base implementations of several more relevant S-boxes. For the 4×4 size, Piccolo [31] has 17.33 GE, Prince [3] has 17 GE, Rectangle [26] has 24 GE, and Midori [16] S-boxes Sb_0 and Sb_1 have 13.67 and 15.33 GE, respectively. For the 5×5 size, Ascon [4] has 30.67 GE and PRIMATEs [14] has 36 GE.

4.4 Reducing the Search Space Size

In the previous section we showed that heuristics can be used to find improved S-boxes when considering affine variant properties. However, still the search space is large and a natural question is whether there is a way to reduce it. It turns out that is possible when considering certain properties as shown next.

Imagine a scenario where one does not know the minimal necessary affine transformation needed to find the optimal value of a property (here, we consider the modified transparency order property). Then, one would need to run all possible combinations of affine transformations in order to find the minimal one. The question is whether this approach can be simplified and automated. In the next experiment we change the objective function to the following one:

$$objective = MT_F * 100 + HW(X), \tag{15}$$

where $HW(X)$ represents the Hamming weight of all matrices and constants in the affine transformation as in Eq. (5). Therefore, with this objective function, we aim to minimize not only the value of the modified transparency order property, but also the number of ones in the matrices and constants from Eq. (5). Note that the main gain in the search space reduction is for the cases in which one or both of the matrices are not necessary (i.e. identity matrix). We add a weight factor to the objective function, since optimizing the modified transparency order is our primary objective. Therefore, a solution that has a good MT_F value and a relatively large $HW(X)$ value will not be replaced with a solution that has a worse MT_F value, but a better $HW(X)$ value. The tuning procedure for the weight factor is based on the observation that if we multiply MT_F with a multiplicand that is higher than the worse case value of $HW(X)$, the results do not change; the only difference is the number of necessary iterations of the algorithm. For S-boxes of size 4×4, the HW is equal to 40 when every matrix and constant consist of all ones. Note that we disregard that the matrices must be invertible; therefore the worst case is an all-ones matrix.

The results for the new objective are given in Table 5. In column *Affine transformation* we give examples of matrix and constant values one needs to use to obtain an S-box with the reported value of the modified transparency order. We denote the identity matrix of dimension n with I_n.

Table 5. Results for GA, Eq. (15).

S-box	MT_F	Affine transformation
PRESENT	1.9	A = I_4, B = [1, 4, 13, 2], a = b = 0
Keccak	2.645	A = I_5, B = [11, 1, 2, 20, 4], a = b = 0
AES	6.88	A = I_8, B = [2, 32, 8, 4, 64, 48, 128, 3], a = b = 0

5 Discussion

When considering the results for the modified transparency order, we see that our approach manages to find S-boxes that outperform the original ones. For the 4×4 and 5×5 sizes that difference is a rather significant one. Furthermore, for the 4×4 size, we can confirm that our method reaches the best possible value compared to previously reported results [17,32]. When looking at the results for 8×8, one can consider them somewhat disappointing since we see only a marginal improvement over the original AES S-box and an even smaller improvement over previously known results. However, we emphasize that here we did not concentrate on the modified transparency property, but on a method to find affine transformations that result in improved properties. Furthermore, it is unknown what is the best possible value of the modified transparency order for the 8×8 S-box having the same cryptographic properties as the AES S-box, so this result could be better than it seems.

When considering the branch number results, for the Keccak case the value is improved, while for the PRESENT and the AES case our method could not find any improved S-boxes. However, we note that it easily finds equivalent S-boxes with the same branch number value, which is quite difficult to do if one would for instance use affine transformations with random values (that still give invertible matrices).

Besides those results, we want to emphasize the results obtained in Table 5. With those experiments we started with an affine transformation of the form given in Eq. (5). The algorithm itself reduced it to an affine transformation as given in Eq. (12). Although we experimented here with a property for which we a priori know the minimal affine transformation, we can easily imagine a scenario where one does not know the minimal transformation. In such a case, our approach can be used for faster evaluation. Indeed, if one observes that the same affine transformation is used in three different S-box sizes, it is a reasonable assumption that that transformation is also the minimal necessary transformation to change a certain affine variant property. Even when it is not the minimal necessary transformation, it will still have a smaller number of terms and therefore the search space size will be reduced. We note that we are not sure whether this approach can be used in cases when there exist only a few possible property values to reach. There, one could improve the objective function to differentiate between matrices A and B, i.e. to first try to minimize the first one, and only then the second one.

When considering the area results, our method is again successful. For smaller sizes it finishes in a short period of time with significantly better area results, as evident from the 4×4 scenario. For the 5×5 size, our best obtained S-box is somewhat worse than the Keccak S-box. Since the size of the Keccak S-box is already quite small, it would be unrealistic to expect a big difference. However, we note that our 5×5 S-box does not have fixed points. A fixed point in the Keccak S-box is also the reason why our heuristics could not output the original S-box (i.e. with constants a and b equal to zero and matrices A and B set as identity matrices), since it can evolve only affine transformations that result in

S-boxes without fixed points. Note that for instance the Ascon S-box (which is an affine transformation of the Keccak S-box, without fixed points and with a branch number equal to 3) has an area of 30.67 GE, which is 30% worse than our S-box. Additionally, since there are usually several clock frequencies of interest, one would need to repeat the experiments for each frequency. With the heuristic approach, one could even combine the search for several frequencies and aim to find one S-box that is good for all settings. Besides optimizing only for the implementation properties, one could at the same time optimize for some affine variant cryptographic properties like the branch number, for example.

On a more general level, we see two possible drawbacks of our approach. The first one is the relevance of the properties we investigate here. However, we believe our approach should be regarded as a more general methodology that works for other properties as well. The second possible drawback is the heuristic nature of our approach where there is no guarantee that the optimal solution is found. However, our results show that the method is quite reliable and consistently producing good solutions. Finally, because of its heuristic nature, our algorithm works well on bigger sizes, which is usually not the case with deterministic algorithms [8,33]. Finally, we believe this approach is substantially better than for instance using heuristics to find completely new S-boxes. This is because already for the 5×5 size it is not easy to find S-boxes with the best possible values for the invariant cryptographic properties (e.g., nonlinearity and δ-uniformity). As already noted, our methodology makes sense only when designing new ciphers and should not be considered as a source of S-boxes that can replace existing ones in ciphers. Indeed, to fully utilize the advantages of this method, one should first find an S-box that is in accordance with his criteria. Only after that, the linear layer can be designed so the design goals of a cipher are met.

In all our experiments, we see that we are interested in more than one property, i.e. we always minimize the number of fixed points alongside some other cryptographic or implementation property. However, the question is whether can we combine even more properties and still obtain good results. To that end, we evolved S-boxes of size 4×4 that are without fixed points, with an as high as possible branch number, and an as low as possible modified transparency order (therefore, our objective functions consists of three parts). As could be expected, our methodology works, but there is no guarantee that it can find S-boxes with all optimal values. Indeed, on the one hand we found S-boxes without fixed points, with branch number 3, and a modified transparency order equal to 2.13. On the other hand, we found S-boxes without fixed points, with branch number 2, and a modified transparency order of 1.9. This shows us a scenario with conflicting objectives, which means it is not possible to obtain a single S-box with all optimal properties.

In future work, we plan to investigate other cryptographic properties as well as the implementation properties of S-boxes of various sizes. Recently, the energy efficient cipher Midori was presented for which the S-box is an involution that is designed with small energy consumption as a goal [16]. However, that S-box has four fixed points which, combined with some other factors, lead in the end to a successful attack on full Midori64 cipher [34]. It would be interesting to see whether it is possible to use our search method in order to obtain involutions that have a smaller number of fixed points, and yet are optimal (with regards to invariant properties). If we consider the 4×4 case, not all optimal classes have involutions, which is only possible when the inverse S-box is a member of the same class as the S-box. Therefore, this helps us to limit our search to only a subset of optimal classes. Accordingly, our search strategy seems to lend itself naturally for that case because it allows us to search only in the relevant classes and to do that much faster than with an exhaustive search method.

6 Conclusions

In this work, we investigate how to find appropriate affine transformations when considering S-box properties that are affine variant. We conduct the analysis for three popular S-box sizes to show our approach scales well even for larger S-boxes. The results show it is possible to efficiently find affine transformations that offer better properties. We also discuss one more possible usage of our methodology, which is the selection of the minimal appropriate affine transformation form. This way, we do not only find better S-boxes (with regards to affine variant cryptographic properties), but also transformations that are easier to enumerate since they consist of a smaller number of terms.

References

1. Daemen, J., Rijmen, V.: The Design of Rijndael. Springer-Verlag New York Inc., Secaucus (2002)
2. Leander, G., Poschmann, A.: On the classification of 4 bit S-boxes. In: Carlet, C., Sunar, B. (eds.) WAIFI 2007. LNCS, vol. 4547, pp. 159–176. Springer, Heidelberg (2007). doi:10.1007/978-3-540-73074-3_13
3. Borghoff, J., et al.: PRINCE – a low-latency block cipher for pervasive computing applications. In: Wang, X., Sako, K. (eds.) ASIACRYPT 2012. LNCS, vol. 7658, pp. 208–225. Springer, Heidelberg (2012). doi:10.1007/978-3-642-34961-4_14
4. Dobraunig, C., Eichlseder, M., Mendel, F., Schläffer, M.: Ascon. AESAR Submission (2014). http://ascon.iaik.tugraz.at/
5. Bertoni, G., Daemen, J., Peeters, M., Assche, G.: Keccak. In: Johansson, T., Nguyen, P.Q. (eds.) EUROCRYPT 2013. LNCS, vol. 7881, pp. 313–314. Springer, Heidelberg (2013). doi:10.1007/978-3-642-38348-9_19
6. Prouff, E.: DPA attacks and S-boxes. In: Gilbert, H., Handschuh, H. (eds.) FSE 2005. LNCS, vol. 3557, pp. 424–441. Springer, Heidelberg (2005). doi:10.1007/11502760_29
7. Chakraborty, K., Sarkar, S., Maitra, S., Mazumdar, D., Mukhopadhyay, D., Prouff, E.: Redefining the transparency order. In: Coding and Cryptography, International Workshop, WCC 2015, Paris, France, 13–17 April 2015 (2015)
8. Ullrich, M., De Cannière, C., Indesteege, S., Küçük, Ö., Mouha, N., Preneel, B.: Finding optimal bitsliced implementations of 4×4-bit S-boxes (2011)

9. Fei, Y., Luo, Q., Ding, A.A.: A statistical model for DPA with novel algorithmic confusion analysis. In: Prouff, E., Schaumont, P. (eds.) CHES 2012. LNCS, vol. 7428, pp. 233–250. Springer, Heidelberg (2012). doi:10.1007/978-3-642-33027-8_14
10. Bogdanov, A., Knudsen, L.R., Leander, G., Paar, C., Poschmann, A., Robshaw, M.J.B., Seurin, Y., Vikkelsoe, C.: PRESENT: an ultra-lightweight block cipher. In: Paillier, P., Verbauwhede, I. (eds.) CHES 2007. LNCS, vol. 4727, pp. 450–466. Springer, Heidelberg (2007). doi:10.1007/978-3-540-74735-2_31
11. Carlet, C.: Vectorial Boolean functions for cryptography. In: Crama, Y., Hammer, P.L. (eds.) Boolean Models and Methods in Mathematics, Computer Science, and Engineering, 1st edn, pp. 398–469. Cambridge University Press, New York (2010)
12. Biham, E., Shamir, A.: Differential cryptanalysis of DES-like cryptosystems. In: Menezes, A.J., Vanstone, S.A. (eds.) CRYPTO 1990. LNCS, vol. 537, pp. 2–21. Springer, Heidelberg (1991). doi:10.1007/3-540-38424-3_1
13. Nyberg, K.: Perfect nonlinear S-boxes. In: Davies, D.W. (ed.) EUROCRYPT 1991. LNCS, vol. 547, pp. 378–386. Springer, Heidelberg (1991). doi:10.1007/3-540-46416-6_32
14. Andreeva, E., Bilgin, B., Bogdanov, A., Luykx, A., Mendel, F., Mennink, B., Mouha, N., Wang, Q., Yasuda, K.: Primates v1 submission to the CAESAR competition (2014). http://competitions.cr.yp.to/round1/primatesv1.pdf
15. Daemen, J., Peeters, M., Assche, G.V., Rijmen, V.: Nessie proposal: the block cipher NOEKEON. Nessie submission (2000) http://gro.noekeon.org/
16. Banik, S., Bogdanov, A., Isobe, T., Shibutani, K., Hiwatari, H., Akishita, T., Regazzoni, F.: Midori: a block cipher for low energy (extended version). Cryptology ePrint Archive, Report 2015/1142 (2015) http://eprint.iacr.org/
17. Picek, S., Mazumdar, B., Mukhopadhyay, D., Batina, L.: Modified transparency order property: solution or just another attempt. In: Chakraborty, R.S., Schwabe, P., Solworth, J. (eds.) SPACE 2015. LNCS, vol. 9354, pp. 210–227. Springer, Heidelberg (2015). doi:10.1007/978-3-319-24126-5_13
18. Saarinen, M.-J.O.: Cryptographic analysis of all 4 × 4-bit S-boxes. In: Miri, A., Vaudenay, S. (eds.) SAC 2011. LNCS, vol. 7118, pp. 118–133. Springer, Heidelberg (2012). doi:10.1007/978-3-642-28496-0_7
19. Osvik, D.A.: Speeding up serpent. In: AES Candidate Conference, pp. 317–329 (2000)
20. de Cannière, C.: Analysis and Design of Symmetric Encryption Algorithms. Ph.D. thesis, Katholieke Universiteit Leuven (2007)
21. Biryukov, A., Cannière, C., Braeken, A., Preneel, B.: A toolbox for cryptanalysis: linear and affine equivalence algorithms. In: Biham, E. (ed.) EUROCRYPT 2003. LNCS, vol. 2656, pp. 33–50. Springer, Heidelberg (2003). doi:10.1007/3-540-39200-9_3
22. Carlet, C., Charpin, P., Zinoviev, V.: Codes, bent functions and permutations suitable for DES-like cryptosystems. Des. Codes Crypt. **15**(2), 125–156 (1998)
23. Picek, S., Ege, B., Papagiannopoulos, K., Batina, L., Jakobovic, D.: Optimality and beyond: the case of 4 × 4 S-boxes. In: IEEE International Symposium on Hardware-Oriented Security and Trust, HOST 2014, Arlington, VA, USA, 6–7 May 2014. IEEE Computer Society, pp. 80–83 (2014)
24. Sarkar, S., Maitra, S., Chakraborty, K.: Differential power analysis in hamming weight model: how to choose among (extended) affine equivalent S-boxes. In: Meier, W., Mukhopadhyay, D. (eds.) INDOCRYPT 2014. LNCS, vol. 8885, pp. 360–373. Springer, Heidelberg (2014). doi:10.1007/978-3-319-13039-2_21

25. Zhang, W., Bao, Z., Rijmen, V., Liu, M.: A new classification of 4-bit optimal S-boxes and its application to PRESENT, RECTANGLE and SPONGENT. In: Leander, G. (ed.) FSE 2015. LNCS, vol. 9054, pp. 494–515. Springer, Heidelberg (2015). doi:10.1007/978-3-662-48116-5_24
26. Zhang, W., Bao, Z., Lin, D., Rijmen, V., Yang, B., Verbauwhede, I.: RECTANGLE: a bit-slice ultra-lightweight block cipher suitable for multiple platforms. IACR Cryptology ePrint Archive 2014, 84 (2014)
27. Bogdanov, A., Knežević, M., Leander, G., Toz, D., Varıcı, K., Verbauwhede, I.: SPONGENT: a lightweight hash function. In: Preneel, B., Takagi, T. (eds.) CHES 2011. LNCS, vol. 6917, pp. 312–325. Springer, Heidelberg (2011). doi:10.1007/978-3-642-23951-9_21
28. Canteaut, A., Roué, J.: On the behaviors of affine equivalent Sboxes regarding differential and linear attacks. In: Oswald, E., Fischlin, M. (eds.) EUROCRYPT 2015. LNCS, vol. 9056, pp. 45–74. Springer, Heidelberg (2015). doi:10.1007/978-3-662-46800-5_3
29. Rotman, J.: An Introduction to the Theory of Groups. Springer, New York (1995)
30. Eiben, A.E., Smith, J.E.: Introduction to Evolutionary Computing. Springer, Heidelberg (2003)
31. Shibutani, K., Isobe, T., Hiwatari, H., Mitsuda, A., Akishita, T., Shirai, T.: *Piccolo*: an ultra-lightweight blockcipher. In: Preneel, B., Takagi, T. (eds.) CHES 2011. LNCS, vol. 6917, pp. 342–357. Springer, Heidelberg (2011). doi:10.1007/978-3-642-23951-9_23
32. Evci, M.A., Kavut, S.: DPA resilience of rotation-symmetric S-boxes. In: Yoshida, M., Mouri, K. (eds.) IWSEC 2014. LNCS, vol. 8639, pp. 146–157. Springer, Heidelberg (2014). doi:10.1007/978-3-319-09843-2_12
33. Courtois, N.T., Hulme, D., Mourouzis, T.: Solving circuit optimisation problems in cryptography and cryptanalysis. Cryptology ePrint Archive, Report 2011/475 (2011). http://eprint.iacr.org/
34. Guo, J., Jean, J., Nikolić, I., Qiao, K., Sasaki, Y., Sim, S.M.: Invariant Subspace Attack Against Full Midori64. Cryptology ePrint Archive, Report 2015/1189 (2015). http://eprint.iacr.org/

Cryptography and Boolean Functions

A Super-Set of Patterson-Wiedemann Functions – Upper Bounds and Possible Nonlinearities

Selçuk Kavut[1(✉)], Subhamoy Maitra[2], and Ferruh Özbudak[3,4]

[1] Department of Computer Engineering, Balıkesir University, 10145 Balıkesir, Turkey
skavut@balikesir.edu.tr

[2] Applied Statistics Unit, Indian Statistical Institute, 203 B. T. Road, Kolkata 700108, India
subho@isical.ac.in

[3] Department of Mathematics and Institute of Applied Mathematics, Middle East Technical University, 06800 Ankara, Turkey
ozbudak@metu.edu.tr

[4] Department of Mathematical Sciences, Aalborg University, Aalborg, Denmark

Abstract. Constructing Boolean functions on odd number of variables with nonlinearity exceeding the bent concatenation bound is one of the most difficult combinatorial problems in the domain of Boolean functions and it has deep implications to coding theory and cryptology. After demonstration of such functions by Patterson and Wiedemann in 1983, for more than three decades the efforts have been channelized in obtaining the instances only. For the first time, in this paper, we try to explore non trivial upper bounds on nonlinearity of such functions which are invariant under several group actions. In fact, we consider much larger sets of functions than what have been considered so far and obtain tight upper bounds on the nonlinearity in several cases. To support our claims, we present computational results for functions on n variables where n is an odd composite integer, $9 \leq n \leq 39$. In particular, our results for $n = 15$ and 21 are of immediate interest given recent research results in this domain. Not only the upper bounds, we also identify what are the nonlinearities that can actually be achieved above the bent concatenation bound for such class of functions.

Keywords: Nonlinearity bound · Patterson-Wiedemann type functions · Covering radius · First order Reed-Muller code

1 Introduction

The maximum achievable nonlinearity of an n-variable Boolean function for n odd and $n > 7$ is a long standing open problem. The problem is directly connected to coding theory, since it corresponds to the covering radius of the

S. Duquesne and S. Petkova-Nikova (Eds.): WAIFI 2016, LNCS 10064, pp. 227–242, 2016.
DOI: 10.1007/978-3-319-55227-9_16

first order Reed-Muller codes of block length 2^n. High nonlinearity is an important property for the Boolean functions used in cryptographic primitives for resisting linear cryptanalysis [10] as well as correlation and fast correlation attacks [11,15] and hence this issue is related to cryptology also. For n even, the functions with provably maximum nonlinearity $2^{n-1} - 2^{\frac{n}{2}-1}$ exist and such functions are called bent, though the complete characterization of such functions is not yet known for $n \geq 8$. Let us consider an n-variable function f constructed by concatenating two $(n-1)$-variable bent functions g and h, i.e., $f(x_0, x_1, \ldots, x_{n-1}) = x_0 g(x_1, \ldots, x_{n-1}) \oplus (x_0 \oplus 1)h(x_1, \ldots, x_{n-1})$ for all $(x_0, x_1, \ldots, x_{n-1}) \in \mathbb{F}_2^n$. One can then easily check that the nonlinearity of f is $2^{n-1} - 2^{\frac{n-1}{2}}$. This is called the bent concatenation bound, which had been conjectured [4] to be the maximum attainable nonlinearity until disproved [13] in 1983.

Solving the question for small number of variables dates back to 1972 when it was shown [1] that the nonlinearity of a 5-variable function is at most 12 which is in fact the bent concatenation bound. Almost a decade later, in 1980, the problem was solved [12] for 7-variable functions noting that the maximum nonlinearity here is also equal to the bent concatenation bound which is 56. The existence of functions on odd number of variables having nonlinearity greater than the bent concatenation bound had remained unknown till Patterson and Wiedemann demonstrated [13] in 1983 two functions on 15 variables achieving nonlinearity $2^{15-1} - 2^{\frac{15-1}{2}} + 20 = 16276$. Both these functions are obtained in a very small class consisting of merely 2^{11} functions that are idempotent (a function f is called idempotent if it is invariant under the action of the group of Frobenius automorphisms, i.e., such a function satisfies the condition $f(\alpha) = f(\alpha^2)$ for all $\alpha \in \mathbb{F}_{2^n}$) and invariant under the action of $\mathbb{F}_{2^5}^* \cdot \mathbb{F}_{2^3}^*$. Over two decades later, in 2006, the 9-variable functions having nonlinearity 241 were found [8] in the class of idempotent functions and shortly after that this result was improved [7] to 242 by defining a generalized class of idempotent functions, in which a function f satisfies the condition $f(\alpha) = f(\alpha^{2^k})$ for all $\alpha \in \mathbb{F}_{2^n}$ where k is a fixed divisor of n.

Consider an n-variable function (n odd) f having nonlinearity $2^{n-1} - 2^{\frac{n-1}{2}} + \mu_n$ ($\mu_n > 0$, integer, i.e., nonlinearity more than the bent concatenation bound) and an m-variable bent function (m even) g. If f and g are functions on different input variables, $f \oplus g$ (known as direct sum) is an $(n+m)$-variable Boolean function having nonlinearity $2^{n+m-1} - 2^{\frac{n+m-1}{2}} + \mu_n \cdot 2^{\frac{m}{2}}$. Thus, if one starts with a 9-variable function with nonlinearity $2^{9-1} - 2^{\frac{9-1}{2}} + 2 = 242$ [7], then it is possible to construct functions of $9+m$ variables having nonlinearity $2^{9+m-1} - 2^{\frac{9+m-1}{2}} + 2 \cdot 2^{\frac{m}{2}}$. For example, by this method, we will have functions on $9 + 6 = 15$ variables with nonlinearity $2^{15-1} - 2^{\frac{15-1}{2}} + 2 \cdot 2^{\frac{15-9}{2}} = 16256 + 16 = 16272$. However, one should note that the functions identified by Patterson and Wiedemann [13] are of nonlinearity $2^{15-1} - 2^{\frac{15-1}{2}} + 20 = 16256 + 16 = 16276 > 16272$. Thus for odd $n > 15$, one should start the construction from such 15-variable functions as available from [13] in direct sum construction with bent functions to have the highest achievable nonlinearity known so far. Thus, we like to motivate the term $\frac{\mu_n}{2^{\frac{n-1}{2}}} = \mu_n \cdot 2^{-\frac{n-1}{2}}$ in such case. As long as we obtain some construction with maximum known nonlinearity beating the bent concatenation bound, we should look at this term. For $n = 9$ [7], this term is $2 \cdot 2^{-4} = \frac{1}{8}$ and for $n = 15$, this is $20 \cdot 2^{-7} = \frac{5}{32}$ [13] which is greater than $\frac{1}{8}$. Naturally, till date the best known $\mu_n \cdot 2^{-\frac{n-1}{2}}$ is for $n = 15$ [13].

Now let us get into the details of the Patterson-Wiedemann functions [13], which we will refer to as PW functions in this document. Let $\phi_2 \in GL_{\mathbb{F}_2}(\mathbb{F}_{2^n})$ be the Frobenius automorphism of $\mathbb{F}_{2^n}$, which is given by $\phi_2(\alpha) = \alpha^2$ for all $\alpha \in \mathbb{F}_{2^n}$. As aforementioned, the class containing only 2^{11} functions considered in [13] is formed by imposing the constraint of being invariant under the action of $\mathbb{F}_{2^5}^* \cdot \mathbb{F}_{2^3}^*$ and the group of Frobenius automorphisms $\langle \phi_2 \rangle$. In fact, it is easy to determine, by performing an exhaustive search, the nonlinearities attained in this class are 16268, 16269, 16275, and 16276. There are quite a few open questions here.

- Can we decide what nonlinearities are possible to achieve without performing the exhaustive search? This question is pertinent as one may like to relax the constraints and try to search a larger class in the hope of better nonlinearity.
- What are the possible nonlinearities above the bent concatenation bound when we consider the action of $\mathbb{F}_{2^5}^* \cdot \mathbb{F}_{2^3}^*$ only and not of the group of Frobenius automorphisms $\langle \phi_2 \rangle$? For this, the search complexity becomes 2^{151} for 15-variable functions.
- Further, we may consider an even larger class when we consider the action of $\mathbb{F}_{2^5}^*$ only. This makes the search spacc incredibly huge containing 2^{1057} functions. How can one obtain the possible nonlinearities greater than the bent concatenation bound in such a large class that cannot be searched exhaustively?

We could answer all these questions for 15-variable PW functions and show that for all these larger classes, the maximum nonlinearity is 16276, that had been achieved long back in [13] in a very small class of 2^{11} functions. This is an important negative result that would save a lot of unsuccessful search in those larger classes had our result been known. There are further implications of our results. Until recently, the PW functions beating the bent concatenation bound were known only for $n = 15 = 5 \cdot 3$. The next possible candidate had been $n = 21 = 7 \cdot 3$ and such functions could be found [6] after a long gap of more than three decades. Each function obtained in [6] are of nonlinearity $2^{21-1} - 2^{\frac{21-1}{2}} + 61$ and the authors left the open question whether there can be functions having higher nonlinearity. Our method shows that the upper bound of nonlinearity in this case could be as high as $2^{21-1} - 2^{\frac{21-1}{2}} + 196$ considering the action of $\mathbb{F}_{2^7}^* \cdot \mathbb{F}_{2^3}^*$ and further the upper bound slightly increases to $2^{21-1} - 2^{\frac{21-1}{2}} + 199$ considering the action of $\mathbb{F}_{2^7}^*$ only. Thus, this is a result in the positive direction that shows one may indeed put further search effort with the expectation of obtaining instances of 21-variable functions with higher nonlinearity values.

Thus, in general framework, we consider n-variable functions[1], where $n = p \cdot q$ such that p and q are distinct odd primes with $p > q$. Then we try to obtain the possible nonlinearity values greater than the bent concatenation bound for the functions which are

- invariant under the action of $\mathbb{F}_{2^p}^* \cdot \mathbb{F}_{2^q}^*$ and also
- invariant under the action of either $\mathbb{F}_{2^p}^*$ or $\mathbb{F}_{2^q}^*$.

We present techniques that involve basic combinatorics and elementary number theoretic techniques. Numerical results are presented for the odd composite integers n where $9 \leq n \leq 39$. To the best of our knowledge, no such result on upper bound of nonlinearity in these larger classes could be explored since the construction of 15-variable PW functions [13] that dates back to 1983. The generic upper bound on nonlinearity for functions on odd number of variables n is $2\lfloor 2^{n-2} - 2^{\frac{n}{2}-2} \rfloor$ [5]. Our results show nontrivial upper bounds for the class of functions we consider here and that is indeed less than the generic upper bound provided in [5]. Next we provide necessary background by reviewing the PW type functions.

2 Background

Let $f : \mathbb{F}_{2^n} \rightarrow \mathbb{F}_2$ be a Boolean function. For any $\omega \in \mathbb{F}_{2^n}$, the Walsh-Hadamard transform $\mathcal{W}_f(\omega)$ of f is defined as $\mathcal{W}_f(\omega) = \sum_{\alpha \in \mathbb{F}_{2^n}} (-1)^{Tr_1^n(\omega\alpha)+f(\alpha)}$, where $Tr_1^n(\alpha) = \alpha + \alpha^2 + \alpha^{2^2} + \ldots + \alpha^{2^{n-1}}$ for all $\alpha \in \mathbb{F}_{2^n}$. From this, the nonlinearity $nl(f)$ can be expressed as $nl(f) = 2^{n-1} - \frac{1}{2} \max_{\omega \in \mathbb{F}_{2^n}} |\mathcal{W}_f(\omega)|$. The distance $d(g, h)$ between g and h is defined as the Hamming distance between 2^n length vectors $(g(\alpha_0), g(\alpha_1), \ldots, g(\alpha_{2^n-1}))$ and $(h(\alpha_0), h(\alpha_1), \ldots, h(\alpha_{2^n-1}))$, where $\{\alpha_0, \alpha_1, \ldots, \alpha_{2^n-1}\}$ are the elements of $\mathbb{F}_{2^n}$. Let $l_\omega(\alpha) = Tr_1^n(\omega\alpha)$ and $h_\omega(\alpha) = l_\omega(\alpha) + 1$. Then the nonlinearity $nl(f)$ can be equivalently defined as the minimum distance of f from all affine functions $\{l_\omega, h_\omega \,|\, \omega \in \mathbb{F}_{2^n}\}$ as $nl(f) = \min_{\omega \in \mathbb{F}_{2^n}} \{d(f, l_\omega), d(f, h_\omega)\}$.

In the following, we briefly revisit the PW construction mostly following [2, 13]. Let $n = p \cdot q$ such that p and q are two distinct odd primes and consider an n-variable Boolean function f having the support $Supp(f) = \{\alpha \in \mathbb{F}_{2^n} \,|\, f(\alpha) = 1\} = \cup_{i=1}^{\ell} \alpha_i \mathbb{F}_{2^p}^*$, where α_i's lie in the different cosets of $\mathbb{F}_{2^p}^*$ in $\mathbb{F}_{2^n}^*$. Then it is clear that

$$d(f, \mathbf{0}) = \ell(2^p - 1),$$
$$d(f, \mathbf{1}) = 2^n - \ell(2^p - 1), \quad (1)$$

where $\mathbf{0}$ and $\mathbf{1}$ are the all-zero and all-one vectors of length 2^n, respectively. Let us define $H_\omega = Supp(h_\omega) = \{\alpha \in \mathbb{F}_{2^n} \,|\, Tr_1^n(\omega\alpha) = 0\}$, which is a hyperplane in $\mathbb{F}_2^n$ when considered as a vector space over $\mathbb{F}_2$. Further, let $H_\omega \restriction_{\alpha_i \mathbb{F}_{2^p}^*}$ be the

[1] In fact we also consider the cases where n is an odd composite integer such as $n = 9$, 25, or 27.

restriction of H_ω to the coset $\alpha_i\mathbb{F}^*_{2^p}$. It can be shown [2,13] that $\left|H_\omega\restriction_{\alpha_i\mathbb{F}^*_{2^p}}\right|$ is either $2^{p-1}-1$ or 2^p-1 and the number of those having cardinality $2^{p-1}-1$ is 2^{n-p}. Suppose $t(\omega) = |\{\alpha_i\mathbb{F}^*_{2^p} \mid \alpha_i\mathbb{F}^*_{2^p} \subseteq Supp(f) \cap H_\omega\}|$, or equivalently $t(\omega)$ is the number of α_i's for which $Tr_1^p(\omega\alpha_i) = 0\ \forall\omega \in \mathbb{F}^*_{2^p}$. Then the number of cosets of $\mathbb{F}^*_{2^p}$ for which both $\left|H_\omega\restriction_{\alpha_i\mathbb{F}^*_{2^p}}\right| = 2^{p-1}-1$ and $\left|Supp(f)\restriction_{\alpha_i\mathbb{F}^*_{2^p}}\right| = 2^p-1$ is found to be $\ell - t(\omega)$. Following this argument, one can get

$$\begin{aligned} d(f,h_\omega) &= (\ell - t(\omega))\cdot 2^{p-1} + (2^{n-p} - \ell + t(\omega))\cdot(2^{p-1}-1) \\ &\quad + \left(\frac{2^n-1}{2^p-1} - 2^{n-p} - t(\omega)\right)\cdot(2^p-1)+1, \\ &= 2^{n-1} - 2^p\cdot t(\omega) + \ell. \end{aligned} \tag{2}$$

Similarly we have

$$d(f,l_\omega) = 2^{n-1} + 2^p\cdot t(\omega) - \ell. \tag{3}$$

As a consequence, from (1)–(3) $nl(f)$ can be rewritten as follows:

$$nl(f) = \min_{\omega\in\mathbb{F}^*_{2^n}} \{\ell(2^p-1), 2^n - \ell(2^p-1), 2^{n-1} - 2^p\cdot t(\omega) + \ell, 2^{n-1} + 2^p\cdot t(\omega) - \ell\},$$

which implies that if $nl(f) > 2^{n-1} - 2^{\frac{n-1}{2}}$ then the following conditions have to be satisfied:

$$\frac{2^{n-1}-2^{(n-1)/2}}{2^p-1} < \ell < \frac{2^{n-1}+2^{(n-1)/2}}{2^p-1}, \tag{4}$$

$$\frac{1}{2^p}\left(\frac{2^{n-1}-2^{(n-1)/2}}{2^p-1} - 2^{(n-1)/2}\right) < t(\omega) < \frac{1}{2^p}\left(\frac{2^{n-1}+2^{(n-1)/2}}{2^p-1} + 2^{(n-1)/2}\right). \tag{5}$$

The condition given by (4) is called the weight condition, as ℓ is the number of cosets in the support of f. In [13], it was computationally shown that there is no function for $n = 3\cdot 3$ satisfying these two conditions. On the other hand, for $n = 5\cdot 3$, there are 1057 cosets of $\mathbb{F}^*_{2^5}$ in $\mathbb{F}^*_{2^{15}}$, which makes an exhaustive search impossible (in the subsequent section we prove that the maximum nonlinearity in this case is 16276). Hence, an additional constraint of being invariant under the action of $\mathbb{F}^*_{2^3}$ and the group of Frobenius automorphism is imposed [13] on f, which provides a very small class of functions leaving only 2^{11} options to search (in fact using the weight condition the number of options can be reduced to $\binom{10}{5} = 252$ as noticed in [13]). Finally, two functions with nonlinearity 16276 are obtained [13] in this class.

At this point, let us recall a more general definition [6] of the aforementioned PW construction:

Definition 1. *Let $n = p\cdot q$, where $p, q > 2$ are prime numbers such that $p > q$. Let the product $\mathcal{R} = \mathbb{F}_{2^p}{}^*\cdot\mathbb{F}_{2^q}{}^*$ be the cyclic group of cardinality $r = (2^p-1)(2^q-1)$ in $\mathbb{F}_{2^n}$. Let $\langle\phi_2\rangle$ be the group of Frobenius automorphisms where $\phi_2 : \mathbb{F}_{2^n} \to \mathbb{F}_{2^n}$ is defined by $\alpha \to \alpha^2$. The function f is called PW type if it is invariant under the action of $\mathcal{R}$ and $\langle\phi_2\rangle$.*

For simplicity, one can view [2] a PW type function as an interleaved sequence [3] which is defined as follows:

Definition 2. *Let $m = dr$, where $d, r > 1$ are integers. The (d, r)-interleaved sequence $A_{d,r}$ corresponding to the binary sequence $A = \{a_0, a_1, a_2, \ldots, a_{m-1}\}$ is defined as the matrix whose $(i, j)^{th}$ entry is equal to $a_{i.d+j}$, where $i = 0, 1, \ldots, r-1$ and $j = 0, 1, \ldots, d-1$.*

Suppose $m = 2^n - 1$. An interleaved sequence $A_{d,r}$ can be associated with the ordered sequence $\{f(1), f(\zeta), f(\zeta^2), \ldots, f(\zeta^{2^n-2})\}$ such that $a_{i \cdot d+j} = f(\zeta^{i \cdot d+j})$, where ζ is a primitive element in $\mathbb{F}_{2^n}$. We call this interleaved sequence the (d, r)-interleaved sequence corresponding to f with respect to ζ. Let $d = (2^p - 1)(2^q - 1)$. Then it follows from Definition 1 that the (d, r)-interleaved sequence of an n-variable PW type function consists of either all 0 or all 1 columns, since the corresponding function f is invariant under the action of $\mathcal{R}$. Further, the invariance under the action of $\langle \phi_2 \rangle$ implies that f is an idempotent function and the ith column has the same value as the jth column if $i \equiv j2^s \mod d$ for some non-negative integer s. This equivalence relation, denoted by ρ_d, is given as:

$$i\, \rho_d\, j \Leftrightarrow \text{there exists an integer } s > 0 \text{ such that } i \equiv j2^s \bmod d. \qquad (6)$$

For $n = 15$, the PW construction can be represented by the $(151, 217)$-interleaved sequence. Using the equivalence relation ρ_{151}, one obtains 11 groups, among which there are 10 groups of size 15 and 1 group of size 1. Notice that any element in a group determines 217 positions in the truth table of f. Since the weight condition gives $524.3871 < \ell < 532.6452$, i.e., $525 \leq \ell \leq 532$, we have to choose 5 groups among the 10 groups of size 15 and we may or may not choose the remaining 1 group of size 1 (observe that ℓ can be either $5 \cdot 15 \cdot 7 = 525$ or $5 \cdot 15 \cdot 7 + 1 = 526$). There are 8 functions with nonlinearities 16268, 16269, 16275, and 16276 exceeding the bent concatenation bound $2^{14} - 2^7 = 16256$ in the corresponding search space. One half of these functions are obtained from the other half by complementing the truth tables except their first bits. Hence, the nonlinearity 16276 (resp., 16268) is obtained from the function with nonlinearity 16275 (resp., 16269) in this way, and vice versa (in the following section, without using an exhaustive search, we show that these nonlinearities are the only possible ones that can be attained by the PW construction and in a much broader class).

3 Nonlinearity of the Functions on $n = p \cdot q$ Variables Invariant Under the Action of $\mathbb{F}^*_{2^p} \cdot \mathbb{F}^*_{2^q}$

In this section, we determine all possible nonlinearities of the PW type functions and their variants for which $n = p \cdot q$ where p, q are odd primes and $p > q$. The functions we consider are invariant under the action of $\mathbb{F}^*_{2^p} \cdot \mathbb{F}^*_{2^q}$. This class is much larger than the class considered in [13] where the action of Frobenius automorphism was considered. In most of the cases, we obtain nonlinearity bounds strictly less than the generic upper bound in [5] that provides the importance of our result. We provide detailed examples with $n = 15, 21$ that answers several open questions that remained unanswered for the 15-variable [13] and 21-variable [6] functions.

In the following theorem we show what are the possible nonlinearities greater than the bent concatenation bound when we consider the action of $\mathbb{F}_{2^p}^* \cdot \mathbb{F}_{2^q}^*$. This is a larger class than what was considered in [13] as we do not consider the action of Frobenius automorphism here.

Theorem 1. *Let f be an n-variable function with $nl(f) = 2^{n-1} - 2^{\frac{n-1}{2}} + \mu_n$ which is invariant under the action of $\mathbb{F}_{2^p}^* \cdot \mathbb{F}_{2^q}^*$, where $\mu_n \in \mathbb{Z}^+$ and $n = p \cdot q$ such that p, q are two distinct odd primes with $p > q$. Then at least one of the values in the following two sets is an integer:*

(i) $\left\{ \frac{2^{n-1}-2^{\frac{n-1}{2}}+\mu_n}{(2^p-1)(2^q-1)}, \frac{2^{n-1}+2^{\frac{n-1}{2}}-\mu_n}{(2^p-1)(2^q-1)} \right\}$,

(ii) $\left\{ \frac{\ell(2^q-1)-2^{\frac{n-1}{2}}+\mu_n}{2^p} \mid \ell \in L^{\mu_n} \right\} \cup \left\{ \frac{\ell(2^q-1)+2^{\frac{n-1}{2}}-\mu_n}{2^p} \mid \ell \in L^{\mu_n} \right\}$,

where $1 \le \mu_n \le 2\lfloor 2^{n-2} - 2^{\frac{n}{2}-2} \rfloor - \left(2^{n-1} - 2^{\frac{n-1}{2}}\right)$ *and*

$$L^{\mu_n} = \left\{ \ell \in \mathbb{Z}^+ \mid \frac{2^{n-1} - 2^{\frac{n-1}{2}} + \mu_n}{(2^p-1)(2^q-1)} \le \ell \le \frac{2^{n-1} + 2^{\frac{n-1}{2}} - \mu_n}{(2^p-1)(2^q-1)} \right\}.$$

Proof. In the following, we consider two cases: either the nonlinearity equals $d(f, \mathbf{0})$ or $d(f, \mathbf{1})$, and then at least one of the values displayed in (i) is an integer, or (non-exclusively) the nonlinearity equals $d(f, h_\omega)$ or $d(f, l_\omega)$ and at least one of the values displayed in (ii) is an integer.

(i) Since f is invariant under the action of $\mathbb{F}_{2^p}^* \cdot \mathbb{F}_{2^q}^*$, its support can be written as $Supp(f) = \cup_{i=1}^{\ell} \alpha_i (\mathbb{F}_{2^p}^* \cdot \mathbb{F}_{2^q}^*)$, where α_i's lie in the different cosets of $\mathbb{F}_{2^p}^* \cdot \mathbb{F}_{2^q}^*$ in $\mathbb{F}_{2^n}^*$. Then it is clear that

$$\begin{aligned} d(f, \mathbf{0}) &= \ell(2^p - 1)(2^q - 1), \\ d(f, \mathbf{1}) &= 2^n - \ell(2^p - 1)(2^q - 1). \end{aligned} \quad (7)$$

Note that if $nl(f) = 2^{n-1} - 2^{\frac{n-1}{2}} + \mu_n$, then either $d(f, \mathbf{0})$ or $d(f, \mathbf{1})$ can be equal to $2^{n-1} - 2^{\frac{n-1}{2}} + \mu_n$, which gives the possible values of ℓ in part (i). Finally, recall that the generic upper bound [5] on nonlinearity for functions on odd number of variables n is $2\lfloor 2^{n-2} - 2^{\frac{n}{2}-2} \rfloor$, from which we get the values of μ_n used to compute ℓ (and $t(\omega)$).

(ii) Clearly, (7) is also obtained from (1) by substituting $\ell(2^q - 1)$ for ℓ, due to the fact that each coset of $\mathbb{F}_{2^p}^* \cdot \mathbb{F}_{2^q}^*$ consists of $2^q - 1$ distinct cosets of $\mathbb{F}_{2^p}^*$ in $\mathbb{F}_{2^n}^*$. Following the same argument, one can get the distances below:

$$\begin{aligned} d(f, h_\omega) &= 2^{n-1} - 2^p \cdot t(\omega) + \ell(2^q - 1), \\ d(f, l_\omega) &= 2^{n-1} + 2^p \cdot t(\omega) - \ell(2^q - 1), \end{aligned} \quad (8)$$

which is obtained from (2) and (3) by using the same substitution, where $l_\omega(\alpha) = Tr_1^n(\omega\alpha)$, $h_\omega(\alpha) = l_\omega(\alpha) + 1$, and $t(\omega)$ is the number of cosets of $\mathbb{F}_{2^p}^*$ totally contained in $Supp(h_\omega) \cap Supp(f)$. Next, as in the proof of part (i), it follows from the definition of nonlinearity that $t(\omega)$ can be either $\frac{\ell(2^q-1)-2^{\frac{n-1}{2}}+\mu_n}{2^p}$ or $\frac{\ell(2^q-1)+2^{\frac{n-1}{2}}-\mu_n}{2^p}$. However, note that $d(f,\mathbf{0}), d(f,\mathbf{1}) \geq nl(f) = 2^{n-1} - 2^{\frac{n-1}{2}} + \mu_n$. This gives, $\frac{2^{n-1}-2^{\frac{n-1}{2}}+\mu_n}{(2^p-1)(2^q-1)} \leq \ell \leq \frac{2^{n-1}+2^{\frac{n-1}{2}}-\mu_n}{(2^p-1)(2^q-1)}$. Thus, we get all the possible values of $t(\omega)$ in part (ii).

□

3.1 Case $n = 15$

The 15-variable PW functions are the most important in the domain of nonlinearity of Boolean functions as for the first time the bent concatenation bound has been defeated in this scenario [13]. There has been efforts to obtain Boolean functions with good cryptographic properties by modifying the PW functions [13] as evident from [9,14]. The search space for 15-variable PW functions [13], considering invariance under $\mathbb{F}_{2^5}^* \cdot \mathbb{F}_{2^3}^*$ as well as Frobenius automorphism, was as little as 2^{11} and thus it was very easy to search and obtain the functions with nonlinearity as high as 16276. However, when we do not consider the Frobenius automorphism, the class becomes much larger. In this case $\frac{2^{15-1}}{(2^5-1)(2^3-1)} = 151$ and hence the search space is as large as 2^{151}. Exhaustive search here is not feasible. However, our result below shows that the maximum nonlinearity in this larger class is again 16276. We present the proof in details and then discuss the step by step description of Theorem 1 in this direction.

Corollary 1. *Consider a 15-variable function f that is invariant under the action of $\mathbb{F}_{2^5}^* \cdot \mathbb{F}_{2^3}^*$. Then $nl(f) \leq 16276 = 2^{15-1} - 2^{\frac{15-1}{2}} + 20$.*

Proof. We have $Supp(f) = \cup_{i=1}^{\ell} \alpha_i(\mathbb{F}_{2^p}^* \cdot \mathbb{F}_{2^q}^*)$. One can write $Supp(f)$ in terms of the cosets $\mathbb{F}_{2^p}^*$ in $\mathbb{F}_{2^n}^*$ as $\cup_{i=1}^{\ell'} \beta_i \mathbb{F}_{2^p}^*$, where $\ell' = \ell(2^q - 1)$. Hence the weight condition (4) can be rewritten as

$$\frac{2^{n-1} - 2^{(n-1)/2}}{2^p - 1} < \ell(2^q - 1) < \frac{2^{n-1} + 2^{(n-1)/2}}{2^p - 1}. \tag{9}$$

Notice that in our case $n = 15$, $p = 5$, and $q = 3$. Substituting these values in (9), we get $74.9124 < \ell < 76.0922$, and thus, we have $75 \leq \ell \leq 76$. Now suppose there exists a 15-variable function f with $nl(f) = 2^{15-1} - 2^{\frac{15-1}{2}} + 20 + c = 16276 + c$, where c is a positive integer. Then one can get the corresponding weight condition for the existence of f as

$$\frac{2^{15-1} - 2^{\frac{15-1}{2}} + 20}{2^5 - 1} + \frac{c}{2^5 - 1} \leq \ell(2^3 - 1) \leq \frac{2^{15-1} + 2^{\frac{15-1}{2}} - 20}{2^5 - 1} - \frac{c}{2^5 - 1},$$

from which, to have nonlinearity $16276 + c$, we get
$75.0046 + \frac{c}{7(2^5-1)} \leq \ell \leq 76 - \frac{c}{7(2^5-1)}$, and thus, there is no solution for ℓ. □

To clarify it further, let us consider the following weight condition to have nonlinearity $2^{15-1} - 2^{\frac{15-1}{2}} + 19 = 16275$:

$$\frac{2^{15-1} - 2^{\frac{15-1}{2}} + 19}{7(2^5 - 1)} = 75 \leq \ell \leq 76.0046 = \frac{2^{15-1} + 2^{\frac{15-1}{2}} - 19}{7(2^5 - 1)}.$$

Note that the lower bound is exactly 75 and it exceeds this value if we replace 19 by 20, which provides nonlinearity 16276. In this case the upper bound is exactly 76, which leaves 76 as the only option for ℓ. However, observe that if we instead replace 19 by 21, then the upper bound becomes less than 76 for which there is no ℓ satisfying the weight condition. Recall from [2,13] that the PW constructions with nonlinearity 16276 belong to a class of very small size in which there exist only 2^{11} functions that are idempotent and invariant under the action of $\mathbb{F}^*_{2^5} \cdot \mathbb{F}^*_{2^3}$. On the other hand, the above corollary shows that there is no function with nonlinearity >16276 in a much larger class of size 2^{151} which is formed by lifting the condition of being idempotent.

We finally discuss what are the possible values of μ_n in the case of $n = 15$. From the above arguments, we deduce that the possible values of μ_{15} are 19 and 20, when we take the integer values of ℓ into account. In other words, if we consider only part (i) of Theorem 1, then the following values of ℓ becomes an integer for $\mu_{15} = 19$ and 20:

$$\ell \in \left\{ \frac{2^{15-1} - 2^{\frac{15-1}{2}} + \mu_{15}}{(2^5 - 1)(2^3 - 1)}, \frac{2^{15-1} + 2^{\frac{15-1}{2}} - \mu_{15}}{(2^5 - 1)(2^3 - 1)} \right\},$$

which yields the nonlinearities $2^{15-1} - 2^{\frac{15-1}{2}} + 19 = 16275$ and $2^{15-1} - 2^{\frac{15-1}{2}} + 20 = 16276$ respectively. To obtain the other possible values of μ_{15} (and the corresponding nonlinearities), let us consider part (ii) of Theorem 1. It is easy to check that $75 \leq \ell \leq 76$ (i.e., $L^{\mu_{15}} = \{75, 76\}$) for all $1 \leq \mu_{15} \leq 18$. Hence, one of the following values of $t(\omega)$ must be an integer to have nonlinearity $nl(f) = 2^{15-1} - 2^{\frac{15-1}{2}} + \mu_{15}$:

$$t(\omega) \in \left\{ \frac{\ell(2^3 - 1) - 2^{\frac{15-1}{2}} + \mu_{15}}{2^5}, \frac{\ell(2^3 - 1) + 2^{\frac{15-1}{2}} - \mu_{15}}{2^5} \right\},$$

where $\ell \in L^{\mu_{15}}$ and $1 \leq \mu_{15} \leq 18$. One can computationally find that $t(\omega)$ is an integer for $\mu_{15} = 12$ and 13, from which we get the nonlinearities $2^{15-1} - 2^{\frac{15-1}{2}} + 12 = 16268$ and $2^{15-1} - 2^{\frac{15-1}{2}} + 13 = 16269$ respectively.

Thus, with these values, we completely solve why such nonlinearities are obtained for the PW functions [13], which could never be answered before. Note that the consideration of Frobenius automorphism does not affect the possible nonlinearity values. As we know, functions invariant under the action of Frobenius automorphism (also called idempotents) are actually rotation symmetric Boolean functions [8,14]. Considering this restriction reduces the search space, but at the same time this space provides a good sample of highly nonlinear functions. It was indeed quite judicious to study this small search space [13], however, why such nonlinearity values could be obtained was not known earlier that we answer here. With our result, now we know that the nonlinearity 16276 is the maximum possible value in a much larger class of size 2^{151}, which was attained in a much smaller sample space of only 2^{11} in [13].

3.2 Case $n = 21$

Now we consider the case $n = p \cdot q = 7 \cdot 3 = 21$. The constraints for 15-variable functions was such that one had to satisfy 11 inequalities for 11 binary variables and it could be done easily by exhaustive search. The situation is not as simple for 21 or more variables. In [13], the choice of orbits for general case could not be explained and also in [2] it has been commented that such search might be infeasible. In a very recent result [6], by heuristic search, the existence of PW functions could be demonstrated for $n = 21$. The nonlinearity of such functions are $2^{21-1} - 2^{\frac{21-1}{2}} + 61$, i.e., $\mu_{21} = 61$ and one may easily note that $61 \cdot 2^{-\frac{21-1}{2}} < 20 \cdot 2^{-\frac{15-1}{2}}$. Thus, even after the discovery of 21-variable PW functions having nonlinearity more than bent concatenation bound, the old maximum achievable nonlinearity using the 15-variable PW functions could not be beaten. Thus, the most natural question is: Can there be the existence of 21-variable functions such that $\mu_{21} > 20 \cdot 2^{\frac{21-15}{2}} = 160$? Our analysis shows that the non-trivial upper bound here corresponds to $\mu_{21} = 196$ and thus there is a hope for improved result with further search effort.

Let us first proceed as in the analysis for the 15-variable case. Here we consider the class of 21-variable functions that are invariant under the action of $\mathbb{F}^*_{2^7} \cdot \mathbb{F}^*_{2^3}$. There are 2359 cosets and hence the search space is of size 2^{2359}. We obtain the following inequality from (9):

$$\frac{2^{21-1} - 2^{\frac{21-1}{2}}}{2^7 - 1} < \ell(2^3 - 1) < \frac{2^{21-1} + 2^{\frac{21-1}{2}}}{2^7 - 1},$$

and thus, we get $1178.3487 < \ell < 1180.6524$, i.e., $1179 \leq \ell \leq 1180$. Suppose there exists a function f with $nl(f) = 2^{21-1} - 2^{\frac{21-1}{2}} + \mu_{21}$. Then, to achieve this nonlinearity, the following condition has to be satisfied:

$$\frac{2^{21-1} - 2^{\frac{21-1}{2}}}{2^7 - 1} + \frac{\mu_{21}}{2^7 - 1} \leq 7\ell \leq \frac{2^{21-1} + 2^{\frac{21-1}{2}}}{2^7 - 1} - \frac{\mu_{21}}{2^7 - 1},$$

and thus, we have $1178.3487 + \frac{\mu_{21}}{7(2^7-1)} \leq \ell \leq 1180.6524 - \frac{\mu_{21}}{7(2^7-1)}$. This inequality has no solution for ℓ only if $\mu_{21} > 580$. However, the gap between the bent concatenation nonlinearity ($1047552 = 2^{21-1} - 2^{\frac{21-1}{2}}$) and the generic upper bound [5] ($1047850 = 2\lfloor 2^{21-2} - 2^{\frac{21}{2}-2} \rfloor$) on the nonlinearity of 21-variable functions is 298 (<580). Thus, what we have obtained so far does not provide any non-trivial upper bound for this class. However, with more detailed analysis we obtain the following non-trivial upper bound.

Corollary 2. *Let us consider a 21-variable function f which is invariant under the action of $\mathbb{F}^*_{2^7} \cdot \mathbb{F}^*_{2^3}$. Then $nl(f) \leq 1047748 = 2^{21-1} - 2^{\frac{21-1}{2}} + 196$.*

Proof. It follows from Theorem 1. Let us consider part (i) of Theorem 1. Then there can be the existence of 21-variable PW functions with $nl(f) = 2^{21-1} - 2^{\frac{21-1}{2}} + \mu_{21}$ if one of the following values of ℓ is an integer:

$$\ell \in \left\{ \frac{2^{21-1} - 2^{\frac{21-1}{2}} + \mu_{21}}{(2^7-1)(2^3-1)}, \frac{2^{21-1} + 2^{\frac{21-1}{2}} - \mu_{21}}{(2^7-1)(2^3-1)} \right\},$$

where $1 \leq \mu_{21} \leq 298 \left(= 2\left\lfloor 2^{21-2} - 2^{\frac{21}{2}-2} \right\rfloor - \left(2^{21-1} - 2^{\frac{21-1}{2}}\right)\right)$. However, it can be computationally checked that no value of μ_{21} makes ℓ an integer.

Next, consider part (ii) of Theorem 1. From the earlier discussion, it is evident that $L^{\mu_{21}} = \{1179, 1180\}$ (i.e., $1179 \leq \ell \leq 1180$) for all $1 \leq \mu_{21} \leq 298$. Then at least one of the following values of $t(\omega)$ must be an integer to have nonlinearity $nl(f) = 2^{21-1} - 2^{\frac{21-1}{2}} + \mu_{21}$:

$$t(\omega) \in \left\{ \frac{\ell(2^3-1) - 2^{\frac{21-1}{2}} + \mu_{21}}{2^7}, \frac{\ell(2^3-1) + 2^{\frac{21-1}{2}} - \mu_{21}}{2^7} \right\},$$

where $\ell \in L^{\mu_{21}}$ and $1 \leq \mu_{21} \leq 298$. We find that only the 8 values of μ_{21} make $t(\omega)$ an integer: $\mu_{21} \in \{60, 61, 67, 68, 188, 189, 195, 196\}$. Hence, the maximum possible nonlinearity is $1047748 = 2^{20} - 2^{10} + 196$. □

Note from the above proof that one of the possible values of μ_{21} is 61, which corresponds to the nonlinearity $1047613 = 2^{20} - 2^{10} + 61$ achieved in [6] as aforementioned. However, there can be the existence of $\mu_{21} \in \{188, 189, 195, 196\}$, which provides $\mu_{21} \cdot 2^{-\frac{21-1}{2}} > 20 \cdot 2^{-\frac{15-1}{2}}$, yielding the best known nonlinearity till date.

3.3 The Algorithm and Numerical Results

Considering Theorem 1, we devise Algorithm 1 to find all possible nonlinearities greater than the bent concatenation bound for the functions on odd number $n = p \cdot q$ of variables that are invariant under the action of $\mathbb{F}^*_{2^p} \cdot \mathbb{F}^*_{2^q}$, where p and q are distinct odd primes such that $p > q$. In this algorithm, for each value of

$$\ell \in \left\{ \ell \in \mathbb{Z}^+ \mid \frac{2^{n-1} - 2^{\frac{n-1}{2}}}{(2^p-1)(2^q-1)} = lb < \ell < ub = \frac{2^{n-1} + 2^{\frac{n-1}{2}}}{(2^p-1)(2^q-1)} \right\}, \quad (10)$$

we store the possible nonlinearities $2^{n-1} - 2^{\frac{n-1}{2}} + \mu_n$ in the array NL whenever the condition given by Theorem 1 is satisfied given the value of $\mu_n \in \{1, 2, \ldots, \mu_n^{max}\}$, where μ_n^{max} (computed using the generic upper bound in [5]) is the maximum possible value of μ_n for odd number n of variables.

We have performed Algorithm 1 and give the maximum values $\mu_{p \cdot q}^{max} (\leq \mu_n^{max})$ of μ_n achievable by the n-variable functions that are invariant under the action of $\mathbb{F}^*_{2^p} \cdot \mathbb{F}^*_{2^q}$ in Table 1 for $15 \leq n = p \cdot q \leq 39$, where p and q are distinct odd primes such that $p > q$. From Table 1, it is seen that $\mu_{p \cdot q}^{max} = \mu_n^{max}$ for only $n = 35$ and $\mu_{p \cdot q}^{max} < \mu_n^{max}$ for all the remaining values of n.

Algorithm 1: Computation of all possible nonlinearities $> 2^{n-1} - 2^{\frac{n-1}{2}}$.

input : n, p, where $n = p \cdot q$ such that p and q are odd primes with $p > q$.
output: Nonlinearities given by the array NL.

$k \leftarrow 0$;
$\mu_n^{max} \leftarrow 2\left\lfloor 2^{n-2} - 2^{\frac{n}{2}-2} \right\rfloor - \left(2^{n-1} - 2^{\frac{n-1}{2}}\right)$;
$lb \leftarrow \frac{2^{n-1}-2^{\frac{n-1}{2}}}{(2^p-1)(2^q-1)}$;
$ub \leftarrow \frac{2^{n-1}+2^{\frac{n-1}{2}}}{(2^p-1)(2^q-1)}$;
for $\ell \leftarrow lb$ *to* ub **do**
  for $\mu_n \leftarrow 1$ *to* μ_n^{max} **do**
    $lbn \leftarrow \frac{2^{n-1}-2^{\frac{n-1}{2}}+\mu_n}{(2^p-1)(2^q-1)}$;
    $ubn \leftarrow \frac{2^{n-1}+2^{\frac{n-1}{2}}-\mu_n}{(2^p-1)(2^q-1)}$;
    if $lbn \leq \ell \leq ubn$ **then**
      $tl \leftarrow \frac{\ell(2^q-1)-2^{\frac{n-1}{2}}+\mu_n}{2^p}$;
      $tu \leftarrow \frac{\ell(2^q-1)+2^{\frac{n-1}{2}}-\mu_n}{2^p}$;
      $\kappa \leftarrow \{lbn, lbu, tl, tu\}$;
      if *any value in* κ *is an integer* **then**
        $nl \leftarrow 2^{n-1} - 2^{\frac{n-1}{2}} + \mu_n$;
        if nl *is not in* NL **then**
          $NL[k] \leftarrow nl$;
          $k \leftarrow k+1$;
return NL;

3.4 Idempotents, i.e., Functions Invariant Under the Action of $\langle\phi_2\rangle$

Next we consider the class of functions that are invariant under the action of $\mathbb{F}_{2^p}^* \cdot \mathbb{F}_{2^q}^*$ and $\langle\phi_2\rangle$ as well. In this case, we need to take only some combinations of the groups (obtained by the equivalence relation (6)) satisfying the weight condition in (10) into account, which reduces the number of possible values of ℓ in Algorithm 1. Thus, modifying Algorithm 1 accordingly, we have computed (see Table 1) the maximum values $\mu_{p\cdot q,\langle\phi_2\rangle}^{max}$ of μ_n which are achievable in this class. In Table 1, $\ell_{p\cdot q}$ denotes the values of ℓ given by (10) and $\ell_{p\cdot q,\langle\phi_2\rangle}$ denotes those of $\ell_{p\cdot q}$ obtained by considering also the action of $\langle\phi_2\rangle$. Comparing the values of $\mu_{p\cdot q,\langle\phi_2\rangle}^{max}$ with μ_n^{max} in Table 1, we find that $\mu_{p\cdot q,\langle\phi_2\rangle}^{max} < \mu_n^{max}$ for all $15 \leq n \leq 39$ whereas $\mu_{p\cdot q,\langle\phi_2\rangle}^{max} < \mu_{p\cdot q}^{max}$ for $n = 33$ and 35 only. Note that the achievable nonlinearities for $n = 15$ and 21, given in the previous subsections, remain the same even after imposing the constraint of being invariant under the action of $\langle\phi_2\rangle$, since the values of $\ell_{p\cdot q}$ are the same as those of $\ell_{p\cdot q,\langle\phi_2\rangle}$ in both cases.

Table 1. The values of $\mu_{p\cdot q}^{max}$ and $\mu_{p\cdot q,\langle\phi_2\rangle}^{max}$ together with those of $\ell_{p\cdot q}$ and $\ell_{p\cdot q,\langle\phi_2\rangle}$ for $15 \leq n \leq 39$, where $n = p \cdot q$ such that p and q are two distinct odd primes with $p > q$.

n	$\ell_{p\cdot q}$	$\ell_{p\cdot q,\langle\phi_2\rangle}$	$\mu_{p\cdot q}^{max}$	$\mu_{p\cdot q,\langle\phi_2\rangle}^{max}$	μ_n^{max}
15	[75, 76]	{75, 76}	20	20	36
21	[1179, 1180]	{1179, 1180}	196	196	298
33	[299735, 299744]	{299739, 299740}	17426	17412	19194
35	[4363663, 4363728]	{4363695, 4363696}	38390	38352	38390
39	[4794067, 4794084]	{4794075, 4794076}	151598	151598	153560

4 Functions Invariant Under the Action of $\mathbb{F}_{2^p}^*$

We start with the following corollary of Theorem 1.

Corollary 3. *Let f be an n-variable function with $nl(f) = 2^{n-1} - 2^{\frac{n-1}{2}} + \mu_n$ which is invariant under the action of $\mathbb{F}_{2^p}^*$, where $1 < p|n$, n is odd, and $\mu_n \in \mathbb{Z}^+$. Then at least one of the values in the following two sets is an integer:*

(i) $\left\{\frac{2^{n-1}-2^{\frac{n-1}{2}}+\mu_n}{2^p-1}, \frac{2^{n-1}+2^{\frac{n-1}{2}}-\mu_n}{2^p-1}\right\}$,

(ii) $\left\{\frac{\ell-2^{\frac{n-1}{2}}+\mu_n}{2^p} \mid \ell \in L^{\mu_n}\right\} \cup \left\{\frac{\ell+2^{\frac{n-1}{2}}-\mu_n}{2^p} \mid \ell \in L^{\mu_n}\right\}$,

where $1 \leq \mu_n \leq 2\left\lfloor 2^{n-2} - 2^{\frac{n}{2}-2}\right\rfloor - \left(2^{n-1} - 2^{\frac{n-1}{2}}\right)$ *and*

$$L^{\mu_n} = \left\{\ell \in \mathbb{Z}^+ \mid \frac{2^{n-1} - 2^{\frac{n-1}{2}} + \mu_n}{2^p - 1} \leq \ell \leq \frac{2^{n-1} + 2^{\frac{n-1}{2}} - \mu_n}{2^p - 1}\right\}.$$

Proof. At least one of the following distances have to be equal to $nl(f) = 2^{n-1} - 2^{\frac{n-1}{2}} + \mu_n$: $d(f, \mathbf{0}) = \ell(2^p - 1)$, $d(f, \mathbf{1}) = 2^n - \ell(2^p - 1)$, $d(f, h_\omega) = 2^{n-1} - 2^p \cdot t(\omega) + \ell'$, $d(f, l_\omega) = 2^{n-1} + 2^p \cdot t(\omega) - \ell'$, where $\ell' \in L^{\mu_n}$. Thus, one or more values given by (i) and (ii) have to be an integer. Recall that the upper bound [5] gives $nl(f) \leq 2\left\lfloor 2^{n-2} - 2^{\frac{n}{2}-2}\right\rfloor$. Hence the proof. □

Using this corollary, we modify Algorithm 1 to find all possible nonlinearities greater than $2^{n-1} - 2^{\frac{n-1}{2}}$ that can be achieved in the class of functions that are invariant under the action of $\mathbb{F}_{2^p}^*$, where n is an odd composite integer and $1 < p|n$. More specifically, in the modified version of Algorithm 1, we replace $\ell(2^q - 1)$ in the numerators of tl and tu (given by lines 10 and 11 of Algorithm 1 respectively) with ℓ and remove the term $(2^q - 1)$ from the denominators of lb, ub, lbn, and ubn (given by lines 3, 4, 7, and 8 of Algorithm 1 respectively).

Performing the modified version, we present in Table 2 the maximum values $\mu_{n,p}^{max} (\leq \mu_n^{max})$ of μ_n attainable in the corresponding classes for the composite integers n, where $9 \leq n \leq 39$ and $1 < p|n$. We have also given the values of

Table 2. The values of $\mu_{n,p}^{max}$ $(= \mu_{n,p,\langle\phi_2\rangle}^{max})$ together with those of $\ell_{n,p}$ for $9 \leq n \leq 39$, where n is an odd composite integer and $1 < p|n$.

(n,p)	$\ell_{n,p}$	$\mu_{n,p}^{max}$ $(= \mu_{n,p,\langle\phi_2\rangle}^{max})$	μ_n^{max}
$(9,3)$	$[35,38]$	4	4
$(15,3)$	$[2323,2358]$	36	36
$(15,5)$	$[525,532]$	20	
$(21,3)$	$[149651,149942]$	298	298
$(21,7)$	$[8249,8264]$	199	
$(25,5)$	$[541069,541332]$	1198	1198
$(27,3)$	$[9585811,9588150]$	2398	2398
$(27,9)$	$[131313,131344]$	2316	
$(33,3)$	$[613557395,613576118]$	19194	19194
$(33,11)$	$[2098145,2098208]$	17432	
$(35,5)$	$[554185101,554193556]$	38390	38390
$(35,7)$	$[135273529,135275592]$	38390	
$(39,3)$	$[39268197523,39268347318]$	153560	153560
$(39,13)$	$[33558465,33558592]$	151598	

ℓ (referred to as $\ell_{n,p}$) used in the same algorithm. It is found that the generic upper bound $2^{n-1} - 2^{\frac{n-1}{2}} + \mu_n^{max}$ remains the same, i.e., $\mu_{n,p}^{max} = \mu_n^{max}$, for all $p \leq \frac{n}{p}$. However, except for $n = 35$, it provides noticeably lower nonlinearities than the generic upper bound for all $p > \frac{n}{p}$. Note that only for $n = 35$, we have $\mu_n^{max} = \mu_{n,p}^{max} = \mu_{n,q}^{max}$. We see from Table 2 that $\mu_{15,5}^{max} = 20$. This means $nl(f) \leq 2^{15-1} - 2^{\frac{15-1}{2}} + 20 = 16276$ in the class of functions that are invariant under the action of $\mathbb{F}_{2^5}^*$ for which the search space becomes 2^{1057} as there are $\frac{2^{15}-1}{2^5-1} = 1057$ cosets of $\mathbb{F}_{2^5}^*$ in $\mathbb{F}_{2^{15}}^*$. On the other hand, we observe from Table 2 that $\mu_{21,7}^{max} = 199$ whereas $\mu_{7\cdot3}^{max} = \mu_{7\cdot3,\langle\phi_2\rangle}^{max} = 196$ (see Table 1). Keeping in mind that there are 115 groups under the equivalence relation $\rho_{\frac{2^{21}-1}{(2^7-1)(2^3-1)}} = \rho_{2359}$ and $\frac{2^{21}-1}{2^7-1} = 16513$ cosets of $\mathbb{F}_{2^7}^*$ in $\mathbb{F}_{2^{21}}^*$, this implies that if we increase the search space from 2^{115} to 2^{16513}, the nonlinearity bound slightly increases from $2^{21-1}-2^{\frac{21-1}{2}}+196$ to $2^{21-1}-2^{\frac{21-1}{2}}+199$. For $n = 9, 15$, and 21, all the achievable values $\mu_{n,p}$ of μ_n are given in Table 3.

Let us now consider the class of functions that are invariant under the action of $\mathbb{F}_{2^p}^*$ and $\langle\phi_2\rangle$. As in the previous section, we expect that some of the possible values of ℓ computed in the aforementioned version of Algorithm 1 are eliminated. Hence, by suitably adapting it, we have computed the maximum values $\mu_{n,p,\langle\phi_2\rangle}^{max}$ of μ_n which are attainable in the corresponding classes; however, we find that, as shown in Table 2, the values of $\mu_{n,p,\langle\phi_2\rangle}^{max}$ are the same as those of $\mu_{n,p}^{max}$.

In Table 3, all the possible values $\mu_{n,p,\langle\phi_2\rangle}$ of μ_n are also given for $n = 9, 15$, and 21. We note that the values of $\mu_{n,p,\langle\phi_2\rangle}$ do not cover all those of $\mu_{n,p}$ for $(n,p) = (9,3)$ and $(21,7)$, although $\mu_{n,p,\langle\phi_2\rangle}^{max} = \mu_{n,p}^{max}$ for all the values of (n,p) in Table 2.

Table 3. The values of $\mu_{n,p}$ and $\mu_{n,p,\langle\phi_2\rangle}$ for $n = 9, 15$, and 21.

(n,p)	$\mu_{n,p}$	$\mu_{n,p,\langle\phi_2\rangle}$	μ_n^{max}
$(9,3)$	$[2,4]$	$[3,4]$	4
$(15,3)$	$[1,36]$	$[1,36]$	36
$(15,5)$	$[12,20]$	$[12,20]$	
$(21,3)$	$[1,298]$	$[1,298]$	298
$(21,7)$	$[56,72] \cup [185,199]$	$[60,68] \cup [71,72] \cup [188,196] \cup [198,199]$	

5 Conclusion

We have presented non-trivial upper bounds on the nonlinearity of PW type functions and their super-sets. Our method can be applied algorithmically for any n-variable function, where n is odd and not prime. It also identifies all the attainable nonlinearities higher than the bent concatenation bounds. Computational results are presented for functions on n-variables where n is odd composite and $9 \leq n \leq 39$. We particularly explain the issues for $n = 15$ and 21 in detail as these are the cases that received serious attention recently. We also provide numerical results for larger variables and more research in this area is necessary to explore the situation further. The results obtained in this paper related to upper bound of nonlinearity couldn't be achieved for more than three decades even after substantial efforts as evident from literature. Towards further research, one may study the Walsh spectra of such functions in more details and apply our strategy for prime $n \geq 11$ considering the factors of $2^n - 1$.

References

1. Berlekamp, E.R., Welch, L.R.: Weight distributions of the cosets of the (32, 6) Reed-Muller code. IEEE Trans. Inf. Theory **18**(1), 203–207 (1972)
2. Gangopadhyay, S., Keskar, P.H., Maitra, S.: Patterson-Wiedemann construction revisited. Discret. Math. **306**(14), 1540–1556 (2006)
3. Gong, G.: Theory and applications of q-ary interleaved sequences. IEEE Trans. Inf. Theory **41**(2), 400–411 (1995)
4. Helleseth, T., Kløve, T., Mvkkeltveit, J.: On the covering radius of binary codes. IEEE Trans. Inf. Theory **24**(5), 627–628 (1978)
5. Hou, X.-D.: On the norm and covering radius of first-order Reed-Muller codes. IEEE Trans. Inf. Theory **43**(3), 1025–1027 (1997)
6. Kavut, S., Maitra, S.: Patterson-Wiedemann type functions on 21 variables with nonlinearity greater than bent concatenation bound. IEEE Trans. Inf. Theory **62**(4), 2277–2282 (2016)
7. Kavut, S., Yücel, M.D.: 9-variable Boolean functions with nonlinearity 242 in the generalized rotation symmetric class. Inf. Comput. **208**(4), 341–350 (2010)
8. Kavut, S., Maitra, S., Yücel, M.D.: Search for Boolean functions with excellent profiles in the rotation symmetric class. IEEE Trans. Inf. Theory **53**(5), 1743–1751 (2007)

9. Maitra, S., Sarkar, P.: Modifications of Patterson-Wiedemann functions for cryptographic applications. IEEE Trans. Inf. Theory **48**(1), 278–284 (2002)
10. Matsui, M.: Linear cryptanalysis method for DES cipher. In: Helleseth, T. (ed.) EUROCRYPT 1993. LNCS, vol. 765, pp. 386–397. Springer, Heidelberg (1994). doi:10.1007/3-540-48285-7_33
11. Meier, W., Staffelbach, O.: Fast correlation attacks on stream ciphers. In: Barstow, D., Brauer, W., Brinch Hansen, P., Gries, D., Luckham, D., Moler, C., Pnueli, A., Seegmüller, G., Stoer, J., Wirth, N., Günther, C.G. (eds.) EUROCRYPT 1988. LNCS, vol. 330, pp. 301–314. Springer, Heidelberg (1988). doi:10.1007/3-540-45961-8_28
12. Mykkeltveit, J.J.: The covering radius of the $(128, 8)$ Reed-Muller code is 56. IEEE Trans. Inf. Theory **26**(3), 358–362 (1983)
13. Patterson, N.J., Wiedemann, D.H.: The covering radius of the $(2^{15}, 16)$ Reed-Muller code is at least 16276. IEEE Trans. Inf. Theory **IT–29**(3), 354–356 (1983). See also correction: IEEE Trans. Inf. Theory **IT-36**(2), 443 (1990)
14. Sarkar, S., Maitra, S.: Idempotents in the neighbourhood of Patterson-Wiedemann functions having Walsh spectra zeros. Des. Codes Crypt. **49**(1–3), 95–103 (2008)
15. Siegenthaler, T.: Decrypting a class of stream ciphers using ciphertext only. IEEE Trans. Comput. **C–34**(1), 81–85 (1985)

A Correction and Improvements of Some Recent Results on Walsh Transforms of Gold Type and Kasami-Welch Type Functions

Ayhan Coşgun[1(✉)] and Ferruh Özbudak[2,3]

[1] Department of Mathematics, Middle East Technical University, Dumlupınar Bul., No:1, 06800 Ankara, Turkey
cosgun@metu.edu.tr

[2] Department of Mathematics and Institute of Applied Mathematics, Middle East Technical University, Dumlupınar Bul., No:1, 06800 Ankara, Turkey
ozbudak@metu.edu.tr

[3] Department of Mathematical Sciences, Aalborg University, Aalborg, Denmark

Abstract. We give explicit evaluations of Walsh transforms of Gold type functions $f(x) = \mathrm{Tr}_K\left(x^{2^a+1} + x^{2^b+1}\right)$, $0 \le a < b$ when $\gcd(b-a,k) = \gcd(b+a,k)$ and Kasami-Welch type functions $f(x) = \mathrm{Tr}_K\left(x^{\frac{2^{ta}+1}{2^a+1}}\right)$, when t is odd, $\gcd(2^k-1, 2^a+1) = 1$, k is even. Therefore we correct a recent result of Roy'2012, we solve an open problem stated in Roy'2012 and we improve and generalize some results of Roy'2012 and Lahtonen-McGuire-Ward'2007.

Keywords: Finite fields · Gold type functions · Kasami-Welch type functions · Walsh transform

1 Introduction

Let $K = \mathbb{F}_{2^k}$ denote the finite field of 2^k elements. We will denote the absolute trace map from a finite field F to $\mathbb{F}_2$ with Tr_F.

Let f be a Boolean function $f : V_k \longrightarrow \mathbb{F}_2$, where V_k is a k-dimensional vector space over $\mathbb{F}_2$. The *Walsh transform* of f at α is the function $f^W : V_k \longrightarrow \mathbb{Z}$ defined by

$$f^W(\alpha) = \sum_{x \in V_k} (-1)^{f(x)+\langle \alpha, x\rangle} \tag{1}$$

where $\langle \alpha, x\rangle$ denotes an (non-degenerate) inner product on V_k. When $V_k = K$, a natural choice for $\langle \alpha, x\rangle$ is $\mathrm{Tr}_K(\alpha x)$. We refer, for example, to [1] for more details on Walsh transform for Boolean functions. Then Eq. (1) becomes

$$f^W(\alpha) = \sum_{x \in K} (-1)^{f(x)+\mathrm{Tr}_K(\alpha x)}. \tag{2}$$

S. Duquesne and S. Petkova-Nikova (Eds.): WAIFI 2016, LNCS 10064, pp. 243–257, 2016.
DOI: 10.1007/978-3-319-55227-9_17

The *Walsh spectrum* of a Boolean function $f : K \longrightarrow \mathbb{F}_2$ is defined to be the set

$$\{f^W(\alpha) : \alpha \in K\}.$$

When the spectrum is precisely $\left\{\pm 2^{\frac{k}{2}}\right\}$, f is called *bent function.* For an integer $0 \leq r \leq k$, a function $f : K \longrightarrow \mathbb{F}_2$ is called *r-plateaued* (*r-partially bent*) if its Walsh spectrum is $\left\{0, \pm 2^{\frac{1}{2}(k+r)}\right\}$. Bent functions have significance due to their applications in cryptography and r-plateaued functions gain interest as they can be used to construct bent functions (see [8,11] for instance).

Among the most famous examples of functions having 3-valued Walsh spectrum, we have Gold functions [4] $f(x) = \mathrm{Tr}_K\left(x^{2^a+1}\right)$, with a is relatively prime to k and k is odd. Gold [4] determined $f^W(\alpha)$ in terms of $f^W(1)$ and $f^W(1)$ is evaluated first in [2] and then in [8]. Furthermore, more general Gold functions are studied in the appendix of [2].

The other famous examples are Kasami-Welch functions [7] (see also [3]) $f(x) = \mathrm{Tr}_K\left(x^{4^a-2^a+1}\right)$. With the same hypothesis that a is relatively prime to k and k is odd, both Gold and Kasami-Welch functions have the spectrum $\left\{0, \pm 2^{\frac{(k+1)}{2}}\right\}$ (i.e. they are 1-plateaued with the given hypothesis).

In this paper we deal with the Walsh transforms of Gold type and Kasami-Welch type functions. By a Gold type function we mean

$$f(x) = \mathrm{Tr}_K\left(x^{2^a+1} + x^{2^b+1}\right), 0 \leq a < b,$$

and by a Kasami-Welch type function we mean

$$f(x) = \mathrm{Tr}_K\left(x^{\frac{2^{ta}+1}{2^a+1}}\right), t \text{ odd}.$$

Gold type functions were studied by various authors in literature. For instance, in [8], Lahtonen, McGuire and Ward give $f^W(0)$ for $f(x) = \mathrm{Tr}_K\left(x^{2^a+1} + x^{2^b+1}\right)$, where $0 \leq a < b$, $\gcd(b-a,k) = \gcd(b+a,k) = 1$ and k odd. Then, using the results of Fitzgerald in [6], Roy [11] evaluated $f^W(\alpha)$

- for any $\alpha \in K$ with k odd,
- for $\alpha \in K$ with $\mathrm{Tr}_K(\alpha) = 0$ and k even,

and stated that the case

- $\mathrm{Tr}_K(\alpha) = 1$ with k even

is open. However, we observed that Roy's result for the case

- $\alpha \in K$ with $\mathrm{Tr}_K(\alpha) = 0$ and k even

does not hold for some α's. We give a counterexample for such an α in Example 1 below in Sect. 3. In Corollary 1 in Sect. 3, we will complete the evaluation of

$f^W(\alpha)$ by fixing the problem in the result of Roy and giving $f^W(\alpha)$ for the remaining open case $\mathrm{Tr}_K(\alpha) = 1$ with k even.

Furthermore, in our main result Theorem 1, we consider a more general function $f(x) = \mathrm{Tr}_K\left(x^{2^a+1} + x^{2^b+1}\right)$, $0 \le a < b$ with the assumption $\gcd(b-a, k) = \gcd(b+a, k)$. For any positive integer n, let $v_2(n)$ denote the highest non-negative exponent v such that 2^v divides n. Two cases occur for the evaluation of $f^W(\alpha)$:

- when "$v_2(b-a) = v_2(b+a) = v_2(k) - 1$" does not hold,

we completed the evaluation.

- When $v_2(b-a) = v_2(b+a) = v_2(k) - 1$

we computed $f^W(\alpha)$ up to its sign, and determined its sign exactly under some extra assumptions (see Theorem 2).

By means of Theorem 1, we will determine the Walsh spectrum of $f(x) = \mathrm{Tr}_K\left(x^{2^a+1} + x^{2^b+1}\right)$, $0 \le a < b$ and $\gcd(b-a, k) = \gcd(b+a, k)$ in Corollary 2 in Sect. 3. We observe that the Walsh spectrum is more complicated than the special case obtained in [11] (see the paragraph before [11, Theorem 7]).

Kasami-Welch type functions, $f(x) = \mathrm{Tr}_K(x^e)$ where $e = \frac{2^{ta}+1}{2^a+1}$ and t is odd, were studied by Niho in his thesis [10]. In [8], Lahtonen, McGuire and Ward evaluated $f^W(1)$ under certain conditions and then Roy in [11] generalized their result up to k odd. We also give generalization of Roy's result for k even with Theorem 3 in Sect. 3 below.

The rest of the paper is organized as follows. We give some background in Sect. 2. We present our results in Sect. 3.

2 Preliminaries

In this section we introduce our notation and present some facts about quadratic forms and linearized polynomials that we use when proving our main results in Sect. 3.

Let n be an arbitrary positive integer. Throughout the paper $v_2(n)$ will denote the highest non-negative exponent v such that 2^v divides n (that is, the 2-adic valuation) and $\left(\frac{a}{n}\right)$ will denote the Jacobi symbol of a modulo n. For finite fields F and E, we will write $\mathrm{Tr}_{E/F}$ for the relative trace from E to F. We also introduce the notation $\chi_E(x)$ for $(-1)^{\mathrm{Tr}_E(x)}$ for any finite field E of characteristic 2.

Let

$$R(x) = \sum_{i=0}^{h} a_i x^{2^i} + \alpha,$$

where $a_i, \alpha \in K$. Let $Q : K \longrightarrow \mathbb{F}_2$ be the quadratic form given by

$$Q(x) = \mathrm{Tr}_K(xR(x)).$$

Then we have

$$\sum_{x\in K}(-1)^{Q(x)} = \Lambda\,(Q)\,2^{\frac{1}{2}(k+r(Q))}. \tag{3}$$

Here $r\,(Q) = \dim rad\,(Q)$ is the dimension of the radical. By $rad\,(Q)$ we mean the radical of the corresponding bilinear form

$$B_Q(x,y) = Q(x+y) + Q(x) + Q(y) \quad \text{for } x, y \in K.$$

More precisely,

$$rad\,(Q) = \{y \in K | B_Q(x,y) = 0 \text{ for all } x \in K\}.$$

Moreover, it is well-known that the invariant $\Lambda\,(Q)$ of the quadratic form Q takes values in the set $\{-1, 0, +1\}$. We refer to [9] for further details.

Combining definition (2) and Eq. (3) above, we have that if $f(x) = \mathrm{Tr}_K\left(x\left(\sum_{i=0}^{h} a_i x^{2^i}\right)\right)$, then

$$f^W(\alpha) = \Lambda\,(Q)\,2^{\frac{1}{2}(k+r(Q))} \tag{4}$$

where $R(x) = \sum_{i=0}^{h} a_i x^{2^i} + \alpha$. Therefore, in order to evaluate $f^W(\alpha)$ it is enough to determine $\Lambda\,(Q)$ and $r\,(Q)$. Furthermore, quadratic functions are $r\,(Q)$-plateaued by Eq. (4).

The following is an another well-known fact and employed in many papers:

$$rad\,(Q) = \log_2\left[deg\left(\gcd\left(R^*(x), x^{2^k} + x\right)\right)\right]$$

where

$$R^*(x) = \sum_{i=0}^{h} a_i \left(x^{2^{h+i}} + x^{2^{h-i}}\right)$$

is the radical polynomial of $R(x)$.

It is easy to observe that $rad\,(Q)$ is independent of the affine part of Q, and this yields:

Lemma 1. *Define $Q_\alpha(x) = \mathrm{Tr}_K\,(xR_\alpha(x))$ and $Q_\beta(x) = \mathrm{Tr}_K\,(xR_\beta(x))$ where $R_\alpha(x) = \sum_{i=0}^{h} a_i x^{2^i} + \alpha \in K[x]$ and $R_\beta(x) = \sum_{i=0}^{h} a_i x^{2^i} + \beta \in K[x]$. Then*

$$r\,(Q_\alpha) = r\,(Q_\beta).$$

Lemma 1 shows that the relation between $f^W(0)$ and $f^W(\alpha)$ depends only on the relation between the invariants $\Lambda\,(Q_0)$ and $\Lambda\,(Q_\alpha)$.

A polynomial of the form

$$L(x) = \sum_{i=0}^{h} a_i x^{q^i} \in \mathbb{F}_{q^m}[x]$$

is called a *linearized polynomial* over $\mathbb{F}_{q^m}$. Its *q-associate* is defined as $l(t) = \sum_{i=0}^{h} a_i t^i \in \mathbb{F}_{q^m}[t]$ and $L(x)$ is called the *inverse q-associate* of $l(t)$.

Let $A(x), B(x) \in \mathbb{F}_{q^m}[x]$ be linearized polynomials and $a(t), b(t) \in \mathbb{F}_{q^m}[t]$ be their q-associates. Then we define the right division "$|_r$" in $\mathbb{F}_{q^m}[x]$ by

$$A(x)|_r B(x) \quad \text{if and only if} \quad B(x) = C(x) \circ A(x)$$

for some linearized polynomial $C(x) \in \mathbb{F}_{q^m}[x]$

When, $m = 1$, in particular, it is a well-known fact that

- q-associate of $A(x) \circ B(x)$ is $a(t)b(t)$ and
- $A(x)|_r B(x)$ if and only if $A(x)$ divides $B(x)$ in ordinary sense.

We will use the following well-known fact about linearized polynomials in the proof of Lemma 3 below in Sect. 3. We provide a proof for completeness (see also [9]).

Proposition 1. *Suppose $L_1(x), L_2(x) \in \mathbb{F}_q[x]$ are two linearized polynomials over $\mathbb{F}_q$, and their q-associates are $l_1(t), l_2(t) \in \mathbb{F}_q[t]$ respectively. Then*

$$\gcd(L_1(x), L_2(x)) = \textit{the inverse q-associate of} \gcd(l_1(t), l_2(t))$$

where $\gcd(L_1(x), L_2(x))$ *is the greatest common divisor of two polynomials $L_1(x)$ and $L_2(x)$ for Euclidean division.*

Proof. Let $\gcd(L_1(x), L_2(x)) = A(x)$, $\gcd(l_1(t), l_2(t)) = b(t)$ and $B(x)$ be the inverse q-associate of $b(t)$. Then we will show that $A(x) = B(x)$.

- $B(x)$ divides $A(x)$:

Let $l_1(t) = c_1(t)b(t)$ and $l_2(t) = c_2(t)b(t)$ for some $c_1(t)$ and $c_2(t)$ in $\mathbb{F}_q[t]$. Then their inverse q-associates are $L_1(x) = C_1(x) \circ B(x)$ and $L_2(x) = C_2(x) \circ B(x)$ where $C_1(x)$, $C_2(x)$ are inverse q-associates of $c_1(t)$ and $c_2(t)$, respectively. So

$$B(x)| \gcd(L_1(x), L_2(x)) = A(x).$$

- $A(x)$ divides $B(x)$:

As $\gcd(L_1(x), L_2(x)) = A(x)$, we have $L_1(x) = D_1(x) \circ A(x)$ and $L_2(x) = D_2(x) \circ A(x)$ for some linearized polynomials $D_1(x), D_2(x) \in \mathbb{F}_q[x]$. Then their q-associates are $l_1(t) = d_1(t)a(t)$ and $l_2(t) = d_2(t)a(t)$ where $d_1(t)$, $d_2(t)$ are q-associates of $D_1(x)$ and $D_2(x)$, respectively. So $a(t)| \gcd(l_1(t), l_2(t)) = b(t)$. That is, $A(x)|B(x)$.

□

3 Main Results

In this section, we will give a counterexample to the result in [11] and then present our main result in Theorem 1. Moreover, Theorem 2 and Corollary 1 solves an open problem of [11] and Theorem 3 generalizes a result of [11].

In the example below, we will see that the result in [11, Theorem 11] does not hold for some $\alpha \in K$.

Example 1. Let $k = 2$, so $K = \mathbb{F}_4$. Also let $f(x) = \mathrm{Tr}_K\left(x^{2^0+1} + x^{2^1+1}\right)$. So, $a = 0$, $b = 1$ and $\gcd(b-a,k) = \gcd(b+a,k) = 1$. Then, by [6, Theorem 2.1] we have $f^W(0) = 0$. Therefore, we would have $f^W(\alpha) = 0$ for all $\alpha \in K$ with $\mathrm{Tr}_K(\alpha) = 0$ according to [11, Theorem 11].

Now, let $\gamma \in K = \mathbb{F}_4$ be the element such that $\gamma^2 = \gamma+1$ (Note that x^2+x+1 is irreducible over $\mathbb{F}_2$). Then $\mathbb{F}_4 = \{0, 1, \gamma, \gamma+1\}$.

For $\alpha = 1$ (so $\mathrm{Tr}_K(1) = 1 + 1^2 = 0$) we have

$$\begin{aligned} f^W(1) &= \sum_{x\in K} (-1)^{\mathrm{Tr}_K(x^2+x^3+x)} \\ &= (-1)^{\mathrm{Tr}_K(0^2+0^3+0)} + (-1)^{\mathrm{Tr}_K(1^2+1^3+1)} \\ &+ (-1)^{\mathrm{Tr}_K(\gamma^2+\gamma^3+\gamma)} + (-1)^{\mathrm{Tr}_K((\gamma+1)^2+(\gamma+1)^3+(\gamma+1))} \\ &= (-1)^{\mathrm{Tr}_K(0)} + (-1)^{\mathrm{Tr}_K(1)} + (-1)^{\mathrm{Tr}_K(0)} + (-1)^{\mathrm{Tr}_K(0)} = 4 \end{aligned}$$

and so $f^W(1) \neq 0$.

The problem in the proof of [11, Theorem 11] is about the image $\mathrm{Im}(L)$ of L where $L(x) = x^{2^a} + x^{2^{-a}} + x^{2^b} + x^{2^{-b}}$. In [11, Theorem 7] it is shown that when $\gcd(b-a,k) = \gcd(b+a,k) = 1$ and k is odd, we have $\mathrm{Im}(L) = K_0$ where K_0 is the set of elements of K with absolute trace 0. The equality "$\mathrm{Im}(L) = K_0$" is assumed also in the proof of [11, Theorem 11], when k is even. However, the equality is not true for even k.

For any integer n dividing k, define the set

$$S_n = \{x \in K : \mathrm{Tr}_{K/\mathbb{F}_{2^n}}(x) = 0\}$$

from now on. In fact, we will see below in Lemma 3 that

$$\mathrm{Im}(L) = \begin{cases} S_d, & \text{if } k/d \text{ is odd,} \\ S_{2d}, & \text{if } k/d \text{ is even.} \end{cases}$$

Therefore, [11, Theorem 11] (where $d = 1$ and k is even) does not necessarily hold for an $\alpha \in K$ such that $Tr_{K/\mathbb{F}_{2^2}}(\alpha) = 1$ as in the Example 1, although we have still $Tr_K(\alpha) = 0$.

Before proving Lemma 3 we will present the following observation which will play a central role in its proof.

Lemma 2. *Assume* $\gcd(b-a,k) = \gcd(b+a,k)$. *Put* $d = \gcd(b-a,k)$ $(= \gcd(b+a,k))$ *and let* $\delta = \gcd(2d,k)$. *Then we have* $\delta \in \{d, 2d\}$ *and*

(i) $\delta = d \iff k/d$ *is odd* $\iff \delta|(b-a)$ *and* $\delta|(b+a)$,
(ii) $\delta = 2d \iff k/d$ *is even* $\iff \delta \nmid (b-a)$, $\delta \nmid (b+a)$ *and* $\delta|2a$, $\delta|2b$.

Proof. We have $\delta = \gcd(2d,k) = \begin{cases} d, & \text{if } k/d \text{ is odd,} \\ 2d, & \text{if } k/d \text{ is even.} \end{cases}$

If $\delta = d$, then $\delta|(b-a)$ and $\delta|(b+a)$ by assumption. So *(i)* is proved.

Assume $\delta = 2d$. So k/d is even and $v_2(k) > v_2(d)$. Then we get $v_2(b-a) = v_2(b+a) = v_2(d)$. Hence, $\delta \nmid (b-a)$ and $\delta \nmid (b+a)$.

Furthermore, $v_2(2b) > v_2(d)$ and $v_2(2a) > v_2(d)$ and this yields $v_2(2b) - v_2(d) = v_2(2b/d) \geq 1$ and $v_2(2a) - v_2(d) = v_2(2a/d) \geq 1$. Then, both $2b/d$ and $2a/d$ are even (note that d divides both $2b$ and $2a$). That is, $\delta = 2d$ divides both $2b$ and $2a$. □

Now we are ready for the next lemma.

Lemma 3. *Let* $L : K \longrightarrow K$ *where* $L(x) = x^{2^a} + x^{2^{-a}} + x^{2^b} + x^{2^{-b}}$. *Under the notation of Lemma 2 we have*

$$\mathrm{Im}(L) = S_\delta.$$

Proof. Clearly $L : K \longrightarrow K$ is linear. We claim:

(1) $\mathrm{Im}(L) \subseteq S_\delta$.
(2) $\mathrm{Ker}(L) = \mathbb{F}_{2^\delta}$.

Proof of (1): We will show that $\mathrm{Tr}_{K/\mathbb{F}_{2^\delta}}(L(x)) = 0$ for all $x \in K$.

$$\begin{aligned}\mathrm{Tr}_{K/\mathbb{F}_{2^\delta}}(L(x)) &= \mathrm{Tr}_{K/\mathbb{F}_{2^\delta}}\left(x^{2^a} + x^{2^{-a}} + x^{2^b} + x^{2^{-b}}\right)\\ &= \mathrm{Tr}_{K/\mathbb{F}_{2^\delta}}\left(x^{2^a} + x^{2^{k-a}} + x^{2^b} + x^{2^{k-b}}\right).\end{aligned}$$

Case (i): If $\delta = d$, $\delta|(b-a)$ by Lemma 2. Then

$$\left[x^{2^a}\right]^{(2^\delta)^{\frac{k+b-a}{\delta}}} = x^{2^{a+k+b-a}} = x^{2^{k+b}} = x^{2^b}$$

and

$$\left[x^{2^{k-a}}\right]^{(2^\delta)^{\frac{k-(b-a)}{\delta}}} = x^{2^{k-b}}$$

similarly. That is, $\mathrm{Tr}_{K/\mathbb{F}_{2^\delta}}\left(x^{2^a}\right) = \mathrm{Tr}_{K/\mathbb{F}_{2^\delta}}\left(x^{2^b}\right)$ and $\mathrm{Tr}_{K/\mathbb{F}_{2^\delta}}\left(x^{2^{k-a}}\right) = \mathrm{Tr}_{K/\mathbb{F}_{2^\delta}}\left(x^{2^{k-b}}\right)$.

Case (ii): If $\delta = 2d$, $\delta|2b$ and $\delta|2a$ by Lemma 2. Then

$$\left[x^{2^a}\right]^{(2^\delta)^{\frac{k-2a}{\delta}}} = x^{2^{k-a}} \quad \text{and} \quad \left[x^{2^b}\right]^{(2^\delta)^{\frac{k-2b}{\delta}}} = x^{2^{k-b}}.$$

Thus, $\mathrm{Tr}_{K/\mathbb{F}_{2^\delta}}\left(x^{2^a}\right) = \mathrm{Tr}_{K/\mathbb{F}_{2^\delta}}\left(x^{2^{k-a}}\right)$ and $\mathrm{Tr}_{K/\mathbb{F}_{2^\delta}}\left(x^{2^b}\right) = \mathrm{Tr}_{K/\mathbb{F}_{2^\delta}}\left(x^{2^{k-b}}\right)$.

Therefore, in both cases we get $\mathrm{Tr}_{K/\mathbb{F}_{2^\delta}}(L(x)) = 0$ for all $x \in K$.

Proof of (2):

$$L(x) = x^{2^a} + x^{2^{-a}} + x^{2^b} + x^{2^{-b}} = 0 \text{ if and only if } x^{2^{a+b}} + x^{2^{b-a}} + x^{2^{2b}} + x = 0.$$

It will be sufficient to show that

$$\gcd\left(x^{2^{a+b}} + x^{2^{b-a}} + x^{2^{2b}} + x, x^{2^k} + x\right) = x^{2^\delta} + x.$$

The linearized polynomial $x^{2^{a+b}} + x^{2^{b-a}} + x^{2^{2b}} + x \in \mathbb{F}_2[x]$ has the 2-associate $x^{a+b} + x^{b-a} + x^{2b} + 1$ which has the following factorization

$$x^{a+b} + x^{b-a} + x^{2b} + 1 = \left(x^{a+b} + 1\right)\left(x^{b-a} + 1\right).$$

Since $\gcd(b-a, k) = \gcd(b+a, k) = d$, we have

$$\gcd\left(x^{a+b} + 1, x^k + 1\right) = x^d + 1 \text{ and } \gcd\left(x^{b-a} + 1, x^k + 1\right) = x^d + 1.$$

Then

$$\begin{aligned}\gcd\left(x^{a+b} + x^{b-a} + x^{2b} + 1, x^k + 1\right) &= \begin{cases} x^d + 1, & \text{if } k/d \text{ is odd,} \\ x^{2d} + 1, & \text{if } k/d \text{ is even} \end{cases} \\ &= x^\delta + 1\end{aligned}$$

and the result follows by Proposition 1.

Hence, $K/\mathrm{Ker}(L) \cong \mathrm{Im}(L)$ implies $|\mathrm{Im}(L)| = 2^{k-\delta}$ and then $\mathrm{Im}(L) = S_\delta$ as $|S_\delta| = 2^{k-\delta}$. □

Now we present the main result of the paper. The evaluation of $f^W(0)$ is already completed in [6]. We find $f^W(\alpha)$ in terms of $f^W(0)$ in some cases of our main result, and we give $f^W(0)$ in absolute value only.

Theorem 1. *Assume that* $\gcd(b-a, k) = \gcd(b+a, k)$ *with* $0 \le a < b$ *and put* $d = \gcd(b-a, k)$ $(= \gcd(b+a, k))$. *Let* $K = \mathbb{F}_{2^k}$, $E = \mathbb{F}_{2^\delta}$ *where* $\delta = \gcd(2d, k)$, *and* $f(x) = \mathrm{Tr}_K\left(x^{2^a+1} + x^{2^b+1}\right)$.

Case 1: *"*$v_2(b-a) = v_2(b+a) = v_2(k) - 1$*" does not hold:*

If $\mathrm{Tr}_{K/E}(\alpha) = 0$, *then we choose* $\beta \in K$ *such that* $\beta^{2^a} + \beta^{2^{-a}} + \beta^{2^b} + \beta^{2^{-b}} = \alpha$ *(see Lemma 3 for existence of such* β*). Then,*

$$f^W(\alpha) = \begin{cases} (-1)^{\mathrm{Tr}_K\left(\beta^{2^a+1} + \beta^{2^b+1} + \alpha\beta\right)} f^W(0), & \textit{if } \mathrm{Tr}_{K/E}(\alpha) = 0, \\ 0, & \textit{otherwise,} \end{cases}$$

where $|f^W(0)| = 2^{\frac{1}{2}(k+\delta)}$.

Case 2: *If $v_2(b-a) = v_2(b+a) = v_2(k) - 1$:*

In this case $[K : E]$ is odd. Put $\tau = \mathrm{Tr}_{K/E}(\alpha)$. Then $\mathrm{Tr}_{K/E}(\alpha + \tau) = 0$ and hence we choose $\beta \in K$ such that $\beta^{2^a} + \beta^{2^{-a}} + \beta^{2^b} + \beta^{2^{-b}} = \alpha + \tau$ (see Lemma 3 for existence of such β). Then,

$$f^W(\alpha) = (-1)^{\mathrm{Tr}_K\left(\beta^{2^a+1} + \beta^{2^b+1} + \alpha\beta\right)} f^W(\tau).$$

Furthermore,

$$f^W(\tau) = \Lambda(g)\, 2^{\frac{1}{2}(k+2d)}$$

where g is the quadratic form $g(x) = \mathrm{Tr}_E\left(x^{2^a+1} + x^{2^b+1} + x\tau\right)$ and $\Lambda(g)$ denotes its invariant.

Remark 1. To avoid a very long and complicated statement in Theorem 1, we will continue the evaluation of $\Lambda(g)$ separately in Theorem 2.

Proof. Firstly,

$$f^W(0) = \sum_{x \in K} (-1)^{f(x)} = \Lambda(f)\, 2^{\frac{1}{2}(k+r)}$$

where

$$r = deg\left(\gcd\left(x^{a+b} + x^{b-a} + x^{2b} + 1, x^k + 1\right)\right) = deg\left(x^\delta + 1\right) = \delta$$

by Proposition 1 and proof of Lemma 3. As the dimension of the radical does not depend on α, we have $f^W(\alpha) = 0$ or $|f^W(\alpha)| = 2^{\frac{1}{2}(k+\delta)}$. So it is left to determine the sign of $f^W(\alpha)$.

By [6, Theorem 2.1],

$$\text{invariant of } f = \Lambda(f) = 0$$

if and only if

$$v_2(b-a) = v_2(b+a) = v_2(k) - 1$$

Thus,

$$f^W(0) = 0 \quad \text{if and only if} \quad v_2(b-a) = v_2(b+a) = v_2(k) - 1$$

where $\chi_K(x) = (-1)^{f(x)}$.

Case 1: When "$v_2(b-a) = v_2(b+a) = v_2(k) - 1$" does not hold.

In this case we are sure that $f^W(0) \neq 0$. Then, by [5, Proposition 3.2],

$$f^W(\alpha) = \begin{cases} (-1)^{f(x_0)} f^W(0), & \text{if } R^*(x) = \alpha^{2^b} \text{ has a solution } x_0 \in K, \\ 0, & \text{otherwise,} \end{cases}$$

where $R^*(x) = x^{2^{b+a}} + x^{2^{b-a}} + x^{2^{2b}} + x$ is the radical polynomial of $R(x) = x^{2^a} + x^{2^b}$.

We have

$$R^*(x) = \alpha^{2^b} \text{ for some } x_0 \in K \;\; \textit{if and only if} \;\; L(x_0) = \alpha \text{ for the same } x_0 \in K$$
$$\textit{if and only if} \;\; \mathrm{Tr}_{K/E}(\alpha) = \tau = 0$$

by Lemma 3.

When $\tau = 0$, let $\beta \in K$ be such that $\alpha = \beta^{2^a} + \beta^{2^{-a}} + \beta^{2^b} + \beta^{2^{-b}}$ and observe that $\mathrm{Tr}_K(\alpha\beta) = \mathrm{Tr}_K\left(\beta^{2^a+1} + \beta^{2^{-a}+1} + \beta^{2^b+1} + \beta^{2^{-b}+1}\right) = 0$ as $\left(\beta^{2^{-t}+1}\right)^{2^t} = \beta^{2^t+1}$ for all integers t. Hence, $f(\beta) = \mathrm{Tr}_K\left(\beta^{2^a+1} + \beta^{2^b+1}\right) = \mathrm{Tr}_K\left(\beta^{2^a+1} + \beta^{2^b+1} + \alpha\beta\right)$ and

$$f^W(\alpha) = \begin{cases} (-1)^{\mathrm{Tr}_K\left(\beta^{2^a+1}+\beta^{2^b+1}+\alpha\beta\right)} f^W(0), & \text{if } \tau = 0, \\ 0, & \text{otherwise.} \end{cases}$$

Case 2: $v_2(b-a) = v_2(b+a) = v_2(k) - 1$.

This is the case when $v_2(k/d) = 1$. So we have $\delta = 2d$.

We will use a similar idea as Roy used in [11]. For any element β of K, we have

$$f^W(\alpha) = \chi_K\left(\beta^{2^a+1} + \beta^{2^b+1} + \alpha\beta\right) \sum_{x\in K} \chi_K\left(x^{2^a+1} + x^{2^b+1} + x\left(L(\beta) + \alpha\right)\right)$$

where $L(\beta) = \beta^{2^a} + \beta^{2^{-a}} + \beta^{2^b} + \beta^{2^{-b}}$.

Now, let $\tau = \mathrm{Tr}_{K/E}(\alpha) \in E$. Then we have

$$\mathrm{Tr}_{K/E}(\alpha + \tau) = \mathrm{Tr}_{K/E}(\alpha) + \tau \mathrm{Tr}_{K/E}(1).$$

The extension degree k/δ is odd in this case, and then $\mathrm{Tr}_{K/E}(1) = 1$. So,

$$\mathrm{Tr}_{K/E}(\alpha + \tau) = \tau + \tau = 0$$

Then, by Lemma 3 there exists $\beta \in K$ such that $L(\beta) = \alpha + \tau$. That is,

$$L(\beta) + \alpha = \tau.$$

Therefore,

$$\begin{aligned} f^W(\alpha) &= \chi_K\left(\beta^{2^a+1} + \beta^{2^b+1} + \alpha\beta\right) \sum_{x\in K} \chi_K\left(x^{2^a+1} + x^{2^b+1} + x\tau\right) \\ &= (-1)^{\mathrm{Tr}_K\left(\beta^{2^a+1}+\beta^{2^b+1}+\alpha\beta\right)} f^W(\tau) \end{aligned}$$

where $\beta \in K$ such that $\alpha + \tau = \beta^{2^a} + \beta^{2^{-a}} + \beta^{2^b} + \beta^{2^{-b}}$.

Let $\Lambda(h_\tau)$ denote the invariant of the quadratic form

$$h_\tau(x) = \mathrm{Tr}_K\left(x^{2^a+1} + x^{2^b+1} + x\tau\right).$$

So,

$$f^W(\tau) = \Lambda(h_\tau)\, 2^{\frac{1}{2}(k+2d)}.$$

It is left to show $\Lambda(h_\tau) = \Lambda(g)$. The equality can be observed by a result in [5].

Apply [5, Theorem 4.2] with $n = 2d$ and $k = 2dp_1p_2...p_s$ (Note that $v_2(k) = v_2(2d)$). Then, by means of [6, Theorem 1.5] we have

$$r\left(Q_\tau^{F_i}\right) = r\left(Q_0^{F_i}\right) = r\left(Q_0^{F_{i-1}}\right) = r\left(Q_\tau^{F_{i-1}}\right) = 2d$$

for all $1 \leq i \leq s$ where $Q_\tau^{F_i}(x) = \mathrm{Tr}_{F_i}\left(x^{2^a+1} + x^{2^b+1} + x\tau\right)$ and $F_i = \mathbb{F}_{2^{(p_1p_2\cdots p_i n)}}$. This yields

$$2^{\frac{1}{2}\left(r\left(Q_\tau^{F_i}\right) - r\left(Q_\tau^{F_{i-1}}\right)\right)} = 2^{\frac{1}{2}(0)} = 2^0 = 1 \equiv (-1)^0 (\mod p_i)$$

for all $1 \leq i \leq s$. Also we have

$$\left(\frac{2}{p_1...p_s}\right)^n = 1$$

as $n = 2d$ is even. Hence,

$$\Lambda(h_\tau) = (-1)^0 \left(\frac{2}{p_1...p_s}\right)^n \Lambda(g) = \Lambda(g).$$

This completes the proof of Theorem 1. □

For the case $v_2(b-a) = v_2(b+a) = v_2(k) - 1$, the evaluation of $f^W(\alpha)$ depends on the evaluation of $\Lambda(g)$ according to Theorem 1.

Next we evaluate $\Lambda(g)$ when $v_2(b-a) = v_2(b+a) = v_2(k) - 1$, $\tau \in \mathbb{F}_{2^2}$ and d is odd.

Theorem 2. *Under the notation of Theorem 1, assume $v_2(b-a) = v_2(b+a) = v_2(k) - 1$, $\tau \in \mathbb{F}_{2^2}$ and d is odd. Then*

$$\Lambda(g) = \begin{cases} +1, & \text{if } \tau = 1, \\ 0, & \text{otherwise.} \end{cases}$$

Proof. By the assumptions we have $0 = v_2(d) = v_2(b-a) = v_2(b+a) = v_2(k) - 1$ and $\delta = 2d$. Also observe that $v_2(2d) = 1 = v_2(2)$. Applying [6, Theorem 1.5] and [5, Theorem 4.2] together, as in the proof of Theorem 1, we get

$$\Lambda(g) = \Lambda(h_\tau)$$

where $\Lambda(h_\tau)$ denotes the invariant of the quadratic form

$$h_\tau(x) = \mathrm{Tr}_{\mathbb{F}_4}\left(x^{2^a+1} + x^{2^b+1} + x\tau\right).$$

Now, we will focus on $\Lambda(h_\tau)$. By Eq. (3),

$$\Lambda(h_\tau)\, 2^{\frac{1}{2}(2+r(h_\tau))} = \sum_{x\in\mathbb{F}_4} (-1)^{h_\tau(x)}.$$

As $\mathbb{F}_4 = \{0, 1, \gamma, \gamma+1\}$ where $\gamma^2 = \gamma + 1$, we are left to deal with 4 cases for τ.

(1) $\tau = 0$:

As $v_2(b-a) = v_2(b+a) = v_2(2) - 1$, we have $\Lambda(h_\tau) = \Lambda(h_0) = 0$ by [6, Theorem 2.1].

(2) $\tau = 1$:

- $h_\tau(0) = \mathrm{Tr}_{\mathbb{F}_4}(0) = 0$,
- $h_\tau(1) = \mathrm{Tr}_{\mathbb{F}_4}(1) = 0$,
- $h_\tau(\gamma) = \mathrm{Tr}_{\mathbb{F}_4}\left(\gamma^{2^a+1} + \gamma^{2^b+1} + \gamma\right) = \mathrm{Tr}_{\mathbb{F}_4}\left(\gamma^{2^a+1} + \gamma^{2^b+1}\right) + 1$,
- and

$$\begin{aligned} h_\tau(\gamma+1) &= \mathrm{Tr}_{\mathbb{F}_4}\left((\gamma+1)^{2^a+1} + (\gamma+1)^{2^b+1} + (\gamma+1)\right) \\ &= \mathrm{Tr}_{\mathbb{F}_4}\left(\gamma^{2^a+1} + \gamma^{2^b+1} + \gamma^{2^a} + \gamma^{2^b} + (\gamma+1)\right) \\ &= \mathrm{Tr}_{\mathbb{F}_4}\left(\gamma^{2^a+1} + \gamma^{2^b+1}\right) + \mathrm{Tr}_{\mathbb{F}_4}(\gamma) + \mathrm{Tr}_{\mathbb{F}_4}(\gamma) + \mathrm{Tr}_{\mathbb{F}_4}(\gamma+1) \\ &= \mathrm{Tr}_{\mathbb{F}_4}\left(\gamma^{2^a+1} + \gamma^{2^b+1}\right) + 1. \end{aligned}$$

Thus,

$$\sum_{x\in\mathbb{F}_4} (-1)^{h_\tau(x)} = 2 - 2(-1)^{\mathrm{Tr}_{\mathbb{F}_4}\left(\gamma^{2^a+1}+\gamma^{2^b+1}\right)}.$$

As $\gamma^2 = \gamma + 1$, we have

$$\gamma^t = \begin{cases} 1 & \text{if } t \equiv 0 \mod 3, \\ \gamma & \text{if } t \equiv 1 \mod 3, \\ \gamma+1 & \text{if } t \equiv 2 \mod 3. \end{cases}$$

Thus,

$$\mathrm{Tr}_{\mathbb{F}_4}\left(\gamma^{2^t+1}\right) = \begin{cases} 0 & \text{if } t \text{ is odd}, \\ 1 & \text{if } t \text{ is even}. \end{cases}$$

In our case, we have $v_2(b+a) = v_2(d) = 0$ and so $b+a$ is odd. Then, one of a and b is odd and the other one is even. That is,

$$\mathrm{Tr}_{\mathbb{F}_4}\left(\gamma^{2^a+1} + \gamma^{2^b+1}\right) = 1 \tag{5}$$

for all such a and b. Finally we deduce that $\sum_{x\in\mathbb{F}_4} (-1)^{h_\tau(x)} = 4$ and $\Lambda(h_\tau) = +1$ for $\tau = 1$.

(3) $\tau = \gamma$**:**

- $h_\tau(0) = \mathrm{Tr}_{\mathbb{F}_4}(0) = 0$,
- $h_\tau(1) = \mathrm{Tr}_{\mathbb{F}_4}(\gamma) = 1$,
- $h_\tau(\gamma) = \mathrm{Tr}_{\mathbb{F}_4}\left(\gamma^{2^a+1} + \gamma^{2^b+1} + (\gamma+1)\right) = 1 + 1 = 0$,
- and

$$\begin{aligned} h_\tau(\gamma+1) &= \mathrm{Tr}_{\mathbb{F}_4}\left((\gamma+1)^{2^a+1} + (\gamma+1)^{2^b+1} + (\gamma+1)\gamma\right) \\ &= \mathrm{Tr}_{\mathbb{F}_4}\left(\gamma^{2^a+1} + \gamma^{2^b+1} + \gamma^{2^a} + \gamma^{2^b} + 1\right) = 1, \end{aligned}$$

using Eq. (5). Then, $\sum_{x\in\mathbb{F}_4}(-1)^{h_\tau(x)} = 0$ and $\Lambda(h_\tau) = 0$ for $\tau = \gamma$.

(4) $\tau = \gamma + 1$**:**

- $h_\tau(0) = \mathrm{Tr}_{\mathbb{F}_4}(0) = 0$,
- $h_\tau(1) = \mathrm{Tr}_{\mathbb{F}_4}(\gamma+1) = 1$,
- $h_\tau(\gamma) = \mathrm{Tr}_{\mathbb{F}_4}\left(\gamma^{2^a+1} + \gamma^{2^b+1} + \gamma(\gamma+1)\right) = 1 + 0 = 1$,
- and

$$\begin{aligned} h_\tau(\gamma+1) &= \mathrm{Tr}_{\mathbb{F}_4}\left((\gamma+1)^{2^a+1} + (\gamma+1)^{2^b+1} + (\gamma+1)^2\right) \\ &= \mathrm{Tr}_{\mathbb{F}_4}\left(\gamma^{2^a+1} + \gamma^{2^b+1} + \gamma^{2^a} + \gamma^{2^b} + \gamma\right) = 0, \end{aligned}$$

using Eq. (5). Then, $\sum_{x\in\mathbb{F}_4}(-1)^{h_\tau(x)} = 0$ and $\Lambda(h_\tau) = 0$ for $\tau = \gamma + 1$.

□

As a consequence, we can complete the evaluation of $f^W(\alpha)$ where f is as given in [11, Theorem 11]. The following corollary completely solves the open problem stated in [11] (see pages 901–903 of [11]), in particular in the paragraph before [11, Theorem 9] and in the Remark in page 903.

Corollary 1. *Under the notation of Theorem 1, with k even and* $d = 1$*, we have*
Case 1: $v_2(k) > 1$

$$f^W(\alpha) = \begin{cases} (-1)^{\mathrm{Tr}_K\left(\beta^{2^a+1}+\beta^{2^b+1}+\alpha\beta\right)} f^W(0), & \textit{if } \tau = 0, \\ 0, & \textit{otherwise}, \end{cases}$$

where $|f^W(0)| = 2^{\frac{1}{2}(k+2)}$.

Case 2: $v_2(k) = 1$

$$f^W(\alpha) = \begin{cases} (-1)^{\mathrm{Tr}_K\left(\beta^{2^a+1}+\beta^{2^b+1}+\alpha\beta\right)} 2^{\frac{1}{2}(k+2)}, & \textit{if } \tau = 1, \\ 0, & \textit{otherwise}. \end{cases}$$

Furthermore, we are able to determine the Walsh spectrum of f which satisfies the assumptions of Theorem 1.

Corollary 2. *Under the notation of Theorem 1 for f, the Walsh spectrum of f is precisely*

$$\begin{cases} \left\{0, \pm 2^{\frac{1}{2}(k+d)}\right\}, & \text{if } k/d \text{ is odd,} \\ \left\{0, \pm 2^{\frac{1}{2}(k+2d)}\right\}, & \text{if } k/d \text{ is even.} \end{cases}$$

Proof. This corollary follows easily from Theorem 1. □

We note that in Corollary 2, the Walsh spectrum of f has two different forms depending on the parity of k/d.

The following is a related but a different result. It gives a generalization of one of the main results of [11] (see [11, Theorem 7]) for k even.

Theorem 3. *Let $K = \mathbb{F}_{2^k}$, k even, a be such that $\gcd\left(2^k - 1, 2^a + 1\right) = 1$. Let t be odd and $e = \frac{2^{ta}+1}{2^a+1}$ with a is a positive integer. If $f(x) = \mathrm{Tr}_K\left(x^e\right)$ on the field K, then*

$$f^W(1) = 2^{\frac{1}{2}(k+r(k))}$$

where $r(k) = \gcd\left((t-1)a, k\right) + \gcd\left((t+1)a, k\right) - \gcd\left(2a, k\right)$.

Proof. Since $\gcd\left(2^k - 1, 2^a + 1\right) = 1$, we have

$$\begin{aligned} f^W(1) = \sum_{x \in K} \chi_K\left(x^{\frac{2^{ta}+1}{2^a+1}} + x\right) &= \sum_{x \in K} \chi_K\left(x^{2^{ta}+1} + x^{2^a+1}\right) \\ &= \Lambda(Q)\, 2^{\frac{1}{2}(k+r(Q))} \end{aligned}$$

where $Q(x) = \mathrm{Tr}_K\left(x^{2^{ta}+1} + x^{2^a+1}\right)$. Denote $r(Q)$ by $r(k)$. Then

$$\gcd\left(2^k - 1, 2^1 + 1\right) = 1 \quad \text{if and only if} \quad v_2(k) \le v_2(a)$$

(see [5, Lemma 5.3]). So

$$v_2(k) \le v_2(a) \le v_2(ta + a)$$

and

$$v_2(k) \le v_2(a) \le v_2(ta - a)$$

as t is odd. Then, by [6, Theorem 1.5] we have

$$r(k) = \gcd\left(ta - a, k\right) + \gcd\left(ta + a, k\right) - \gcd\left(s, k\right)$$

where $s = \gcd\left(ta + a, ta - a\right) = 2a$.

Now, it is left to determine $\Lambda(Q)$. Combining [6, Theorem 3.7] and [6, Theorem 4.9] we obtain $\Lambda(Q) = 1$. □

Acknowledgements. We would like to thank the anonymous reviewers for their insightful and helpful comments that improved the presentation of this paper.

References

1. Carlet, C.: Boolean functions for cryptography and error correcting codes (chap. 8). In: Crama, Y., Hammer, P.L. (eds.) Boolean Models and Methods in Mathematics, Computer Science, and Engineering. Encyclopedia of Mathematics and Its Applications, vol. 134, pp. 257–397. Cambridge University Press, Cambridge (2010)
2. Dillon, J.F., Dobbertin, H.: New cyclic difference sets with Singer parameters. Finite Fields Appl. **10**, 342–389 (2004)
3. Dobbertin, H.: Another proof of Kasami's theorem. Des. Codes Cryptogr. **17**, 177–180 (1999)
4. Gold, R.: Maximal recursive sequences with 3-valued recursive cross-correlation functions. IEEE Trans. Inform. Theory **14**, 154–156 (1968)
5. Hou, X.D.: Explicit evaluation of certain exponential sums of binary quadratic functions. Finite Fields Appl. **13**, 843–868 (2007)
6. Fitzgerald, R.: Invariants of trace forms over finite fields of characteristic two. Finite Fields Appl. **15**, 261–275 (2009)
7. Kasami, T.: The weight enumerators for several classes of subcodes of the second order binary Reed-Muller codes. Inf. Control **18**, 369–394 (1971)
8. Lahtonen, J., McGuire, G., Ward, H.N.: Gold and Kasami-Welch functions, quadratic forms, and bent functions. Adv. Math. Commun. **1**(2), 243–250 (2007)
9. Lidl, R., Niederreiter, H.: Finite Fields. Encyclopedia of Mathematics and its Applications, vol. 20, 2nd edn. Cambridge University Press, Cambridge (1997)
10. Niho, Y.: Multi-valued cross-correlation functions between two maximal linear recursive sequences. Ph.D. thesis, University of Southern California, Los Angeles (1972)
11. Roy, S.: Generalization of some results on Gold and Kasami-Welch functions. Finite Fields Appl. **18**, 894–903 (2012)

A Practical Group Signature Scheme Based on Rank Metric

Quentin Alamélou[1,2](✉), Olivier Blazy[1], Stéphane Cauchie[2], and Philippe Gaborit[1]

[1] Université de Limoges, XLIM-DMI, Limoges, France
{quentin.alamelou,olivier.blazy,philippe.gaborit}@xlim.fr
[2] R&D Department, Worldline, Seclin, France
{quentin.alamelou,stephane.cauchie}@worldline.com

Abstract. In this work, we propose the first rank-based group signature. Our construction enjoys two major advantages compared to concurrent post-quantum schemes since it is both practicably instantiated with public key and signature sizes logarithmic in the number of group members, and dynamic in a relaxation of the reference BSZ model. For such a result, we introduce a new rank-based tool, referred as the *Rank Concatenated Stern's protocol*, enabling to link different users to a common syndrome. This protocol, which could be of independent interest, can be seen as a Stern-like protocol with an additional property that permits a verifier to check the weight of each part of a split secret. Along with this work, we also define two rank-based adaptations of Hamming-based problems, referred as the *One More Rank Syndrome Decoding* and the *Decision Rank Syndrome Decoding* problems for which we discuss the security. Embedded into Fiat-Shamir paradigm, our authentication protocol leads to a group signature scheme secure in the Random Oracle Model assuming the security of rank-based systems (namely RankSign and LRPC codes) and the newly introduced problems. For a 100 bits security level, we give an example of parameters which lead to a signature size of **550 kB** and **5 kB** for the public key.

Keywords: Group signature · Post-quantum cryptography · Rank metric · Zero-knowledge

1 Introduction

A group signature scheme allows members of a group to anonymously issue signatures on behalf of the group while an opener may revoke anonymity. It turned out to be very useful in real-life applications such as, for instance, e-voting or company access policy. While current practical group signatures schemes are still based on number theory, it is worth looking for constructions facing the quantum computer.

Related Work. Since its introduction in [1], group signature has been extensively studied and two main formalization works [2,3] have been proposed (BMW and BSZ models). In the latter case, Bellare et al. consider the case where new

S. Duquesne and S. Petkova-Nikova (Eds.): WAIFI 2016, LNCS 10064, pp. 258–275, 2016.
DOI: 10.1007/978-3-319-55227-9_18

users can be added during the lifetime of the group. Numerous efficient pairing-based group signatures have tended to fit or extend these aforesaid models among which the following non-exhaustive works [4–9]. To address the quantum threat, Gordon et al. designed the first lattice-based group signature scheme [10] and in spite of recent progress in this area [11–16] or even in code-based cryptography [17, 18], post quantum schemes still suffer parameters inefficiency and/or lack of properties compared to number-theoretic based constructions previously mentioned. In parallel to this strong attention for group signature and its subsequent improvements, the field of rank-based cryptography has also gained a lot of interest due to, notably, the recent design of efficient and post quantum cryptosystems [19, 20] with strong security reductions [21].

Our Contributions. Our rank-based construction constitutes the first post quantum group signature scheme to both enable enrollment of new users and enjoy practical parameters. Indeed, our scheme benefits from public keys sizes logarithmic in the number of group members, leading to an instantiation with signatures and public key, respectively of sizes **550 kB** and **5 kB** for a 100 bits security level. For such a purpose, we propose a novel approach while designing a (rank-based) Stern-like authentication protocol, referred as the Rank Concatenated Stern's Protocol. The key idea is to enable a verifier to check the weights of both parts of a split secret. This protocol is then turned into a group signature via Fiat-Shamir (FS) paradigm [22] to constitute a new tool in the growth of rank-based cryptography. We describe a generic scheme that we instantiate with the LRPC cryptosystem and RankSign scheme so that the practical security of our scheme relies on these aforesaid schemes and the two rank-based problems introduced along with this work, referred as the *One More Rank Syndrome Decoding* (OMSD) and the *Decision Rank Syndrome Decoding* (D-RSD) problems.

Road Map. In the following, Sects. 2 and 3 are respectively concerned with preliminaries, notably about rank based cryptography, and our group signature model. Section 4 introduces our new ZK authentication protocol while Sect. 5 describes the ensuing group signature scheme. Finally, Sect. 6 provides a security analysis and Sect. 7 gives parameters for instantiating our scheme.

2 Preliminaries

We first define notation and then give some basic background.

2.1 Notation

All through this work, we use the following notation.

h denote a random oracle. λ denotes a security parameter; $\mathcal{H}_\lambda$ denotes a random oracle whose output depends on λ. Given a prover-verifier protocol, l_λ denotes the number of rounds to run to achieve security related to λ. Except stated otherwise, *log* denotes the binary logarithm. We denote by $[\![n]\!]$ the set $\{0, \ldots, n\}$. $\mathcal{A}(z; \mathcal{O})$ denotes that entity $\mathcal{A}$ has knowledge of z and access to oracle $\mathcal{O}$.

Rank Metric. Let q be a power of a prime p, m an integer and let V_n be an n dimensional vector space over a finite field $GF(q^m)$. Let β denote a basis $(\beta_1, \ldots, \beta_m)$ of $GF(q^m)$ over $GF(q)$. Let $\mathcal{F}_i$ be the map from $GF(q^m)$ to $GF(q)$ where $\mathcal{F}_i(x)$ is the i–th coordinate of x in the basis β. To any $v = (v_1, \ldots, v_n) \in V_n$, we associate $\overline{v} \in \mathcal{M}_{m,n}(GF(q))$ defined by $\overline{v}_{i,j} = \mathcal{F}_i(v_j)$. For a basis β, we denote ψ_β the inverse of the application $V_n \to \mathcal{M}_{m,n}(GF(q)) : x \to \overline{x}$ computed with the basis β.

2.2 Background on Rank Metric and Cryptography

Rank Metric Codes. We first recall some definitions.

Definition 1 (Matrix Code). *A linear matrix code C of length $m \times n$ over $GF(q)$ is a subspace of matrices space of size $m \times n$ over $GF(q)$. If C is of dimension K, we say that C is a $[m \times n, K]_q$ matrix code, or $[m \times n, K]_q$ if there is no ambiguity.*

The difference between such a $[m \times n, K]_q$ matrix code and a code of length mn and of dimension K is that we can define a natural metric through the matrix rank function.

Definition 2. *For any $v \in V_n$, the rank weight of v, denoted $\omega t(v)$, is defined as the rank of the associated matrix $\overline{v}$. We can now define the rank metric between two vectors x and y such as $d_r(x, y) := \omega t(x - y) = rank\ (\overline{x} - \overline{y})$. From now, $\mathcal{B}(S, \omega) := \{v \in S : \omega t(v) = \omega\}$.*

Definition 3 (Linear Rank Code). *A $[n, k]_{q^m}$ rank code C of length n and dimension k over $GF(q^m)$ is a linear subspace of dimension k of $GF(q^m)^n$ viewed as a rank metric space. Each word $c = (c_1, \ldots, c_n)$ of C can be associated to a $m \times n$ matrix over $GF(q)$ by representing each coordinate c_i by a column vector with respect to a basis β.*

Defined as the rank of its associate matrix $\overline{x}$, the weight of a word x does not depend on the choice of the basis β.

Definition 4. *Let $x = (x_1, \ldots, x_n) \in GF(q^m)^n$ be a vector of rank ω. We denote E the $GF(q)$-subvector space of $GF(q^m)$ generated by $x_1, x_2, \ldots, x_n$. The vector space E is called the **support** of x.*

Remark 1. The notion of support of a codeword for the Hamming metric and for the rank metric are different but share a common principle: in both cases, given a syndrome s for which it exists a low weight vector x such that $H.x^t = s$, then, if the support of x is known, it is possible to recover all the coordinates values of x by solving a linear system.

Definition 5. *Let e be an error vector of rank r and error support space E. We denote by generalized erasure of dimension t of an error e, a subspace T of dimension t of its error support E.*

Similarly to the Hamming case where an erasure corresponds to knowing the position of an error, this rank erasure notion is the knowledge of a subspace T of the error support E.

Rank-Based Cryptography. The main interest of rank-based cryptography is that for hard problems with same size of parameters, the computational complexity is higher than problems based on Hamming metric. It is then possible to generate instances of problems, with high computational complexity and with small size of keys (a few thousand bits) when such sizes are only reached with additional structure (like cyclicity) for Hamming (code-based cryptography) or Euclidean (lattice-based cryptography) distances. We now recall Syndrome Decoding problem and Gilbert-Varshamov bound analogues in rank metric. For more details, we refer the reader to [23, 24] and references therein.

Syndrome Decoding Problem (RSD). As in the Hamming case, the problem consists in finding a weighted constrained antecedent to a random syndrome by a dual matrix.

Definition 6 (Rank Syndrome Decoding). *Let H be a $(n-k) \times n$ matrix over $GF(q^m)$ $(k < n)$, $s \in GF(q^m)^{n-k}$ and ω an integer. The RSD problem consists in finding a vector $x \in GF(q^m)^n$ verifying $H.x^T = s$ and $\omega t(x) \leq \omega$.*

The hardness of this problem was proven in [21] while the complexity of the best known attacks can be found in [25]. The RSD problem can be seen as a rank adaptation of the well-known Syndrome Decoding (SD) problem which relies on Hamming metric and was proven to be NP-complete in [26].

Gilbert-Varshamov Bound (GVR). The number of elements $S(m,q,\omega)$ of a sphere of radius ω in $GF(q^m)^n$, is equal to the number of $m \times n$ q-ary matrices of rank weight t. For $t = 0$, $S_0 = 1$ and for $\omega \geq 1$, we have (see [23]):

$$S(n,m,q,\omega) = \prod_{j=0}^{\omega-1} \frac{(q^n - q^j)(q^m - q^j)}{q^t - q^j}$$

From this, we then deduce the volume of a ball $B(n,m,q,\omega)$ of radius t in $GF(q^m)$ to be:

$$B(n,m,q,\omega) = \sum_{i=0}^{\omega} S(n,m,q,i)$$

In the (frequent) linear case, the rank Gilbert-Varshamov bound $GVR(n,k,m,q)$ for a linear code $[k,n]_{q^m}$ is then defined as the smallest integer ω such as $B(n,m,q,\omega) \geq q^{m(n-k)}$ where $\mathcal{B}(n,m,q,\omega)$ denotes a ball of radius ω in in $GF(q^m)$.

For a rank code C with dual matrix H, the GVR bound is the smallest rank weight ω for which, for any syndrome s, there exists on average one word x solving the RSD instance (H,s,ω).

Rank Stern-like Protocol. Introduced by Goldwasser, Micali and Rackoff [27], Zero-Knowledge (ZK) protocols fast aroused interest. Stern then first proposed such a scheme based on coding theory [28]. Fixing the attempt of Chen [29],

Gaborit et al. designed a 3-pass prover-verifier protocol constituting a rank alternative to Stern's protocol [30]. To fulfill such a goal, they had to define, in rank metric, an equivalent notion of permutation used in the Hamming metric setting. More precisely, they came up with an operation that, without leaking any information about its support, can associate any word of rank ω to any particular word of same rank ω. We now recall this operation.

Definition 7. *Let $Q \in GL_m(q)$, $v \in V_n$ and a basis β. We define the product $Q * v$ such that $Q * v := \psi_\beta(Q\overline{v})$.*

For any $x, y \in V_n$ such that $rk(x) = rk(y)$, it is possible to find $P \in GL_n(q)$ and $Q \in GL_m(q)$ such that $x = Q * yP$.

New Rank-Based Problems. We now introduce two rank-based problems referred as the One-More Rank Syndrome (OMRSD) Decoding problem and the Decision Rank Syndrome Decoding (D-RSD) problem. We prove that the D-RSD problem is a hard problem and we justify the very likely difficulty of the OMRSD problem.

One More Rank Syndrome Decoding Problem. We first discuss the situation where one is given some solutions of a RSD instance and is asked to find a new one.

Definition 8 (OMRSD Problem). *Given an RSD instance $sd = (H, s, \omega)$ and l solutions to sd, denoted $x_1, \ldots, x_l$, the OMRSD$(sd, x_1, \ldots, x_l)$ problem consists in finding x_{l+1} solution of sd such as : $\forall i = 1 \ldots l, x_i \neq x_{l+1}$.*

Assumption 1: the OMRSD problem is hard.

Discussion on assumption 1: There is no known reduction to the RSD problem for the OMRSD problem. This problem is an adaptation in a coding context of a similar problem which exists for classical cryptography. At the difference of the classical RSD problem where an attacker knows only a syndrome and wants to find a small weight vector, in that case the attacker knows l small weight vectors of weight ω and search for a new one. It is of course possible to consider linear combinations of small weight vectors to find another small weight vector, meanwhile because of the properties of the metric, adding two random small weight vectors of weight ω leads in general to a vector of weight close to 2ω which is of no use for our problem. In particular in the case of rank metric, if ω is greater than the rank Gilbert-Varshamov bound (which has to be the case in general, if more than one preimage of a syndrome does exist), the problem of finding a pre-image of weight more than twice the GVR bound is always easy. Hence this means that using linear combinations of known solutions is not likely to be of any help. This type of result is not true for instance for Hamming metric or for Euclidean norm, for which in some cases finding preimage of weight twice the Gilbert-Varshamov bound can be difficult. Moreover the number of linear independent such solutions is upper bounded by the dimension of the code. Overall, although there is no known reduction for this problem, the problem is considered difficult by the communauty and no attack exploiting the l known vectors are known, so that the best attack for the problem consists in directly attacking the RSD problem.

Decision Rank Syndrome Decoding Problem. We now define the D-RSD problem which consists in distinguishing a random syndrome from a syndrome issued from a small weight vector.

Definition 9 (D-RSD Problem). *Given a random $H \in \mathcal{M}_{n-k,n}(GF(q^m))$, a word $x \in \mathcal{B}(GF(q^m)^n, \omega > 0)$ a random syndrome $s \in GF(q^m)^{n-k}$, is it possible to distinguish $H.x^T$ from s?*

Once again, this problem can be seen as a rank adaptation of the Decision Syndrome Decoding problem defined in Hamming-based cryptography.

Proposition 1. *The D-RSD problem is hard.*

Proof. Decision problems are very important in cryptography; in the case of Hamming-based cryptography, the Decision Syndrome Decoding problem has been proven equivalent to the search problem in Theorem 2 [31], based on the Goldreich-Levin theorem. The result is presented in term of indistinguishability of a pseudo-random generator based on the SD problem. Recently a transformation from a binary code to a q-ary code was proposed in [21] which permits to obtain a randomized reduction from the SD problem to the RSD problem. This transformation was used in [32] to adapt the result of Fisher-Stern [31] for rank metric, but with a reduction to the computational SD problem. These results hence show that there is a randomized reduction from the binary computational SD problem to the D-RSD problem, and hence that the D-RSD problem is hard. In practice the best attacks for this problem are attacks towards the RSD problem.

2.3 LRPC Codes and Related Cryptosystems

We now introduce LRPC codes [19] and ensuing cryptosystems.

Definition 10. *A Low Rank Parity Check (LRPC) code of rank d, length n and dimension k over $GF(q^m)$ is a code defined by a $(n-k) \times n$ parity check matrix $H = (h_{i,j})$, such that all its coordinates $h_{i,j}$ belong to the same $GF(q)$-subspace of dimension d of $GF(q^m)$. We denote by $\{F_1, F_2, \dots, F_d\}$ a basis of F.*

An efficient decoding algorithm was provided in a way close to the classical decoding procedure of BCH. Indeed, the general idea for decoding a word y is as follows: when the parity matrix H has rank weight small enough, the space generated by the coordinates of a syndrome $s = H.y^T$ enables to recover the product space $P = \langle E.F \rangle$. Then, knowledge of both P and F enables to deduce E the support of the error e and finally the error e contained in y by solving a linear system.

LRPC Cryptosystem. Contrary to Gabidulin codes [33], LRPC codes enjoy a poor structure so that they end up to be a well-suited candidate for rank-based cryptography. When embedded into either McEliece or Niederreiter cryptographic setting, they enable the design of an LRPC-based encryption scheme. In Sect. 7), we will focus on the Niederreiter setting to instantiate our scheme.

RankSign or Decoding a Random Syndrome Beyond GVR. LRPC codes cannot decode up to the GVR bound so to circumvent the impossibility of applying the CFS methodology, Gaborit et al. [20] then proposed to decode random syndromes above GVR. Given a syndrome, the idea is to randomly fix some subspace of the error support (Definition 4) which leads to increasing the size of decoding balls. The RankSign signature scheme was then deduced from this result applying a methodology close to CFS where, given a secret key, one is then able to output a small weight vector solving an RSD instance relatively to a public matrix. The RankSign public key can then be seen as a trapdoor matrix.

3 Definition and Security Model

3.1 Definition

Under the existence of a PKI for exchanges between users and authorities, we propose the following definition where two authorities, a group manager (also called issuer) and an opener, are involved.

Definition 11. *A group signature $\mathcal{GS}$ scheme is a sequence of protocols (KeyGen, Join, Sign, Verif, Open) such as:*

- $KeyGen(1^\lambda)$*: it generates the group public key gpk and the private keys: the group manager secret key gmsk and the opener secret key skO containing some tracing table tr which could be publicly revealed;*
- $Join(\mathcal{U}_i, gmsk, gpk)$*: interactive protocol between a user* $\mathcal{U}_i$ *and the group manager. In the end, the user gets a secret key usk[i] and the issuer contacts the opener to update tr;*
- $Sign(usk[i], gpk, m; \mu)$*: to sign a message m, the user uses his secret key usk[i] and some randomness* μ *to output a signature* σ *valid under the group public key;*
- $Verif(gpk, m, \sigma)$*: anybody can check the validity of a signature* σ *on the message m with respect to gpk. It outputs 1 if the signature is valid, and 0 otherwise;*
- $Open(skO, gpk, m, \sigma)$*: for a valid signature* σ *with respect to gpk, the opener can provide signer's identity: it thus outputs the user* $\mathcal{U}_i$ *when it succeeds and 0 otherwise.*

3.2 Security Model

Compared to classic BMW related static models, our scheme enjoys the possibility of adding group members (protocol *Join*) during lifetime of the group but does not fulfill the non-frameability security property required by the dynamic BSZ model.

Remark 2. Informally, non-frameability guarantees that, even if both the group manager and the opener are corrupted, no honest user could be accused of having generated a signature if he did not. Even if non-frameability appears as a nice property, many real life applications, such as authentication systems, assume issuer integrity so that the interest of our model does hold in numerous contexts.

We then require our scheme to fulfill properties of correctness, anonymity and traceability.

Correctness. This guarantees that honest users should be able to generate valid signatures, and the opener should then be able to revoke anonymity of the signers.

In the following experiments, we denote the set of corrupted users by CU, made of users for which an adversary $\mathcal{A}$ knows their secret keys in opposition to honest users referred as the set HU. $\mathcal{A}$ is granted some oracles:

- $\mathcal{O}^{join}(\mathcal{U}_i)$, a new user $\mathcal{U}_i$ is added to HU;
- $\mathcal{O}^{sign}(\mathcal{U}_i, m)$, if $\mathcal{U}_i \in HU$, returns $Sign(gmsk, sk[i], m)$ and adds i to $S[m]$, the list of users for which a signature on message m exists;
- $\mathcal{O}^{corrupt}(\mathcal{U}_i)$, if $\mathcal{U}_i \in HU$, provides user's secret key $usk[i]$ and moves $\mathcal{U}_i$ to CU;
- $\mathcal{O}^{open}(m, \sigma)$, returns $Open(skO, gpk, m, \sigma)$.

Anonymity and Traceability. Informally, the anonymity notion requires that signatures issued by two users are computationally indistinguishable to an adversary $\mathcal{A}$. Traceability ensures that no group member or coalition of group members and the opener can produce a valid signature that cannot be opened or for which the opening process might accuse an honest user.

Anonymity. The anonymity security game (Fig. 1(a)) consists in a challenger randomly choosing a bit $b \in \{0, 1\}$ while the adversary $\mathcal{A}$ is asked to guess this value. More precisely, $\mathcal{A}$ targets two users i_0 and i_1 and the challenger issues a signature on behalf of i_b. Granted aforesaid oracles, the adversary wins the game if it outputs $b' = b$.

Traceability. Concerning traceability (Fig. 1(b)), the adversary aims at producing a valid signature for which the opening procedure either fails or accuses an honest user. More precisely, algorithm *Verif* must output 1 on inputs a cople (m, σ) generated by the adversary and *gpk* while procedure *Open* should not output the identity of a corrupted user because it would simply mean that $\mathcal{A}$ signed m with a secret key he already knew. On the contrary, if $\mathcal{A}$ is able to produce a valid signature that cannot be opened or that traces back to an honest user, we consider that he has succeeded in attacking the security of the scheme.

(a) Experiment $Exp_{GS,\mathcal{A}}^{anon-b}(\lambda)$
1. $(gpk, gmsk) \leftarrow KeyGen(1^\lambda)$
2. $(m, i_0, i_1) \leftarrow \mathcal{A}(gpk, tr : \mathcal{O}^{join}, \mathcal{O}^{corrupt}, \mathcal{O}^{sign}, \mathcal{O}^{open})$
3. $\sigma_b \leftarrow Sign(usk[i_b], gpk, m; \mu)$
4. $b' \leftarrow \mathcal{A}(gpk, \sigma_b : \mathcal{O}^{join}, \mathcal{O}^{corrupt}, \mathcal{O}^{sign}, \mathcal{O}^{open})$
5. If $i_0 \notin HU$ or $i_1 \notin HU$, Return 0.
6. Return b'.

$$Adv_{GS,\mathcal{A}}^{anon}(\lambda) = Pr[Exp_{GS,\mathcal{A}}^{anon-1}(\lambda) = 1] - Pr[Exp_{GS,\mathcal{A}}^{anon-0}(\lambda) = 1]$$

(b) Experiment $Exp_{GS,\mathcal{A}}^{tr}(\lambda)$
1. $(gpk, gmsk, skO) \leftarrow KeyGen(1^\lambda)$
2. $(m, \sigma) \leftarrow \mathcal{A}(gpk, skO : \mathcal{O}^{join}, \mathcal{O}^{Corrupt}, \mathcal{O}^{sign})$
3. If $Verif(gpk, m, \sigma) = 0$, Return 0.
4. If $Open(skO, gpk, m, \sigma) = \perp$, Return 1.
5. If $\exists j \notin CU \cup S[m]$,
$Open(skO, gpk, m, \sigma) = j$, Return 1.
6. Else Return 0.

$$Adv_{GS,\mathcal{A}}^{tr}(\lambda) = Pr[Exp_{GS,\mathcal{A}}^{tr}(\lambda) = 1]$$

Fig. 1. Security notions

Definition 12. *A group signature scheme fulfilling correctness and for which advantages related to anonymity and traceability (Fig. 1) are negligible, is said to be dynamic.*

4 Rank Concatenated Stern's Protocol

For H a public matrix, x a small weight vector of weight w_x, and the syndrome $s = H.x^T$, Stern's authentication protocol [28] permits a prover to convince a verifier that he knows a small weight vector of weight w_x, such that $H.x^T = s$.

Stern's authentication protocol and its variations have been widely used to design group signatures through FS paradigm. A rank-based alternative was first proposed by [29] that was later broken and repaired by Gaborit et al. in [30]. We rely on this latter, referred as the *Rank Stern's protocol* to propose a new rank-based ZK authentication protocol.

4.1 Problematic and Overview of Our Protocol

The problematic is to design an authentication protocol enabling a verifier to check the weight of each part of a split secret. More precisely, let us consider $k \times n$ and $k \times n'$ random matrices Q and R over $GF(q^m)$, a syndrome s, some weights ω_x and ω_y leading to the SD instance depicted in Fig. 2.

$$[Q \mid R] \cdot \binom{x}{y} = s.$$

Fig. 2. High level overview

While (Rank) Stern's protocol only allows to prove knowledge of a secret z of weight $\omega_x + \omega_y$, our goal is to prove knowledge of a split secret $z = (x, y)$ such as $\omega t(x) = \omega_x$ and $\omega t(y) = \omega_y$.

4.2 Rank Concatenated Stern's Protocol

Similarly to [17], Fig. 3 introduces our authentication protocol, from now referred as *Rank Concatenated Stern's Protocol* ($\mathcal{RCSP}$), whose goal is to prove knowledge of a secret (x, y) with weight constraints on both x and y. The idea is somewhat to run, in parallel, 2 instances of Rank Stern's protocol on x and y while linking these two values through commitments. We denote here $V_n = GF(q^m)^n$ and $V_{n'} = GF(q^m)^{n'}$.

RSD instance $((Q|R), s, \omega_Q, \omega_R)$

$\mathcal{P}$'s secret: $(x, y) \in V_n \times V_{n'}$ such as $(Q|R).(x, y)^T = s$ with $\omega t(x) = \omega_x$ and $\omega t(y) = \omega_y$

1. [Commitment step] $\mathcal{P}$ chooses $(v_1, v_2) \xleftarrow{\$} \in V_n \times V_{n'}$, $r_1, r_2, r_3 \xleftarrow{\$} 1^\lambda$, $(P_1, P_2) \xleftarrow{\$} GL_n(GF(q)) \times GL_{n'}(GF(q))$ and $Q_1, Q_2 \xleftarrow{\$} GL_m(q)$ and He then sends c_1, c_2, c_3 where:
   $c_1 = h(Q_1|P_1|Qv_1^T + Rv_2^T|Q_2|P_2|r_1)$,
   $c_2 = h(Q_1 * v_1 P_1|Q_2 * v_2 P_2|r_2)$,
   $c_3 = h(Q_1 * (v_1 + x)P_1|(Q_2 * (v_2 + y)P_2|r_3)$
2. [Challenge step] $\mathcal{V}$ sends $ch \xleftarrow{\$} \{0, 1, 2\}$ to $\mathcal{P}$.
3. [Response step] There are three possibilities:
   $ch = 0$: $\mathcal{P}$ responds v_1, $(Q_1|P_1)$, v_2, $(Q_2|P_2)$, r_1, r_2.
   $ch = 1$: $\mathcal{P}$ responds $v_1 + x$, $(Q_1|P_1)$, $v_2 + y$, $(Q_2|P_2)$, r_1, r_3.
   $ch = 2$: $\mathcal{P}$ responds $Q_1 * v_1 P_1$, $Q_1 * xP_1$, $Q_2 * v_2 P_2$, $Q_2 * yP_2$, r_2, r_3.
4. [Verification step] There are three possibilities:
   $ch = 0$: $\mathcal{V}$ checks c_1, c_2.
   $ch = 1$: $\mathcal{V}$ checks c_1, c_3.
   $ch = 2$: $\mathcal{V}$ checks c_2, c_3 and
   $\omega_t(Q_1 * xP_1) = \omega_x$, $\omega t(Q_2 * yP_2) = \omega_y$.
5. [Final step] $\mathcal{V}$ outputs *Accept* if all checks passed, $\perp$ otherwise.

Fig. 3. Rank Concatenated Stern's Protocol ($\mathcal{RCSP}$).

Remark 3. As pointed out in [17], original version of Stern's authentication protocol [28] (in both Hamming and Rank cases) suffers a witness distinguishability. A simple randomization (roles of seeds r_1, r_2, r_3) addresses this issue to ensure ZK property.

Theorem 1. *$\mathcal{RCSP}$ (Fig. 3) is an honest prover verifier ZK protocol with cheating probability* $2/3$ *thus verifying properties of completeness, soundness and zero-knowledge.*

Proof. Lying on rank version of Stern's protocol [30], the proof is straightforward. Completeness property is straightforward and we only stress out that in the case where $ch = 1$, the verifier can check validity of c_1 by computing $h(Q_1|P_1|(Q(v_1 + x)^T + R(v_2 + y)^T - s|Q_2|P_2|r_1)$. Soundness and ZK properties directly come from those (of the randomized) Rank Stern's protocol.

5 Our Rank-Based Group Signature Scheme

Before going into details, we introduce matrices H_s and H_c, indistinguishable from random ones, verifying:

- H_s is a public *trapdoor matrix* i.e. given a (trapdoor) secret key sk_s, a random syndrome s and an integer ω_s, one can output y solving the RSD instance (H_s, s, ω_s);
- H_c is the public key of a Rank-based Public Key Cryptosystem (R-PKC) with associated secret key sk_c.

5.1 High Level Overview of Our Scheme

The main idea is to instantiate $\mathcal{RCSP}$ with particular matrices Q and R so that a user will be given a small weight split secret (x, y_i) such as:

- y_i is user's signing key committed into a group public syndrome s relatively to H_s;
- x is a random vector committed through a R-PKC ciphertext c enabling to further revoke anonymity.

These secrets are then linked via a syndrome r, leading to the situation depicted in Fig. 4, where A and B are random matrices, H_s is a trapdoor matrix and H_c, a R-PKC public matrix.

To sign a message, $\mathcal{U}_i$ then makes a proof of knowledge on (x, y_i) through Rank Concatenated Stern's protocol (Fig. 3). When the opener, given the R-PKC secret key, wants to revoke anonymity, he first recovers x from c and then computes $r - Ax^T$. This value must appear in its tracing table tr containing all the $B.y_i^T$s from which he can finally deduce signer's identity.

$$\begin{bmatrix} A & B \\ 0 & H_s \\ H_c & 0 \end{bmatrix} . \begin{pmatrix} x \\ y_i \end{pmatrix} = \begin{pmatrix} r \\ s \\ c \end{pmatrix}.$$

Fig. 4. A particular instantiation of $\mathcal{RCSP}$.

5.2 Algorithms KeyGen, Join and Sign

To begin, the algorithm *KeyGen*, according to λ, generates the following data:

- a RSD instance (H_s, s, ω_s) where H_s is a trapdoor matrix with associated secret key sk_s given to the group manager;
- a R-PKC key pair (H_c, sk_c) with sk_c given to the opener;
- an integer ω and two random matrices A, B.

When contacted by user $\mathcal{U}_i$, the group manager uses its trapdoor key sk_s to compute user's secret key $usk[i] = y_i$ as a solution of the aforesaid RSD instance. The opener is then given the syndrome $B.y_i^T$ to update tr.

Now, to authenticate himself, $\mathcal{U}_i$ first chooses a random x of weight ω and computes the syndromes $r = A.x^T + B.y_i^T$ and $c = H_c.x^T$. By instantiating $Q = \begin{bmatrix} A \\ 0 \\ H_c \end{bmatrix}$ and $R = \begin{bmatrix} B \\ H_s \\ 0 \end{bmatrix}$, he makes, through $\mathcal{RCSP}$, a ZK proof on the secret (x, y_i) with $\omega t(x) = \omega$ and $\omega t(y_i) = \omega_s$, solving the RSD instance depicted in Fig. 4.

(a) $\boldsymbol{KeyGen(1^\lambda)}$
1. According to λ, generate:
   1.1. a RSD instance $rsd = (H_s, s, \omega_s)$:
      – H_s a trapdoor matrix;
      – sk_s its related secret key.
   1.2. a R-PKC instance:
      – H_c the public matrix key;
      – sk_c its related secret key.
   1.3. two random matrices A and B and an integer ω.
4. $gpk := (H_s, H_c, A, B, s, \omega_s, \omega)$.
5. $gmsk := sk_s$, $skO := (sk_c, tr = [\,])$.
6. Return $(gpk, gmsk, skO)$.

(b) $\boldsymbol{Join(\mathcal{U}_i, gmsk = sk_s, gpk)}$
1. Use the trapdoor sk_s on H_s to output y_i solving rsd.
2. If $\exists j \leq i-1$, $y_i = usk[j] \vee B.y_i^T = tr[j]$, go to 1.
3. Return $usk[i] = y_i$, $tr[i] = B.y_i^T$.

(c) $\boldsymbol{Sign(usk[j] = y_i, gpk, m; \mu)}$
1. Choose a random x such as $\omega t(x) = \omega$.
   1.1. $r := A.x^T + B.y_i^T$.
   1.2. $c := H_c.x^T$.
2. For $l = 1 \ldots l_\lambda$
   2.1. Set $c_1, c_2, c_3, d_1, d_2, d_3$ according to Figure 3, step 1.
   2.2. $cmt[l] := \{c_1, c_2, \ldots, d_3\}$.
3. $ch := \mathcal{H}_\lambda(m, cmt, r, c) \in [\![2]\!]^{l_\lambda}$.
4. For $i = 1 \ldots l_\lambda$
   Generate $rsp[l]$ according to $ch[l]$ and Figure 3, step 3.
5. Set $\Pi = (cmt, ch, resp)$.
6. Return $\sigma = (\Pi, (r, c))$.

Fig. 5. *KeyGen, Join, Sign* algorithms

Finally, by turning this process non-interactive through FS paradigm, we get the signing algorithm. Algorithms *KeyGen*, *Join* and *Sign* are described in Fig. 5.

5.3 Algorithms *Verif* and *Open*

The verification algorithm relies on the verification step of $\mathcal{RCSP}$ (Fig. 3).

To revoke anonymity, the opener uses sk_c to recover x from the R-PKC ciphertext $c = H_c.x^T$. He can then computes $A.x^T$ and subtracts this value to r transmitted in the signature along with c. The result $r - A.x^T$ must be equal to some $tr[i]$, from which the signer's identity is learnt. These two algorithms appear in Fig. 6.

(a) ***Verif*(*gpk*, *m*, σ)**	(b) ***Open*(*skO*, *gpk*, *m*, σ)**

1. Parse $\sigma = (\Pi, (r, c))$
2. Parse $\Pi = (cmt, ch, rsp)$
3. $\tilde{ch} := \mathcal{H}_\lambda(m, cmt, r, c) \in [\![2]\!]^{l_\lambda}$.
   If ($\tilde{ch} \neq ch$), Return 0.
4. For $l = 1 \ldots l_\lambda$
   3.1. Check $rsp[l]$ according to $cmt[l]$, $ch[l]$ and Figure 3.
   3.2. If a verification fails, Return 0.
5. Return 1.
	1. If ($Verif(gpk, m, \sigma) = 0$)
	   Return $\perp$.
	2. Parse $\sigma = (\Pi, (r, c))$
	3. Parse $skO = (sk_c, tr)$.
	4. Use sk_c to recover x from c.
	5. Set $z = r - Ax^T$
	6. For $i = 1 \ldots N$
	   If ($tr[i] = z$)
	     Return $\mathcal{U}_i$.
	7. Return $\perp$.

Fig. 6. *Verif* and *Open* algorithms

6 Security Analysis

Since correctness directly comes from $\mathcal{RCSP}$, we focus on anonymity and traceability requirements defined in Fig. 1. We begin with the anonymity property.

Theorem 2. *If there exists an adversary $\mathcal{A}$ that can break the anonymity property of the scheme, then there exists an adversary $\mathcal{B}$ that can either break the Zero-Knowledge property of $\mathcal{RCSP}$ or the Decision Rank Syndrome Decoding (D-RSD) problem.*

Proof. Through a sequence of games, we will exhibit that an adversary against our anonymity property would be able to either break the ZK property of our scheme or the D-RSD problem.

$\mathcal{G}_0$ $\mathcal{B}$ runs $KeyGen(1^\lambda)$ and acts honestly as described in Fig. 1(a).

$\mathcal{G}_1$ Now, to answer the opening query, the simulator uses the ROM observability to extract some y, and then compares the value to the $B.y_i^T$ contained in tr. Under the ZK soundness, this is similar to the previous game.

$\mathcal{G}_2$ The simulator now supersedes part of the KeyGen by setting $\begin{bmatrix} A \\ H_c \end{bmatrix} = C$. This game is identical to the previous one.

$\mathcal{G}_3$ The simulator now simulates the proof when answering the challenge queries, by not using the value x, y_i. This game is identical to the previous one under the ZK property.

$\mathcal{G}_4$ Now, he sends randoms $c = s_1$, $r = s_2 + B.y_i^T$. This game is indistinguishable from the previous one under the D-RSD problem. (This is seen, by splitting the challenge s in s_1, s_2 as it was done for the matrix C)

$\mathcal{G}_5$ This last game only displays random values, hence the adversary has no advantage, which terminates the proof.

Concerning traceability, we have the following theorem.

Theorem 3. *If there exists an adversary $\mathcal{A}$ that can break the traceability of the scheme, then there exists an adversary $\mathcal{B}$ that can break either break the Soundness of the Zero-Knowledge proof or the OMRSD problem.*

Proof. The proof is straightforward, the simulator starts by simulating the ZK proofs on every honest signing queries. Then, he picks an identity at random expecting it to be the targeted honest user (this happens with non-negligible probability) and sets his tracing key $B.y_*^\top$ as s, for the other identities the simulator sends a request to the RSD oracle, and forwards the answer.

Receiving the adversary answers, the simulator uses the ROM, to extract the value y_* solution to the challenge.

7 Instantiation

Our scheme is generic and can be used with any trapdoor matrix and public key encryption scheme. In our rank-based context, RankSign and LRPC embedded into Niederreiter setting (see Sect. 2) constitute well-suited candidates for respectively instantiating matrices H_s and H_c introduced in Sect. 5.

According to Sects. 5 and 6, correctness and security of our scheme generically rely on matrices H_s and H_c meant to be indistinguishable from random ones. With such an instantiation, the security is maintained through the putative indistinguashibilty of LRPC and RankSign public matrices with random matrices [19].

Parameters. As it will exhibited below, it is sufficient in practice to consider matrices A and B with only one row and then according to previous section, the security of the protocol relies on the D-RSD problem or the OMRSD problem associated to matrices H_s and H_c. We now give parameters to obtain an overall security of 2^{100}.

Following [20], we can consider parameters $n' = 23$, $k' = 10$, $t = 3$, $m = 24$ and $q = 2^8$ to design a RankSign public matrix H_s of dimension $(n' - k') \times (n' + t)$ with coordinates lying in $GF(q^m)$. In particular, t denotes the number of generalized erasures handled by such an instantiation. The size of the group then consists in the number of potential antecedents to a common public syndrome s; namely it corresponds to the number of possibilities to form t independent vectors lying over $GF(q^m)$ which roughly leads to q^{tm} users ($2^{8\times 24}$ here). We refer the reader to [20] for more details on this point. On the other hand, the matrix H_c can be instantiated with the cyclic LRPC cryptosystem embedded in Niederreiter setting; following [19], we consider parameters $n = 74$, $k = 37$, $q = 2^8$ and the working field $GF(q^m)$. Now, matrices A and B are only used to differentiate the $B.y_i^T$ (procedure *Open*) and since $q = 2^8$ and $m = 24$, there are 2^{192} possibilities for $B.y_i^T$ by simply taking A and B with one row. Finally, with A and B made of one row and H_c cyclic, the size of the public key is mainly due to H_s. By considering the systematic form with aforesaid parameters, it leads to H_s of size $(n' - k') \times (n' + t - n' + k') = (23 - 10) \times (10 + 3)$ over $GF(2^{8\times 24})$. Adding contributions of A, B and one line of H_c, it finally leads to a public key of around 5 kB.

The signature size depends both on the security level and the length of a proof of knowledge in $\mathcal{RCSP}$ (Fig. 3). Let us first notice that random matrices involved in the protocol can be sent through seeds from which they could be regenerated. Hence, the preponderant data sent during the protocol consists in the vectors belonging to V_n and $V_{n'}$: thus we get on average 4/3 elements of the ambient space $GF(q^m)^{n+n'}$ representing $4/3 \times (74 + 23) \times 8 \times 24$ bits. When targeting a 100 bits security level for which $100/log_2(3/2)$ repetitions of the protocol are required, this leads to a signature of 550 kB. One should notice that these parameters are versatile and it would be easy to find parameters to fit another security level.

Asymptotic Complexity. To study the asymptotic complexity, we first recall that:

- the number of users N is roughly $q^{mt} = 2^{mt\,log(q)}$;
- in practice and as exhibited above, parameters t, k' and m are set to $O(n')$ so that the public key is approximated by $O(n'^3 \times log(q))$.

For a given security level, it is possible to increase the number of users by increasing the size of q. In that case we consider all parameters except q as fixed. From the previous recalls, the number of users is then $N = 2^{O(log(q))}$ when the size of parameters is in $O(log(q)) = O(log(N))$.

Finally, in terms of computation time, the protocol is very efficient since in themselves the LRPC and RankSign cryptosystems are very fast (a few milliseconds for encryption/decryption or signature).

Concurrent Works. Our dynamic scheme compares very well with code-based concurrent works such as [17,18]. Indeed, it features public key and signature sizes logarithmic in N while those of the static scheme presented in [18] are linear in N. Furthermore, when considering at maximum 2^{24} users, the latter one leads to a public key of size 1.16 GB with the advantage of only relying on the SD-problem. Even if dynamic and with public key and signature sizes in $N^{1/\sqrt{log(N)}}$, the group signature of [17] leads to signatures of size 20 MB and a public key of 2.5 MB. In parallel, despite recent progress and satisfying asymptotic performances [11,13–16] (public keys and signatures logarithmic in the number of group members), lattice-based constructions still suffer great sizes of parameters. Indeed, the most efficient one due to [16], improving works of [13,14], proposes signatures and a public key respectively of size 61.5 MB and 4.9 MB for a group made of only 1024 users and an overall security of 2^{80}.

8 Conclusion

This work proposes the first rank-based group signature scheme that is dynamic in a relaxation of the BSZ model and compares very well with concurrent post-quantum schemes. By introducing a rank-based ZK authentication protocol, which could be of independent interest, we obtain a signature scheme via Fiat-Shamir paradigm. Its security in the ROM, relies on LRPC related cryptosystems (RankSign and Niederreiter), the RSD problem, and rank-based computational problems introduced along with this work (OMRSD and D-RSD problems). With an asymptotic complexity better than code-based constructions and similar to best lattice-based results, our scheme features public key and signature sizes logarithmic in the number of group members. Last but not least, with well chosen parameters, we obtain an instantiation with public key and signature sizes respectively of **550 kB** and **5 kB** so that our scheme appears fairly practical and as the most efficient post-quantum group signature protocol up to date.

References

1. Chaum, D., Heyst, E.: Group signatures. In: Davies, D.W. (ed.) EUROCRYPT 1991. LNCS, vol. 547, pp. 257–265. Springer, Heidelberg (1991). doi:10.1007/3-540-46416-6_22
2. Bellare, M., Micciancio, D., Warinschi, B.: Foundations of group signatures: formal definitions, simplified requirements, and a construction based on general assumptions. In: Biham, E. (ed.) EUROCRYPT 2003. LNCS, vol. 2656, pp. 614–629. Springer, Heidelberg (2003). doi:10.1007/3-540-39200-9_38
3. Bellare, M., Shi, H., Zhang, C.: Foundations of group signatures: the case of dynamic groups. In: Menezes, A. (ed.) CT-RSA 2005. LNCS, vol. 3376, pp. 136–153. Springer, Heidelberg (2005). doi:10.1007/978-3-540-30574-3_11
4. Boneh, D., Boyen, X., Shacham, H.: Short group signatures. In: Franklin, M. (ed.) CRYPTO 2004. LNCS, vol. 3152, pp. 41–55. Springer, Heidelberg (2004). doi:10.1007/978-3-540-28628-8_3
5. Camenisch, J., Lysyanskaya, A.: Signature schemes and anonymous credentials from bilinear maps. In: Franklin, M. (ed.) CRYPTO 2004. LNCS, vol. 3152, pp. 56–72. Springer, Heidelberg (2004). doi:10.1007/978-3-540-28628-8_4
6. Boneh, D., Shacham, H.: Group signatures with verifier-local revocation. In: Proceedings of CCS 2004, pp. 168–177. ACM Press (2004)
7. Kiayias, A., Tsiounis, Y., Yung, M.: Traceable signatures. In: Cachin, C., Camenisch, J.L. (eds.) EUROCRYPT 2004. LNCS, vol. 3027, pp. 571–589. Springer, Heidelberg (2004). doi:10.1007/978-3-540-24676-3_34

8. Groth, J.: Fully anonymous group signatures without random oracles. In: Kurosawa, K. (ed.) ASIACRYPT 2007. LNCS, vol. 4833, pp. 164–180. Springer, Heidelberg (2007). doi:10.1007/978-3-540-76900-2_10
9. Libert, B., Yung, M.: Efficient traceable signatures in the standard model. In: Shacham, H., Waters, B. (eds.) Pairing 2009. LNCS, vol. 5671, pp. 187–205. Springer, Heidelberg (2009). doi:10.1007/978-3-642-03298-1_13
10. Gordon, S.D., Katz, J., Vaikuntanathan, V.: A group signature scheme from lattice assumptions. In: Abe, M. (ed.) ASIACRYPT 2010. LNCS, vol. 6477, pp. 395–412. Springer, Heidelberg (2010). doi:10.1007/978-3-642-17373-8_23
11. Laguillaumie, F., Langlois, A., Libert, B., Stehlé, D.: Lattice-based group signatures with logarithmic signature size. In: Sako, K., Sarkar, P. (eds.) ASIACRYPT 2013. LNCS, vol. 8270, pp. 41–61. Springer, Heidelberg (2013). doi:10.1007/978-3-642-42045-0_3
12. Langlois, A., Ling, S., Nguyen, K., Wang, H.: Lattice-based group signature scheme with verifier-local revocation. In: Krawczyk, H. (ed.) PKC 2014. LNCS, vol. 8383, pp. 345–361. Springer, Heidelberg (2014). doi:10.1007/978-3-642-54631-0_20
13. Ling, S., Nguyen, K., Wang, H.: Group signatures from lattices: simpler, tighter, shorter, ring-based. In: Katz, J. (ed.) PKC 2015. LNCS, vol. 9020, pp. 427–449. Springer, Heidelberg (2015). doi:10.1007/978-3-662-46447-2_19
14. Nguyen, P.Q., Zhang, J., Zhang, Z.: Simpler efficient group signatures from lattices. In: Katz, J. (ed.) PKC 2015. LNCS, vol. 9020, pp. 401–426. Springer, Heidelberg (2015). doi:10.1007/978-3-662-46447-2_18
15. Libert, B., Mouhartem, F., Nguyen, K.: A lattice-based group signature scheme with message-dependent opening. In: Manulis, M., Sadeghi, A.-R., Schneider, S. (eds.) ACNS 2016. LNCS, vol. 9696, pp. 137–155. Springer, Cham (2016). doi:10.1007/978-3-319-39555-5_8
16. Libert, B., Ling, S., Nguyen, K., Wang, H.: Zero-knowledge arguments for lattice-based accumulators: logarithmic-size ring signatures and group signatures without trapdoors. In: Fischlin, M., Coron, J.-S. (eds.) EUROCRYPT 2016. LNCS, vol. 9666, pp. 1–31. Springer, Heidelberg (2016). doi:10.1007/978-3-662-49896-5_1
17. Alamélou, Q., Blazy, O., Cauchie, S., Gaborit, P.: A code-based group signature scheme. In: Charpin, J.-P.T.P., Sendrier, N. (eds.) Proceedings of the 9th International Workshop on Coding and Cryptography 2015, WCC2015, France, Paris (2015)
18. Ezerman, M.F., Lee, H.T., Ling, S., Nguyen, K., Wang, H.: A provably secure group signature scheme from code-based assumptions. In: Iwata, T., Cheon, J.H. (eds.) ASIACRYPT 2015. LNCS, vol. 9452, pp. 260–285. Springer, Heidelberg (2015). doi:10.1007/978-3-662-48797-6_12
19. Gaborit, P., Murat, G., Ruatta, O., Zémor, G.: Low Rank Parity Check codes and their application to cryptography. In: WCC 2013, Bergen, Norway, April 2013
20. Gaborit, P., Ruatta, O., Schrek, J., Zémor, G.: RankSign: an efficient signature algorithm based on the rank metric. In: Mosca, M. (ed.) PQCrypto 2014. LNCS, vol. 8772, pp. 88–107. Springer, Cham (2014). doi:10.1007/978-3-319-11659-4_6
21. Gaborit, P., Zémor, G.: On the hardness of the decoding and the minimum distance problems for rank codes. CoRR, abs/1404.3482 (2014)
22. Fiat, A., Shamir, A.: How to prove yourself: practical solutions to identification and signature problems. In: Odlyzko, A.M. (ed.) CRYPTO 1986. LNCS, vol. 263, pp. 186–194. Springer, Heidelberg (1987). doi:10.1007/3-540-47721-7_12
23. Loidreau, P.: Properties of codes in rank metric. CoRR, abs/cs/0610057 (2006)

24. Gaborit, P., Ruatta, O., Schrek, J., Zémor, G.: New results for rank-based cryptography. In: Pointcheval, D., Vergnaud, D. (eds.) AFRICACRYPT 2014. LNCS, vol. 8469, pp. 1–12. Springer, Cham (2014). doi:10.1007/978-3-319-06734-6_1
25. Gaborit, P., Ruatta, O., Schrek, J.: On the complexity of the rank syndrome decoding problem. IEEE Trans. Inf. Theory **62**(2), 1006–1019 (2016)
26. Berlekamp, E., McEliece, R.J., Van Tilborg, H.C.A.: On the inherent intractability of certain coding problems (corresp.). IEEE Trans. Inf. Theory **24**(3), 384–386 (1978)
27. Goldwasser, S., Micali, S., Rackoff, C.: The knowledge complexity of interactive proof systems. SIAM J. Comput. **18**(1), 186–208 (1989)
28. Stern, J.: A new identification scheme based on syndrome decoding. In: Stinson, D.R. (ed.) CRYPTO 1993. LNCS, vol. 773, pp. 13–21. Springer, Heidelberg (1994). doi:10.1007/3-540-48329-2_2
29. Chen, K.: A new identification algorithm. In: Dawson, E., Golić, J. (eds.) CPA 1995. LNCS, vol. 1029, pp. 244–249. Springer, Heidelberg (1996). doi:10.1007/BFb0032363
30. Gaborit, P., Schrek, J., Zémor, G.: Full cryptanalysis of the chen identification protocol. In: Yang, B.-Y. (ed.) PQCrypto 2011. LNCS, vol. 7071, pp. 35–50. Springer, Heidelberg (2011). doi:10.1007/978-3-642-25405-5_3
31. Fischer, J.-B., Stern, J.: An efficient pseudo-random generator provably as secure as syndrome decoding. In: Maurer, U. (ed.) EUROCRYPT 1996. LNCS, vol. 1070, pp. 245–255. Springer, Heidelberg (1996). doi:10.1007/3-540-68339-9_22
32. Gaborit, P., Hauteville, A., Tillich, J.-P.: RankSynd a PRNG based on rank metric. In: Takagi, T. (ed.) PQCrypto 2016. LNCS, vol. 9606, pp. 18–28. Springer, Cham (2016). doi:10.1007/978-3-319-29360-8_2
33. Ernst, M.: Gabidulin: theory of codes with maximum rank distance. Probl. Peredachi Inf. **21**(1), 3–16 (1985)

Author Index

Printed in the United States
By Bookmasters